Embodied Intelligence Conference 2022 (EI-2022)

IOP Conference Series: Materials Science and Engineering
Volume 1292

Cambridge, United Kingdom
22 – 25 March 2022

ISBN: 978-1-7138-8185-8
ISSN: 1757-8981

Printed from e-media with permission by:

Curran Associates, Inc.
57 Morehouse Lane
Red Hook, NY 12571

Some format issues inherent in the e-media version may also appear in this print version.

Institute of Physics
Dirac House, Temple Back
Bristol BS1 6BE UK

Phone: 44 1 17 929 7481
Fax: 44 1 17 920 0979

techtracking@iop.org

Additional copies of this publication are available from:

Curran Associates, Inc.
57 Morehouse Lane
Red Hook, NY 12571 USA
Phone: 845-758-0400
Fax: 845-758-2633
Email: curran@proceedings.com
Web: www.proceedings.com

TABLE OF CONTENTS

Preface

Intelligence emerging from interactions between the body and the environment at the material and morphological levels attracts the attention of many different research communities. Researchers with engineering, social sciences, computer sciences, biology, and materials science backgrounds all study and understand embodied intelligence from their own perspectives. Therefore, providing a common definition of 'Embodied Intelligence' remains challenging.

Furthermore, the intricate connection between the body and mind remains an unresolved philosophical question. Exploring this relationship not only deepens our understanding of animals and humans but also serves as a crucial element in the development of embodied artificial intelligence. By unraveling this connection, we can pave the way for creating robots endowed with practical intelligence. The long-standing philosophical debates surrounding consciousness have gained utmost significance in the realm of AI-powered machines.

This book presents a comprehensive selection of papers authored by renowned researchers who participated in the 2nd International Conference on Embodied Intelligence held on 23-25 March, 2022, and the Embodied Intelligence Workshop, held at the RoboSoft 2022 Conference. This collection features reviews and perspectives articles, as well as the original research. The collection aims to stimulate interdisciplinary discussions and presenting novel ideas on the topic of Embodied Intelligence. By exploring the latest concepts and future directions in this field, these papers offer fresh perspectives and help paving the way for breakthroughs in both understanding and practical implementation.

The central focus of this year's conference revolved around two key topics: the significance of modeling in the design of soft robots and the pivotal role of the brain in embodied agents. The design of soft robots can follow a modeling-based approach using formal models and optimization techniques, similar to traditional rigid robotics. However, accurately modeling the behavior of soft robots is challenging. An exploratory approach inspired by biological organisms involves an iterative process, drawing insights from their locomotion and other behaviors. The role of the brain and central nervous system in bio-inspired robotics is significant, with self-organization and embodied intelligence allowing problem-solving without extensive centralized control. Researchers emulate cognitive processes observed in animals to develop intelligent robots, requiring advances in computational models and algorithms. Finding a good balance between exploiting the physical body and relying on computational capabilities remains an open topic of discussion.

The book consists of a wide range of topics surrounding Embodied Intelligence and is loosely structured in two parts. The first set of chapters discuss the principles and perspectives on Embodied Intelligence and the links to robotics and artificial life. In the second half of the book, there is discussion on the connection between humans and Embodied Intelligence and the surrounding philosophy, psychology and cognitive science which ground and situate the research.

We express our sincere gratitude to all the participants of the Conference and Workshop participants, the authors, and the reviewers for their outstanding contributions to this volume.

Arsen Abdulali,
Josie Hughes,
Fumiya Iida.

IOP Conf. Series: Materials Science and Engineering 1292 (2023) 011002 doi:10.1088/1757-899X/1292/1/011002

Peer Review Statement

All papers published in this volume have been reviewed through processes administered by the Editors. Reviews were conducted by expert referees to the professional and scientific standards expected of a proceedings journal published by IOP Publishing.

- **Type of peer review:** Single Anonymous
- **Conference submission management system:** Morressier
- **Number of submissions received:** 28
- **Number of submissions sent for review:** 28
- **Number of submissions accepted:** 28
- **Acceptance Rate (Submissions Accepted / Submissions Received × 100):** 100
- **Average number of reviews per paper:** 1.93
- **Total number of reviewers involved:** 26
- **Contact person for queries:**
  Name: Dr. Fumiya Iida
  Email: fi224@cam.ac.uk
  Affiliation: University of Cambridge

IOP Conf. Series: Materials Science and Engineering 1292 (2023) 012001 doi:10.1088/1757-899X/1292/1/012001

Modeling embodied intelligence: can we capture its essence by modeling internal and external interactions?

Cecilia Laschi

National University of Singapore, Department of Mechanical Engineering
9, Engineering Drive 1, EA-05-04, 117575 Singapore

mpeclc@nus.edu.sg

Abstract. In soft robotics, embodied intelligence leverages on the compliance of the physical body to let sensory-motor behaviour emerge from the interaction with the environment. Soft robot modeling often focuses on the effect of actuation on the deformable body. Can we model and describe mathematically the deformations of a soft robot body under external interaction forces? Would that capture the essence of embodied intelligence? We argue that addressing this challenge, in an interdisciplinary effort, would make embodied intelligence and soft robotics take a leap forward and transition from a trial-and-error pioneering phase to a model-informed discipline era.

1. Introduction

Embodied intelligence is the emergence of sensory-motor behaviour from the complex interaction of the body with the environment [1]. In soft robotics, such external interactions are received by the compliant body as deformations, which are combined with the deformations generated by actuators.

Modeling these internal and external interactions into a unified representation would allow to grasp embodied intelligence, for design and control purposes. A mathematical description of the internal and external interaction forces is crucial for soft robots to achieve controllable and programmable performance [2]. Indeed, we are at a stage where the field of soft robotics, after an impressive growth [3], faces the challenge of evolving into a well-rooted and principled scientific discipline that provides tangible societal benefits, thereby thriving over the next few decades [4]. At the same time, computational modeling has progressed towards techniques and tools that make complex models and simulations affordable.

On the other hand, keeping pursuing prototype-based design and control methodologies, without a supporting engineering modeling framework would inevitably prevent soft robotics from becoming the expected thriving scientific field with clear societal benefits. Prototype-driven design (also referred to as trial-and-error) was likely the fastest and most efficient way to proceed, especially in the exploratory phase. Today, we are at a stage where computational modeling within a model-driven umbrella can be achieved and can enable (i) scaling up soft-robot design and control in response to application needs, and (ii) replicating the behaviour of a soft robot by leveraging simulation and data-driven modeling, opening the path towards soft-robot digital twins.

2. Bibliometric ground

Figure 1 shows the percentage of computational modeling adopted from 1950 to 2020 in the aerospace sector, where it is prominent, in soft robotics, where it is emerging, and in the entire research community, including the humanities and other sectors where computational modeling is almost non-existent. We note how computational modeling kept growing in the aerospace field, bringing a series of tangible breakthroughs that include the Apollo program and space flight, along with the exponential growth of civil aviation, and possibly space tourism. We believe that soft robotics might be at a stage similar to the "Apollo years", where the community must adopt a model-driven workflow for design and control, moving forward from prototype-driven and often trial-and-error methodologies. We argue that this evolution can be achieved by a more pervasive use of rigorous computational mathematical modeling to describe internal and external interactions, and their coupling. This can lead to a flexible hierarchy of improved models, from high- to low-dimensional (or fundamental models) that can thoroughly capture embodied intelligence and can be used for design and control purposes, thereby reducing the sim-to-real gap.

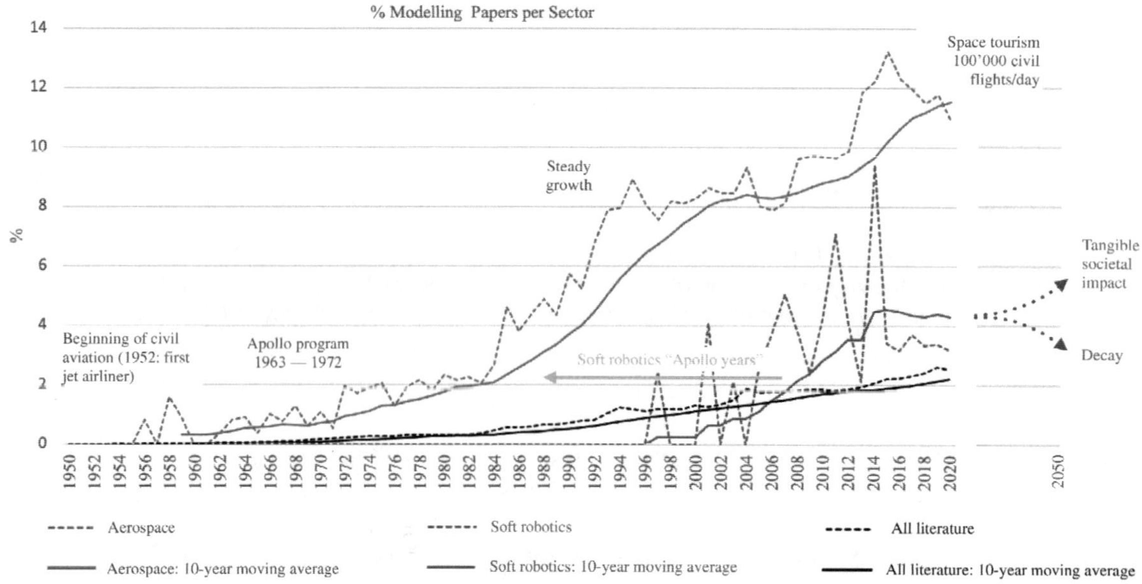

Figure 1. Growth of modeling in aerospace (blue), soft robotics (red) and the whole research community (black) in terms of number of publications with computational modeling, normalized by number of publications per sector; the y axis is the percentage of publications including computational modeling with respect to a given sector. Dashed lines correspond to actual publications with computational modeling per year, while the solid lines are ten years moving averages. Ten years batches are commonly considered research cycles and represent a reasonable approximation to estimate a trend in a given sector. We note that the common exponential growth shadowing the trend of every field in the past few years is not present thanks to the normalization adopted, that emphasize the percentage of computational modeling in a given discipline. The search for publications was done in Scopus, and the search details are as follows for each field: "All literature": (TITLE-ABS-KEY ((fig* OR tab* OR par* OR an* OR the* OR for* OR wor* OR pape*); "Aerospace": (TITLE-ABS-KEY ((ALL) AND (aeronautic* OR aerospace* OR aerodyn*))); "Soft robotics": (TITLE-ABS-KEY ((ALL) AND ("soft robot*" OR "soft bodi*" OR "soft body"))); "Aerospace Computational Modeling" (TITLE-ABS-KEY((ALL) AND (AEROSPACE) AND model* AND (computation* OR digital*))); "Soft robotics Computational Modeling": (TITLE-ABS-KEY((ALL) AND (SOFT ROBOTICS) AND model* AND (computation* OR digital*))).

3. Computational modeling for soft robotics

Computational modeling involves using computer simulations of physical principles translated into mathematical models, to provide accurate, efficient, and flexible proxy solutions of the underlying real-world.

In other fields, mature computational modeling has enabled engineering tasks that were previously unimaginable – the aerospace and biomedical fields (e.g., cardiovascular dynamics modeling) constitutes two notable examples [5]. The computational modeling challenge in soft robotics is considerable, as the range of morphologies (e.g., arms, fingers, legs, and fins), physics (e.g., materials, fluids, etc.), scales (from few mm to few m), abilities (e.g., reaching, grasping, walking, morphing, growing, swimming, jumping, crawling, and digging) and intended applications (e.g., healthcare, manufacturing, underwater sensing and manipulation, scientific exploration, entertainment, and more) is extremely diverse.

Computational modeling in the context of soft robotics inherently leads to multiscale and multiphysics problems, given the diverse configurations that can be adopted. Multiscale and multiphysics are notoriously challenging but have successfully been achieved in some fields, such as aerospace engineering. It is challenging due to a number of reasons: i) multiscale dynamics in space and time often necessitates high-dimensional computational representations, which quickly become intractable (e.g., in turbulence, resolving all relevant scales requires supercomputer simulations); ii) nonlinearity inherent in the dynamics, for example convective nonlinearity in fluid dynamics; iii) partially known interfacial physics and unmodeled dynamics (nonlinear friction, turbulent drag, etc.); iv) ill-conditioned systems of equations. Therefore, it is not surprising that interdisciplinary efforts are required.

Modeling and computational techniques that are used in other sectors can be applied in soft robots, with mutual benefit for soft robotics progress and research opportunities for a range of communities, including applied mathematics, computer science, and computational physics, to name a few. Specifically, soft robotics progress would benefit from further research efforts in 1) modeling the internal interactions of soft robot components that lead to efficient movements and deformation; 2) modeling the robot interactions with the surrounding environment; and 3) modeling their coupling with internal interactions, or embodied intelligence. These computational models could then be used for instance in the context of automatic robot design and optimization [6], where material, geometrical configuration, and actuation/sensing, can all constitute the parameter space that one optimizes to efficiently achieve the desired tasks.

While the first point, i.e., modeling the internal interactions of soft robots, has been addressed in soft robotics, especially for control purposes, modeling external interaction is under-investigated. We main envisage three broad families of approaches: (i) based on continuum mechanics; (ii) based on lumped parameters; (iii) data-driven. They can all be considered both for the case of interaction with fluids (relevant for swimming, flying, and more) and solids (relevant for locomotion and manipulation). Despite the differences among these approaches, they can be formalized into one general equation, as proposed in [7] and reported here to the reader's convenience, whose terms vary with the approach taken.

$$\mathcal{D}_{\mathbf{q}_{sb}} = \mathcal{N}_{sb} + \mathcal{C}_{int} + \mathcal{C}_{ext} \; in \; \Omega_{sb}$$

where $\mathcal{D}$ is a differential operator, $\mathbf{q}_{sb}$ describes the soft body (sb) in the soft body domain Ω_{sb}, $\mathcal{N}_{sb}$ is a non-linear term that describes the soft-body mechanics, $\mathcal{C}_{int}$ and $\mathcal{C}_{ext}$ are coupling terms that accounts for internal and external interactions, respectively.

This unified approach may allow significant improvements in soft robot performance for tasks including grasping, walking, crawling, underwater and underground operations, and morphological adaptation, among others. It might also open novel and sought-after robot abilities and applications, such as next-generation industry and services and human-robot interaction.

4. Conclusions

We highlighted the need and the challenges in modeling internal and external interactions in soft robotics. All this in the quest for mathematically describing the complex interaction of a soft robot with its environment and capturing the underlying principles of emergent behaviour in embodied intelligence. Is such an effort worth? It is, to the extent that it enables model-informed design and control of soft robots. Whether a computationally obtained design could be physically realized should be considered under two perspectives: on one side, ready-to-build soft robots could be designed with available sensors and actuators only. On the other hand, a particularly effective design might steer the development of novel actuation and sensing technologies.

Soft robots that can fulfil given design inputs, fully leveraging on embodied intelligence, can deliver unprecedented abilities, and can respond to unmet needs. Furthermore, computational models can be used to construct digital twins, for real-time monitoring and control of the soft robot asset during a mission or throughout its life cycle.

Soft robotics is at a critical stage of development, and it is the right time for this field to evolve towards becoming a discipline, with a computational modeling framework and structured design methodologies.

References

[1] R. Pfeifer, M. Lungarella, F. Iida, "Self-Organization, Embodiment, and Biologically Inspired Robotics", *Science* 318:1088, 2007.

[2] H. Hauser, A.J. Ijspeert, R.M. Füchslin, R. Pfeifer, W. Maass, "Towards a theoretical foundation for morphological computation with compliant bodies", *Biological Cybernetics* 105:355–370, 2011.

[3] G. Bao, H. Fang, L. Chen, Y. Wan, F. Xu, Q. Yang, L. Zhang, "Soft Robotics: Academic Insights and Perspectives Through Bibliometric Analysis", *Soft Robotics* 5(3), 2018.

[4] E.W. Hawkes, C. Majidi, M.T. Tolley, "Hard questions for soft robotics", *Science Robotics* 6(53), 2021.

[5] S.A. Niederer, M.S. Sacks, M. Girolami, K. Willcox, "Scaling digital twins from the artisanal to the industrial", *Nature Computational Science* 1(5):313–320, 2021.

[6] S.G. Nurzaman, L. Wang, F. Iida, J. Lipton, D. Floreano, D. Rus, "Design Optimization of Soft Robots", *IEEE Robotics & Automation Magazine* 27(4):10-11, 202.

[7] G. Mengaldo, F. Renda, S.L. Brunton, M. Baecher, M. Calisti, C. Duriez, G.S. Chirikjian, C. Laschi, "A concise guide to modeling the physics of embodied intelligence in soft robotics", *Nature Reviews Physics*, 4(9):595-610, 2022.

How to measure embodied intelligence?

Marie Vihmar[1], **Kadri-Ann Valdur**[1], **Saoni Banerji**[1], **Indrek Must**[1]

[1] Institute of Technology, University of Tartu, 60540, Estonia

E-mail: indrek.must@ut.ee

Abstract. Embodied intelligence (EI) summarizes design approaches that give robots efficient physical interaction with their surrounding environment. EI has, to date, found an extensive descriptive treatment yet lacks universal metrics. Could we quantify EI? What would be the EI analog to intelligence quotient (IQ) in humans? We first suggest an intelligent unit (intel-unit) as the smallest entity that displays program execution beyond simple stimuli-responsiveness. The emergence of EI by situatedness of encoded physical agents in an environment is explored in three case studies. First, an awn of a wild oat – a non-living structure in nature that satisfies all robot's descriptions – is analyzed in terms of material-level intelligence. Next, a jocular quantitative scale found with a popular gadget – the 'fortune teller fish' – is given treatment in the EI context. Finally, the learned and summarized EI principles are tested on a well-known object with perceived intelligence yet lacking a consensual robotic description – a knitted woollen sweater. We confirm the necessity of treating environment interactions in expressions and definitions of EI, as situatedness is central in defining functionality. Universal metrics would identify non-intuitive representations of EI and lead from empirical to model-based design of unconventional robots. Robots that engage situatedness provide a physical interface to artificial intelligence, similar to our bodies communicating with our brains.

1. Introduction

Intelligence, either 'biological' or 'artificial,' appears as a nebulous entity without size, shape, or weight. Modern 'cloud' computation further disarrays the idea of the physical embodiment of intelligence – almost rightfully so. First, let's ask how intelligence manifests itself in the physical world. Any machine computation in today's convention entails physical electron transactions in arranged field-effect transistors to algorithmically process information. The code and information are stored similarly: as an increased electron density in gates of tiny electronic capacitors forming memory arrays. Thus, even artificial intelligence (AI) is conditionally embodied, as it does have a traceable physical representation. We could localize the AI, yet the users barely know (and it hardly even matters) if a particular AI algorithm is physically being executed on their smartphone or a server cluster across the ocean. Even probabilistic machine learning algorithms reach a comparable outcome in any syntactically compatible computation device, regardless of their physical location or working principle. Indeed, **AI is inevitably situated, but it very seldom engages situatedness**. Remarkably, synaptic transactions in the brain appear to emerge from hard-wired neurons; yet only to be connected to a situated network of neurons throughout the body, constituting a certain level of EI.

All fields in science benefit from having objective traceable metrics that allow reliable comparison between the works of different scientists, may they be as fundamental as the meter for length or as complex as IQ. Quantification (or quantitation), central to the scientific method, entails the act of measuring (including counting) to transform human sensory observations and experiences into numerical values [1]. Inevitably, EI, as any intelligence, is highly contextual and arguably subjective, yet systematic metrological approaches are essential for comparison or benchmarking. A reproducible (i.e., precise) and valid (i.e., accurate) quantification process must be operationally definable.

Indeed, When considering intelligence, the path of quantification is far from straightforward.

Human intelligence, for example, generally attributed to cognitive ability, is measured by one's ability to perform a given task based on reasoning, quantified by a standardized score, i.e., intelligence quotient (IQ) [2]. Although a single underlying construct, i.e., the general intelligence factor (g), has been used to quantify abstract thinking, theories on the existence of multiple intelligences have been postulated [3]. As a result, scientific consensus on the standardization of measuring human intelligence is still an open question.

Quantification of artificial intelligence (AI) has shown the importance of having an adequate understanding of the quantifiable subject. The quantification of AI has changed in time due to a better understanding of objectives and human intelligence as the main comparing element, creating an iterative loop [4]. Turing test [5] was the first quantification of machine intelligence by evaluating if a human evaluator can reliably distinguish a human and a machine in a natural language conversation. The Turing test has shortcomings as the list of possible expressions of AI has expanded (e.g., problem-solving, decision making, learning [6]), and other tests (such as ImageNets test [4]) have been devised to comply with these expressions.

Although intelligence quantification is already challenging in the biological domain, let's ask an even intriguing yet practically essential question: **could we *measure* embodied intelligence?**

In an attempt to isolate embodied intelligence (EI), let's exercise our brains to define the smallest physical unit (with a finite size and shape, thus embodied), **the intel-unit**, that we recognize and accept as physically intelligent. Isolation of intel-units gives us the first practical tool for measuring EI - by counting.

In the quest for an intel-unit, let's first observe robots that we currently associate with intelligence. In the literature, we are first obstacled by the terms 'smart' or 'intelligent' materials that actually refer to stimuli-responsiveness. As a representative example, would material's swelling (i.e., increase in volume and dimensions) upon exposure to water vapor suffice for intelligent behavior? As an argument in favor, a piece of hygroscopic material displays the highest possible degree of multi- and combined functionality (a cornerstone in soft robotics narrative), as it detects humidity and reacts to it to a defined extent at a molecular level. It appears like a perfect closed-loop control! Yet, treating swelling as an act of computation is barely convincing to anyone.

A higher swelling rate hardly makes one hygroscopic material more intelligent than another. In a similar line, extensive parameters such as weight are hardly any better candidates as intelligence metrics. Yet, sometimes we perceive 'material intelligence,' whereas the connection of this concept to measurable material parameters is not straightforward. The structure in which the material is arranged gets us warmer. Yet, a physical parameter such as fiber density, analogous to transistor count in a microprocessor, barely directly scales with intelligence. If a parameter scales in meters, it probably does not in intelligence. For intelligence comparison, we should look at the program code instead, not trivial in material-encoded robots.

Although anything that could be called a conventional EI robot is still to be established, the most unconventional robots are probably the best to be isolated for exploring the extent of the validity of theoretical assumptions. We consider it safe to assume that **a structure with EI also shares the minimum defining requirements for any robot: sensing, actuation, and execution of code.** Vice versa, the presence of an executable code of any description could define a finite level of intelligence.

IOP Conf. Series: Materials Science and Engineering 1292 (2023) 012002 doi:10.1088/1757-899X/1292/1/012002

2. Case studies

2.1. Case study 1: awns of wild oat

The **awns of wild oat** appear to satisfy all requisites of a robot. As they detach from the plant, the seed-carrying awns are non-living structures that execute a program aimed at better seed dispersal and plantation [7, 8]. The awns react to humidity by twisting their helical structure made of hygroscopic fibers of cellulose. Indeed, the swelling of awns upon exposure to a humid environment and the kinematic amplification of movement by a twisted structure is generally well-understood from a physics and material science point of view at the structural and molecular level. Yet, the microarrangement of linear actuators for joint rotational actuation barely attests to any description of intelligence. The movement per se governs sensing and actuation, but apparently not the code. So, where is intelligence in an awn?

Indeed, when an awn would be observed in an isolated condition, for example, fixed by its base in a humidity chamber, we could easily observe its complex kinematic choreography. The code gets apparently deciphered, but it does not make much sense. However, when exposed to a highly diverse natural environment at the soil surface, the twisting awns interact with the surrounding structures (such as plant debris or roots) so that the likelihood of burying the seed would statistically increase [9]. **Situatedness** defines the mission plan of awn-robots. The program code is next to impossible to decipher without considering the environment. Situatedness appears to be a defining aspect of embodied intelligence, manifested by algorithms that produce environment-dependent results. **The program code for EI entails, per minimum, three interconnected levels: materials + structure + situatedness in an environment.**

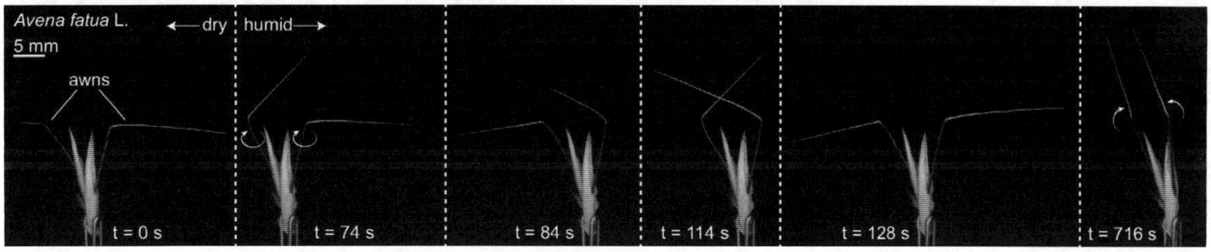

Figure 1. Hygrochoreography of wild oat's awns. Awns made of twisted hygrpscopic fibers respond to humidity change by fast and extensive movement. In isolated conditions, the choreography, although extremely elaborate for its simple structure, is deterministic. Thus, awns fixed on a lamp do not suffice for deciphering of the encoded seed dispersal mission.

One pair of awns, or actually an attached pair of awn-carrying florets (Figure 1), appears a good candidate as an **intel-unit**. Upon disassembly of an awn-robot, all its components would react to stimuli (humidity) (using energy harvested from the environment for movement). Yet, it would break the program code to an undecipherable level. The robot's parts apparently make calculations, yet not in a strategically meaningful way. One pair of awns is **a minimum unit that embodies intelligence.**

The awn-bearing wheat seeds deploy in pairs, allowing the individual awns to temporarily cross-entangle and suddenly release the energy collected during the entanglement phase, enabling fast movements. Such function is not apparent ex-situ but becomes evident when random interactions with the environment necessitate stimuli in a broad frequency spectrum. Equally important is the inter-robot, i.e., between two awns of a seed pair, interaction: the awns do not have any interconnection than a rigid structural link that holds them together. The awns experience a rich interaction mediated by the random interactions at the robot's surface. Consequently, the environment not only contributes to the code and acts **as a randomizer**. The awn's code exemplifies the engagement of intrinsically random interactions for defined action.

Two situated awn pairs (each intel-unit, as suggested above) pose an intriguing perspective. In addition to facilitating random interactions, the environment appears to **interconnect different intel-units for intelligent cooperation**. Thus, for intelligent cooperation, the intel-units may interact internally or externally. In this example, information transfer between intel-units occurs via the robot-environment interface. Randomized interaction between intel-units is much richer and more potent than being strictly 'connected' would allow.

Remarkably, in a wild oat, we could not evidence **intra-robot information transfer** in any more sophisticated form as simple delivery of force through mechanically bonded units, as the same fibers that sense humidity express their swelling response locally. Indeed, we isolated an intelligent agent with only local expression of responsiveness but displaying intelligence as a whole.

The awn's code does not involve decision-making by combining information from multiple inputs. Indeed, a practical embodied computer can be highly robust, well-evidencing Occam's razor concept for the simplest solution to win!

2.2. Case study 2: the 'fortune teller fish'

The awns gave us the first evidence of EI quantification: by counting intel-units. Can we identify and isolate an example for continuous scale as well? Unexpected occurrences may teach us more than established ones!

We note that our example of a continuous-scale EI is familiar to most readers - yet probably unrecognized as such: the commercially available 'novelty' product - "The Fortune Teller Fish" (Figure 2). Under this name, this popular magician requisite dates back at least two centuries. Today, similar products, probably by several manufacturers, are available in numerous retail stores. For those not yet come across one, a "miracle fish" exhibits different morphologies upon exposure to a wet palm. The more sweat on the palm, the more pronounced the morphological effect is. The 'miracle fish' is made from regenerated cellulose - a hygroscopic polymer sodium polyacrylate that bonds and retains water molecules. The swelling and drying cause changes in the shape of the molecules, which in turn bring the particular fish-shaped film to move its 'head', 'tail,' sides or curl up entirely. Remarkably, the product's packaging is accompanied by a **scale** that associates different movements with the participant's temperament, behavior, or mood. For example, held in a palm, the moving tail is for indifference, the moving head indicates jealousy, turning over is interpreted as falsehood, etc. The process is probably slightly empirical to be called a proper 'measurement,' yet an explicit scale provides more than a binary result.

Just like fish in nature needs water to survive, the 'fortune teller fish' needs some water to serve some insights into how to measure embodied intelligence. The polymer (the subject) is not intelligent per se. A sensing and actuating feedback loop is formed upon exposure to a wet surface (i.e., situated in an object, the environment). Through sensing, the perceiving subject and perceived object conclude a primary form of intelligence. Notably, the material of the 'fish' and its specific shape is essential to the code: the thickness and morphological features of higher (the tail) or lower (the head) determine curling patterns. The fish material, shape features, and the situatedness on top of a wet palm, could be seen as an intel-unit.

Intriguingly, instead of a palm, similar curling could be achieved above any humid surface, allowing us to observe the repetitive closed-loop interaction between (in)animate objects, similar to connected intel-units described in the first case study. It looks like the description of its behavior could be readily accepted and even fully described based on simple control theory. Yet, a loop may form as well elsewhere. When the 'fish' is held on a palm of an anxious participant (we can assume a palm more sweaty than in a relaxed state), the curling of the 'fish' visually announces anxiety, thus generating even more moisture on the palm (the environment). In this case, the temperament of the participant dictates the state of the natural environment situated on the palm forming a closed-loop information exchange with the 'fish.'

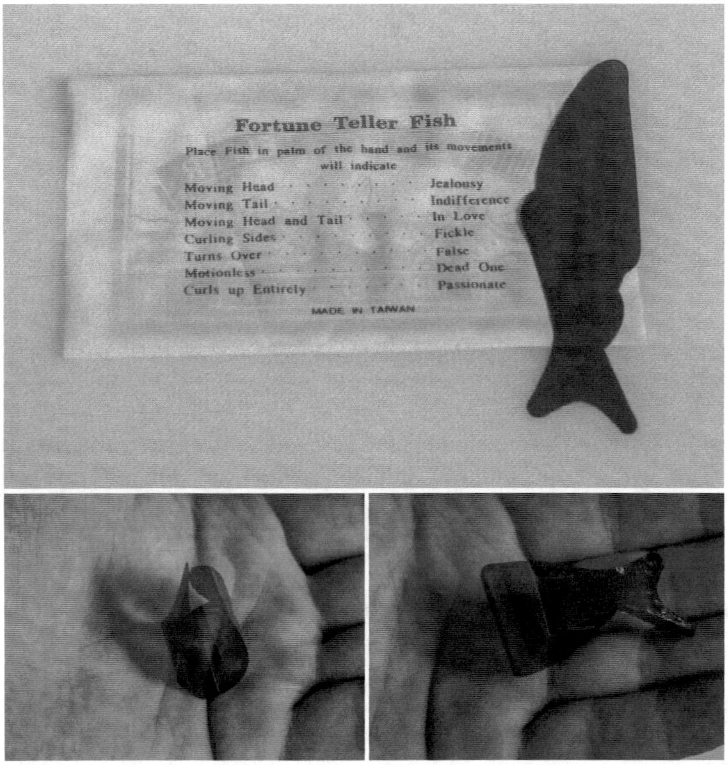

Figure 2. Non-binary scale for EI. The gadget 'Fortune Teller Fish' is accompanied with a non-binary lookup table for interpretation of water-mediated information exchange on a wet palm

The nature of the environment, whether natural or simulated, determines the number of curls the fish would make, which plays an essential role in recognizing the nature of the loop, open or closed.

2.3. Case study 3: a woollen sweater

A woollen sweater is a well-comprehensible everyday item; yet, in our quest for unconventional robotics, let's assume here (provocatively) that a sweater also has the qualities of a robot! Among known robot descriptions, the presumed intelligence of a sweater is safe to be qualified as embodied. Arbitrary at first sight, applying robotic conventions to an unexpected item may bring light to the boundaries of EI.

To understand the nature of its intelligence, let us first zoom into a single woollen fiber - the essential bits being the cuticle and the cortex. The outermost surface of the fiber consists of overlapping cuticle cells that repel water, thanks to a chemically-bonded waxy coating. The cortex comprises two types of cells – ortho-cortical and para-cortical, each with a different chemical composition. The expansion of cells is moisture-dependent. The difference in expansion causes the fiber to 'crimp' or curl, allowing the wool to trap millions of tiny still air pockets, providing insulation. Consequently, a lightly-damped woollen sweater becomes crimper and, thus, warmer. An interplay between humidity-responsive materials and their hierarchical structure gives rise to complex behavior when situated in an environment.

A 'single' fiber on its own probably does not manifest the sweater's embodied intelligence. The cross-interactions between adjacent fibers, the arrangement of entangled fibers into bundles, and the number of loops made thereof already give us sufficient input to quantify the degree of warmth. How is it, however, connected to the goal of intelligence? For example, in coarser fibers, the para-cortical and ortho-cortical cells are arranged more randomly, providing less crimp, hence less warmth. A loosely-knit sweater with finger-sized holes would fail to give warmth to the wearer compared to that of a tightly-knit sweater. In both cases, the sweater still fails to decipher its final goal of intelligence. The presence of and **experience by a wearer** is a central consideration here. Unless coupled with the environment (i.e., in part the wearer's body), the sweater's qualification and quantification of intelligence are inconspicuous. For example, depending on the environment (dry or moist air; calm or sweating), the sweater executes a relevant code, keeping the wearer warm or warmer. Slightly similar to the 'fish' in the previous case study, human emotions and subjective perceptions are a part of the environment descriptions. The environment, in turn, constitutes an essential component of EI, as discussed above.

So, where is the intel-unit of a sweater located? Questioning the uncertainty, one might still argue the sweater as a whole represents a single intel-unit, and most rightfully so. As a single entity, a woollen sweater establishes a specific fiber arrangement and bridges the information exchange between the wearer and the environment. If this entity is cut in half - a piece of woollen cloth would merely display responsiveness to stimuli (humidity) locally by the fiber's inherent properties and arrangement. However, it would fail to decipher its mission of intelligence, a.k.a execution of the program code.

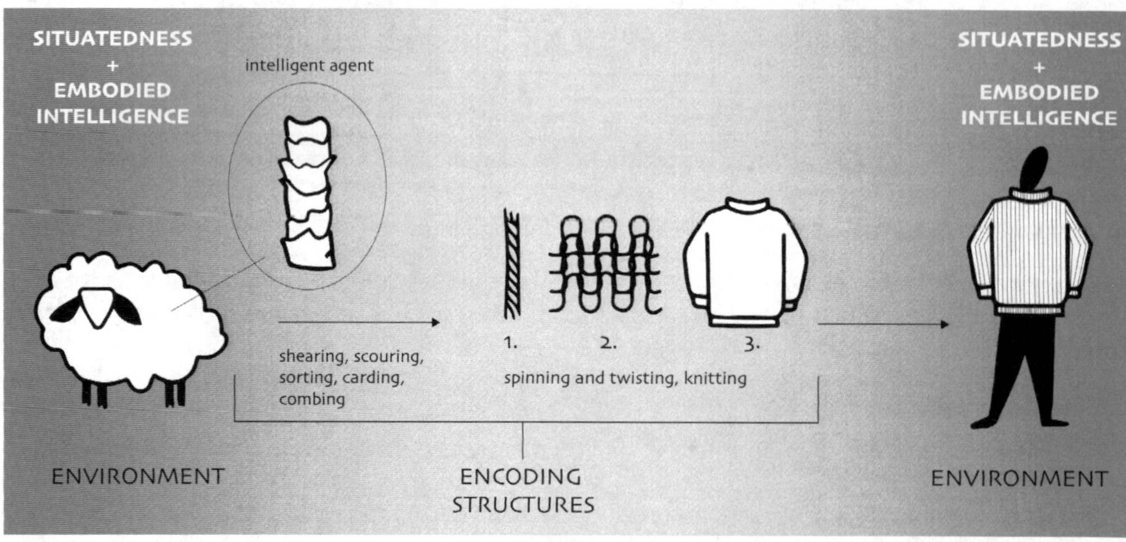

Figure 3. Encoded wool. Reshaping and re-situating a responsive biomaterialwoolas a new intelligent agent. In a new situatedness, the wool macrostructure is changed beyond recognition, yet encodes a remarkably similar program.

The sweater code is written into the material by arranging the thread by knitting and sewing (Figure 3). The hierarchical nature of textiles implies a combination of defined and repeating elementary units for total code formulation. Specific knitting patterns define building blocks yet do not suffice as intel-units. One can practically program any form of code into the sweater to deploy pre-defined functions. However, the style of programming the code (encoding) is, obviously, much different than what is conventionally understood in digital coding [10, 11]. To understand this, let us look into the making of a woollen sweater from a hierarchical perspective. Starting from the origin - wool, while situated in a specific environment, such as on a sheep, expresses material-level intelligence necessary to regulate humidity and keep the sheep warm. Once the fleece is scoured off from the sheep, followed by carding and combing the felt and twisting it into yarn, one can add different levels of 'material intelligence' that define the responsiveness of the material (strength, friction, etc.). Finally, when the fibers are knit into a sweater, a program consisting of loops as mesoscale building blocks is executed. Knitting determines the final set of properties - shape, size, and specific interaction with the wearer. One can alter the code without changing the core material (fiber). Similar to both previous case studies, the material, structure, and environment work in unison in defining the sweater code. By taking into account - the environmental variables, the wearer's body, and the arrangement of fibers to interact and exchange information from and to the wearer, the intended program can be indeed encoded (at least altered) by modifying only the fibers or only the arrangement of fibers. The processing steps applied throughout the wool's journey from sheep to sweater bring changes to its macroscopic structure beyond recognition, yet the reshaped (and resituated!) intelligent agent performs a remarkably similar mission in negotiating the cold weather. A changed situatedness thus may require an unrecognizably different structure to encode a similar behaviour.

Indeed, different wool types do create an empirically felt difference. Well-pronounced in a sweater, the **adaptive properties** can assume a central role in material-level EI. As an observation, the sweater's interaction with the wearer's body allows it to conform to its shape with extensive wearing and learning, manifesting bidirectional, environment-mediated communication between intel-units (one of which being the sweater, the other may be the wearer). The intelligent loop between the wearer and the environment could be fragmented into functional feedback keeping the wearer warm and emotional feedback that may be determined by protective heritage signs, symbols, and colors that could provide emotional support and protection. The fiber properties can measure the former, functional intelligence, for instance, by analyzing the sheep breed and the climatic variables that influence the fiber to be either more adaptive to the climatic fluctuations or not. Additionally, one could measure the yarn and its mechanical properties, quantify the degree of warmth dependent on the knit and seater patterns, and measure felting that occurs with wearing. Traditional craft techniques appear to have definite signs of intuitive EI engagement.

How could we measure a sweater's embodiment of intelligence? In countable intel-units, the previous discussion reached a unity result. Could we quantify the sweater's emotional intelligence as a function of time? Could we measure emotional support expressed by wearing protective colors and heritage symbols [12]? Yet a compiler for implementing specific tasks in a thread is still to be developed.

3. Conclusion

We explored the origin and essence of EI in three case studies (Figure 4) at (or probably exceeding) the consensual boundary of robotics. In all cases, the elementary (bio)material (micro)structure, chemistry, and responsiveness are generally well-accessible to scientists to characterize quantitatively. We intuitively assume some degree of EI in each of the three examples. Yet, even with the complete (mechanical and chemical) blueprint in our hands, we struggle to decipher any meaningful EI code. In two of the latter case studies, the macroscopic arrangement of biomaterial is entirely defined by humans, so the blueprint is original. Consideration of situatedness in and interaction with the environment provided the necessary link. In none of the analyzed cases, we could identify more than one intel-unit, suggesting the environment as a universal relay between individual intel-units. For consistency, it probably makes sense to consider humans as individual intel-units in communication with other intelligent agents, be it even a sweater around the body or a 'fish' on the palm. Randomizing the effect of the environment may reason difficulty in establishing analytical, causal relations between the code features and the product, especially when the outcome is of emotional or otherwise subjective nature. The naturally occurring robot – the wheat awn – confirmed the universal nature of EI and the inability to decipher its code ex-situ. After identifying, a broad and systematic understanding of EI principles enables us to approach design at an entirely new level, functionally combining robotics, biology, and cultural heritage.

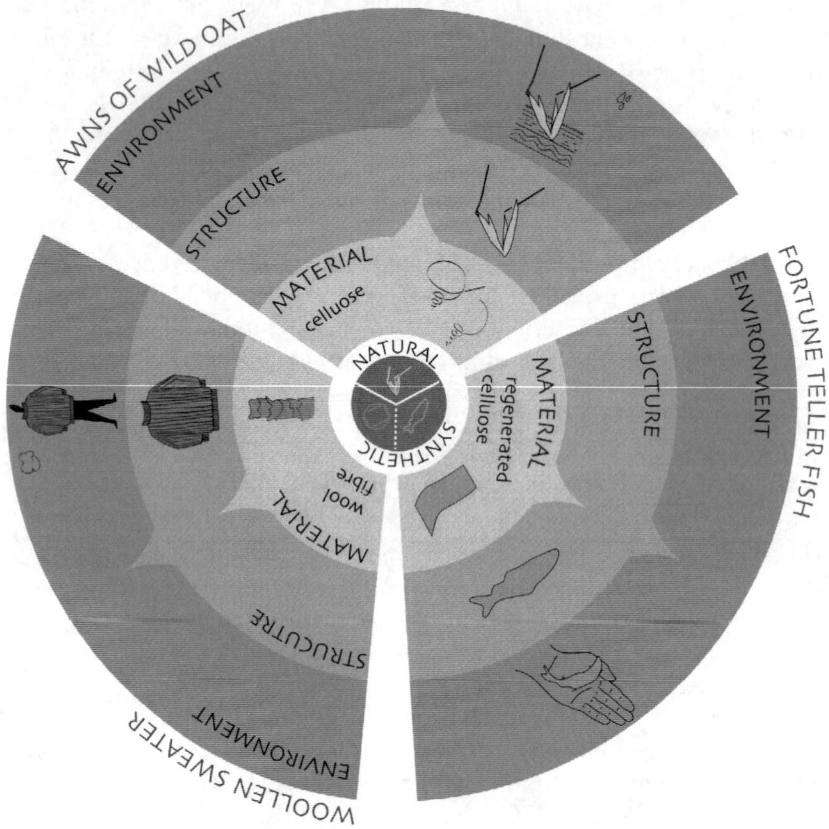

Figure 4. Case studies for EI. Synthetically or naturally encoded biomaterials express their mission when situated in an environment. Three levels - materials, structure, and environment considered together form an intel-unit.

4. Acknowledgments

This research was supported by the Estonian Research Council grant PRG1498, PRG1084, H2020 project TWINNIMS (Grant agreement 857263), and the Estonian Centre of Excellence in ICT Research.

5. References

[1] Wilks S S 1961 *Isis* **52** 135–142
[2] Gottfredson L S 1997 *Intelligence* **24** 13–23
[3] Gardner H E 2000 *Intelligence reframed: Multiple intelligences for the 21st century* (Hachette UK)

IOP Conf. Series: Materials Science and Engineering 1292 (2023) 012002 doi:10.1088/1757-899X/1292/1/012002

[4] Li L, Lin Y L, Zheng N N, Wang F Y, Liu Y, Cao D, Wang K and Huang W L 2018 *Artificial Intelligence Review* **50** 441–465
[5] TURING A M 1950 *Mind* **LIX** 433–460 ISSN 0026-4423 URL https://doi.org/10.1093/mind/LIX.236.433
[6] Back T, Fogel D B and Michalewicz Z 1997 *Handbook of Evolutionary Computation* 1st ed (GBR: IOP Publishing Ltd.) ISBN 0750303921
[7] Lindtner T, Uzan A Y, Eder M, Bar-On B and Elbaum R 2021 *Acta Biomaterialia* **135** 483–492
[8] Luthfikasari R, Patil T V, Patel D K, Dutta S D, Ganguly K, Espinal M M and Lim K T 2022 *Small* **18** 2201417
[9] Mancuso S 2018 *The revolutionary genius of plants: a new understanding of plant intelligence and behavior* (Simon and Schuster)
[10] Underwood J and Zilka L 2022 *Journal of Textile Design Research and Practice* **10** 1–19
[11] Li X 2022 *Random beauty - a curated journey to encounter randomness in knitting* Master's thesis Aalto University. School of Arts, Design and Architecture URL http://urn.fi/URN:NBN:fi:aalto-202205022980
[12] Jones S 2022 *TEXTILE* **0** 1–13

A Task-to-Intelligence Mapping: When is embodied intelligence worth designing?

Aja Carter, Wei-Hsi Chen, Shivangi Misra, Cynthia Sung*

* Authors listed alphabetically

General Robotics, Automation, Sensing & Perception (GRASP) Laboratory, University of Pennsylvania, Philadelphia, PA, USA

E-mail: {caja, weicc, shivangi, crsung}@seas.upenn.edu

Abstract. While there has been much work in the space of embodied intelligence, we as a field have struggled to define what exactly embodied intelligence is and how it should be used. In this paper, we propose that there are multiple types of embodied intelligence, and that these different types of embodied intelligence are suited to different types of tasks. We introduce a method for classifying tasks according to their objective and occurrence, and we describe how existing work in embodied intelligence fits into this framework. We hope that this proposed framework will initiate a discussion to more formally think about the role that embodied intelligence plays and the value that it brings to engineering and robotics.

1. The Role of Embodied Intelligence

In the last several years, there has been a large growth in the use of the term "embodied intelligence" [1] and, at the same time, a debate over what exactly embodied intelligence is. Research on "physically intelligent," "morphologically intelligent," and "embodied" systems display a large spectrum of autonomy, from systems with behaviors "hard-coded" into the physical structure [2] to systems that respond to environmental stimuli in ways that approximate computational planning strategies [3]. The range of definitions have been just as broad, including attempts such as

> "The behavior of a system is not merely the outcome of some internal control structure... but it is also shaped by the environment in which the system is physically embedded" (Pfeifer, Lungarella, Iida, 2012 [4])

> "The exchange of energy and information between a machine and its environment toward some specified set of goals" (Koditschek, 2021 [5])

> "Tight coupling between the body and the brain" (Sitti, 2021 [6])

These definitions emphasize to varying degrees the role of the physical. The first definition, for example, describes for the most part a one-way relationship where the body results in certain types of interactions with the physical world, and the role of intelligence is to accommodate and account for those interactions in determining what actions, behaviors, and adaptions occur. In contrast, Sitti's definition implies a two-way relationship where decision-making is shared between a physical structure and a more abstract computational one. And definitions such as those proposed by Koditschek acknowledge both options, requiring a physical body and a decision maker, but not requiring a specific interaction between them. A common thread, however, is that intelligence constitutes a system's ability to *adapt* and respond to stimuli in the environment [7]. Thus we see that decade ago, notions of including physics, actuator and sensor models, and environment interactions in control algorithms produced powerful systems capable of reactive behavioral adaptation [8]. More recently, the explosion of soft robotics, compliant mechanisms, and minimalist design has further demonstrated how a well-designed body can simplify computation and reduce energetic cost [9]. Witnessing this broad range of capabilities that we have managed to achieve, as we look forward to the future role the idea of embodied intelligence will play in engineering, the major question before us seems to be not what embodied intelligence is, but rather where these different types of intelligence most shine.

1.1. Costs and Rewards for Embodied Intelligence

The problem of designing a system with embodied intelligence is, at its core, a coupled optimization problem balancing the system's physical body and its computational brain. In determining how this balance should be achieved, it is therefore important to consider the trade-offs between the value and cost to the system. In this sense, we consider the idea of embodied intelligence always in the context of a particular task that the system aims to accomplish.

For any task, higher levels of intelligence bring greater flexibility and the potential for more effective or efficient task performance. The ability to adapt and learn new behaviors to approach a task more effectively, to change bulk structure to fit task constraints, or to adjust sensorimotor capabilities to enhance information flow and processing all provide a system with options to improve compared to a nonadaptive system. At the same time, the computational and physical structure required to implement such capabilities incurs a cost. Junge et al. [10] identify, for example, cost of design (actuators, mass, volume), cost of control complexity (tuning parameters), and cost of manufacturing and assembly as a few design-time factors to balance. Further execution-time factors, such as additional energy required for task performance when the number of actuators is too high or too low, computational cost associated with planning and control, and memory and bandwidth required to execute desired behaviors, must also be considered. These factors must be balanced against each other in the context of the desired task to produce an overall efficient design.

This strategy of prioritized down-selection is evidenced in biological systems under Darwinian evolution, where physical traits that enhance an organism's capability to accomplish tasks within one's life (i.e., Whole-Organism Performance Capacity) would be selected for and passed on to the next generation (note: many forms may satisfy a performance need) [11, 12]. For example, locomotion that is directly related to an animal's ability to capture food, avoid becoming food, and disperse into new environments is often a large driver in evolutionary pressures in biological systems [13]. The outside world that biological systems inhabit is a complex, non-linear space. Systems inhabiting this space exhibit complicating features such as joint redundancies, time delays between external excitation and internal response, and environmental noise [14, 15, 16]. The nervous system alone cannot successfully enact wanted behaviors with all this internal and external complexity [17]. Thus, within biological systems, the interactions between the nervous system, the muscular (and within vertebrates skeletal) system, and the environment, we find emerging adaptive (in some literature "embodied intelligence") behavior [15, 18]. Without embodied intelligence, biological systems could not successfully respond to tasks inherent to their survival, which would increase individual risk and reduce species fitness.

Turning to engineered systems, we are inspired to understand whether similar trends appear in robotics.

IOP Conf. Series: Materials Science and Engineering 1292 (2023) 012003 doi:10.1088/1757-899X/1292/1/012003

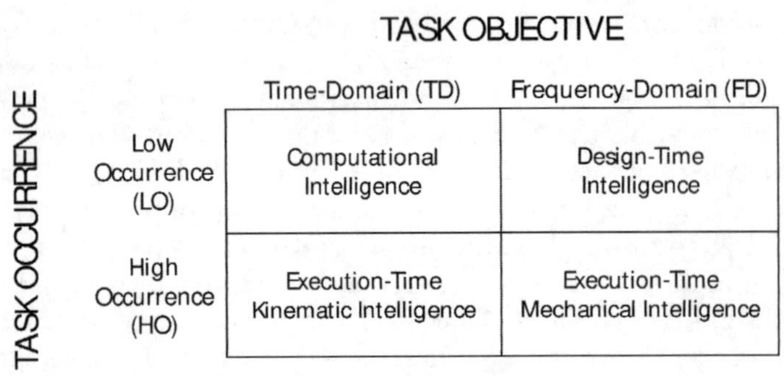

Figure 1. Types of embodied intelligence found in the literature, categorized by potential task applications. Tasks are broken down along task objective and task occurrence axes, which affect the value that different modes of intelligence bring to the system.

1.2. Task-Intelligence Framework for Design

We thus propose a taxonomy of tasks and of embodied intelligence, which we hope will provide a starting point for understanding the relationship between the two and inform future engineering design decisions.

1.2.1. Types of Tasks We propose that tasks can be classified into four different categories along two axes: task objective and task occurrence (Figure 1).

Along the *objectives* axis, we observe that task specifications are typically cast either as a *time-domain* (TD) objective or as a *frequency-domain* (FD) objective. Time-domain objectives are those tasks where the primary aim is to achieve a particular trajectory or a state at a given point in time. For example, most manipulation tasks involving moving an object from one location to another fall into this category. Task specifications involving minimizing time to completion or total path distance would also be of the time-domain type. Frequency-domain objectives are those tasks where the primary aim is to achieve a particular frequency response. For example, cancellation of noise or vibrations would fall into this category. Task specifications involving minimizing energy consumption would also generally be frequency-domain tasks. While in most cases, the complete task specification will include elements from both categories (e.g., minimize time to completion subject to a total energy budget), typically one will dominate. For example, we would call the previous time minimization example a TD task. We also note that periodic trajectories (e.g., repeated pick-and-place tasks), although performed at some frequency, would under these definitions be considered TD tasks unless the frequency of the trajectory was one of the objectives.

Along the *occurrence* axis, we categorize tasks by how often they are encountered by the system. *High occurrence* (HO) tasks are encountered repeatedly over the course of the system's life, while *low occurrence* (LO) tasks are encountered only infrequently. The task occurrence dictates the total cost that completing the task incurs to the system. For the same task on the same system, a higher occurrence will incur a higher task-specific cost since the task must be completed multiple times.

1.2.2. Task-to-Intelligence Mappings We suggest that each of these axes influences the value that different modes of intelligence bring to the system.

In particular, the task objective informs what design characteristics have the greatest impact on the task performance. TD tasks relate to the system's state or configuration over time and will be most strongly influenced by design geometry and kinematic structure. In contrast, FD tasks relate to the system's dynamical response and will be most strongly influenced by design mechanics.

At the same time, the task occurrence informs to what extent physical changes in the body of the system will influence task performance. Based on the premise that reactive computation is cheap relative to physical change, LO tasks where tasks change infrequently should be addressed primarily through computational planning and control. In contrast, HO tasks occur with great enough frequency that the benefits of execution-time morphological change are worth the additional costs of actuation and control complexity. Note that, as observed by [19], physically-enacted embodied intelligence presents a value only when the task takes sufficient time to significantly reduce the cost of long-term computational control and improve overall efficiency. We assume for all cases that this threshold has been reached and thus HO tasks, while frequently occurring, are not necessarily short in duration.

1.2.3. Types of Embodied Intelligence Based on this reasoning, we observe in the literature four major types of embodied intelligence, constituting different degrees of computational versus bodily adaptation exhibited by the system and occupying the four quadrants of the task space. These are:

- Computational intelligence: Computationally-implemented embodied intelligence resides in a physical body but, for the most part, does not require a particular body response to complete a task. Instead, planning and control occurs at the computational level and accounts for body dynamics. We propose that this strategy is most suited to TD-LO tasks, that is, tasks that must be completed infrequently and primarily consist of the ensuring that the system reaches a particular configuration or follows a particular trajectory.
- Design-time intelligence: In design-time intelligence, also sometimes referred to as "physical inteligence" [6], elements of the task objective are designed into the robot body, simplifying control complexity or enabling the system to complete tasks open-loop with no feedback at all. We propose that this strategy is most suited to FD-LO tasks, that is, tasks where the goal is to achieve a particular steady-state behavior, which changes infrequently over the course of the system's lifetime.
- Execution-time kinematic intelligence: Execution-time kinematic intelligence implies kinematic change. Self-reconfigurable systems focusing on shape, geometry, and topology fall into this category. We propose that this strategy is most suited to TD-HO tasks, where a number of specific trajectories or configurations must be achieved repeatedly in time.
- Execution-time mechanical intelligence: Execution-time mechanical intelligence refers to self-reconfigurable systems with the ability to change their fundamental mechanics, including mass distribution, stiffness, damping, or other properties that affect their dynamical response. We propose that this strategy is most suited to FD-HO tasks, particular cases where the ultimate goal is to reduce energy consumption over a number of repeatedly occurring tasks.

In the remainder of this paper, we consider the individual quadrants of the task grid, providing examples of existing robotic systems that address these tasks using different types of embodied intelligence. We focus primarily on physically-implemented embodied intelligence – that is, on design-time intelligence, execution-time kinematic intelligence, and execution-time mechanical intelligence – since there is already a great deal of work in the space of computationally-implemented embodied intelligence and refer the interested reader to [20] for more information on this last type.

2. Type FD-LO: Design-Time Intelligence

The frequency domain, low occurrence quadrant involves studying tasks where a useful frequency response or steady state behavior is required, and the task specification changes only rarely. For example, a change in environment is one of many factors that could entail a shift in task objective. If it happens that the robot is designed with the sole intention to operate without any qualitative variation in movement, with changes in environmental features occurring rarely in its lifetime, then the task domain in consideration would belong to the FD-LO quadrant. One way to think about FD-LO is that these tasks require "specialists." For these tasks, it would be reasonable to optimize the design of the system for a sole task with an associated natural frequency [21], thus producing design-time intelligence.

One common example of this category is soft swimmers. For example, a flexible bionic flipper was optimized in [22] to generate thrust in a periodic paddling motion by changing the flipper flexibility. The authors found that rigid flippers led to unstable motion while flexible flippers, which were designed to be similar in compliance to real sea lion flippers, were much more effective in propelling the underwater vehicle forward. Similarly, for undulatory motion using segment-wise deformation to mimic serpentine motion [23], the natural dynamics of the compliant robot body was harnessed in half of the actuation cycle to generate sinusoidal waves of deformation and simplify computational control. Since segment stiffness and segment coupling stiffness has a large affect on the natural frequency of motion in segmented robots [24], design-time optimization results in large improvements in energy efficiency when the desired frequency is known.

Similar gains have been achieved in other systems. For example, using different material properties to achieve different mechanics in the C-shape leg of a RHex machine changes the robot's gait and its ability to traverse different types of terrain [25]. In our own work, we have also demonstrated that programmable stiffness achieved through origami [26] can be used to control energy storage in a hopping [27] or jumping [28] task. These types of metamaterial are shown to be effective power cascading devices [29] and are easy to design and implement, with very high resilience, as long as the intended task and environmental conditions are known.

3. Type TD-HO: Execution-Time Kinematic Intelligence

The time-domain, high occurrence quadrant involves studying tasks where the robot's shape or mechanism is important over some duration, and the task must be performed very frequently throughout the lifespan of the system. These tasks often require a specific morphological design [30]. For example, mobile robots with wheels are efficient when navigating over smooth terrains. However, on uneven surfaces, the cost of locomotion with wheels becomes high (and sometimes infinite, unable to overcome the obstacles at all). This is when having a second set of kinematic type, or mode, will be helpful, such as having legs or morphing wheels. Ideally, one can incorporate all types of kinematic mechanisms into a single robot. However, the cost of this strategy is high due to the volume and mass of the additional actuators, the limited strength of materials, and financial constraints. Thus, embedding a specific morphological change in a robot design is only cost-effective when the corresponding task is high in occurrence.

As one example, we previously built TurboQuad [31], a robot with the ability to switch between legged and wheeled modes with one set of actuators. The energy consumptions of these three modes are evaluated using an index called the "specific resistance", defined as the averaged power consumption per weight over the averaged forward speed [32]. Overall, the specific resistance of the robot is significantly lower in the wheeled mode compared to the legged mode under every speed, thus is the recommended mode of operation on even terrain. However, in the wheeled mode, the robot cannot climb stairs and overcome large obstacles, resulting in an infinite cost. Thus when steering the robot from an uneven terrain to an even terrain, TurboQuad's morphology changes from wheeled to legged, demonstrating execution-time kinematic intelligence.

In addition to locomotion, kinematically changing the shape of the system can enhance the robot's workspace, adding other functionality such as transportability, protection, and viable transmission [33]. Similar shape changes have been achieved in manipulation, for example, where changing finger geometries results in a system that can transition between different grasps repeatably [34] or perform reliable perching in aerial vehicles [35]. We observe multiple methods of implementing shape change, such as

(i) dedicated actuators. These types of robot change shape using an additional set of actuator on top of the driving actuator. One example is the Quattroped [36], a leg-wheel transformable robot that changes "limb" morphology with the help of a servo motor.
(ii) shared actuators. These types of robot use the same pair of actuators for morphing and the main task. The previously described Turboquad is of this type.
(iii) a control sequence. These types of robot use a particular sequence of motions to reconfigure. Examples include using different controlling jerks for manipulation [34] or momentum in flight [37].
(iv) connection and disconnection. These types of robot change topology by disconnecting and reconnecting parts. Examples include self-reconfigurable modular robots [38] and trusses [39].

In each of these examples, shape change enables the robotic system to access regions of the task space inaccessible to any one of their individual morphologies, but increases their flexibility at the cost of additional design or control complexity.

4. Type FD-HO: Execution-Time Mechanical Intelligence

The frequency-domain, high occurrence quadrant involves studying tasks where a robot's frequency response must change, and such an action must be performed multiple times throughout the lifespan of the robot, often with qualitative changes in behavior. Locomotive gaits (walking, hopping, swimming, flying), as well as vibration damping, lie in the sphere of frequency domain or cyclic actions. For these systems, their stability and efficiency stem from the ability to vary the mechanical properties (such as stiffness, damping, mass, and shape) of their constituent parts according to task requirements [13]. In biological systems, this ability is achieved locally through the neural system as control of low-level mechanics [4]. In robots, this strategy corresponds to hierarchical control, where whole-body motions are stimulated as oscillations via central pattern generators (CPGs) and environmental adaptation is delegated to lower-level mechanics (bodily) change.

In this space, researchers have produced a wealth of new materials and actuators capable of changing mechanical properties [40, 41, 42]. By taking advantage of stiffness-tuning capabilities enabled by these actuators, robotic systems have demonstrated improved energetic efficiency in swimming [43] and hopping [44]. For example, [19] introduces the very interesting concept of a control-morphology space where the authors lay special emphasis on the variability of mechanics (joint stiffness) of their eel-like robot to simplify the control problem, bringing more efficiency to the robots movements or effecting smooth transitions between two qualitatively different oscillatory behaviors. These systems can match the compliance of humans for safe manipulation in prosthetics and rehabilitative therapy [45]. We have previously also shown that these mechanics changes can be computed locally within the body itself without the need for a centralized planner [46]. Other mechanics changes such as tunable damping or added transmission systems can additionally simplify oscillatory control [47] and aid in disturbance rejection [48].

Along with the great performance benefits witnessed by execution-time mechanics intelligence, this type of embodied intelligence presents perhaps the greatest integration cost out of all the types, since combining heterogeneous components into an integrated functional system requires tight coupling between materials, mechanisms, sensing, computation, and control, both at the local and the global level [49]. Compared to the two other types described previously, work in this space is relatively limited, although we expect to see a large growth along with our rapidly increasing technological capabilities.

5. Conclusion

We have presented here a framework for thinking about embodied intelligence and how the level at which intelligence is designed into an engineered system relates to a desired task. We have proposed that tasks can be categorized into four quadrants and identified for each quadrant a type of intelligence that would suit the task type. Our framework is based on examples found in the literature, but we acknowledge that a full analysis is yet incomplete. The concept of embodied intelligence is a rich area for exploration, and we have already seen many incredible technological advancements. However, in the end, all design is an optimization between the value to the system and the cost of implementation, and thus our ability to characterize the benefits of embodied intelligence have been hampered by the lack of concrete and detailed efforts to analyze the cost and reward structure of these existing works. We believe that for future research, this type of analysis will bring new insights, expose existing gaps, and open new possibilities in the design of coupled brain-body systems. Now is the time that we should step back and think carefully about when and where different types of embodied intelligence shine, in order to inform future steps in our field that can produce real impact.

Acknowledgments

This work was supported in part by the National Science Foundation under Grant No. 1845339 and 2121887, and by the Army Research Office under the SLICE Multidisciplinary University Research Initiatives Program Grant #W911NF1810327. A.C. was supported by the University of Pennsylvania through the Vice Provost Fellowship Program. We are also grateful to Jessica Weakly, Gabriel Unger, Erynn H. Johnson, and Daniel Feshbach for helpful feedback and discussions.

References

[1] Hadi Sadati S M, EiDiwiny M, Nurzaman S, Iida F and Nanayakkara T 2021 *Proceedings of the Embodied Intelligence Conference*
[2] Jin L, Forte A E, Deng B, Rafsanjani A and Bertoldi K 2020 *Advanced Materials* **32** 2001863
[3] Zhao Y, Chi Y, Hong Y, Li Y, Yang S and Yin J 2022 *Proceedings of the National Academy of Sciences* **119** e2200265119
[4] Pfeifer R, Lungarella M and Iida F 2012 *Communications of the ACM* **55** 76–87
[5] Koditschek D E 2021 *Annual Review of Control, Robotics, and Autonomous Systems* **4** 1–33
[6] Sitti M 2021 *Extreme Mechanics Letters* **46** 101340
[7] Sternberg R J 2000 *Handbook of intelligence* (Cambridge University Press)
[8] Pfeifer R and Iida F 2004 *Embodied artificial intelligence* 1–26
[9] Rus D and Tolley M T 2015 *Nature* **521** 467–475
[10] Junge K and Hughes J 2021 *Proceedings of the Embodied Intelligence Conference*
[11] Wainwright P C 2007 *Annual Review of Ecology, Evolution, and Systematics* 381–401
[12] Husak J F and Lailvaux S P 2017 *Integrative and comparative biology* **57** 333–343
[13] Dickinson M H, Farley C T, Full R J, Koehl M, Kram R and Lehman S 2000 *science* **288** 100–106
[14] Franklin D W and Wolpert D M 2011 *Neuron* **72** 425–442
[15] Chiel H J and Beer R D 1997 *Trends in neurosciences* **20** 553–557
[16] Hochner B 2012 *Current biology* **22** R887–R892
[17] Bernstein N 1966 *The co-ordination and regulation of movements*
[18] Chiel H J, Ting L H, Ekeberg Ö and Hartmann M J 2009 *Journal of Neuroscience* **29** 12807–12814
[19] Garrad M, Rossiter J and Hauser H 2018 *IEEE Robotics and Automation Letters* **3** 2056–2062
[20] Pfeifer R and Scheier C 2001 *Understanding Intelligence* (MIT press)

[21] Xu F and Wang H 2021 *IEEE/CAA Journal of Automatica Sinica* **8** 1500–1522
[22] Liu Y, Li H, Deng S, Wang S, Liu S and Wang Z 2022 *IEEE/ASME Transactions on Mechatronics* 1–11
[23] Onal C D and Rus D 2013 *Bioinspiration & Biomimetics* **8** 026003
[24] Daltorio K A, Boxerbaum A S, Horchler A D, Shaw K M, Chiel H J and Quinn R D 2013 *Bioinspiration & biomimetics* **8**
[25] Galloway K C, Clark J E and Koditschek D E 2013 *Journal of Mechanisms and Robotics* **5**
[26] Chen W H, Misra S, Gao Y, Lee Y J, Koditschek D E, Yang S and Sung C R 2020 *IEEE Robotics and Automation Letters* **5** 2131–2137
[27] Chen W H, Misra S, Caporale J D, Koditschek D E, Yang S and Sung C R 2020 *2020 3rd IEEE International Conference on Soft Robotics (RoboSoft)* (IEEE) pp 373–380
[28] Carlson J, Friedman J, Kim C and Sung C 2020 *2020 IEEE International Conference on Robotics and Automation (ICRA)* (IEEE) pp 10089–10095
[29] Ilton M, Bhamla M S, Ma X, Cox S M, Fitchett L L, Kim Y, Koh J s, Krishnamurthy D, Kuo C Y, Temel F Z *et al.* 2018 *Science* **360** eaao1082
[30] Russo M and Ceccarelli M 2020 *Robotics* **9** 32
[31] Chen W H, Lin H S, Lin Y M and Lin P C 2017 *IEEE Transactions on Robotics* **33** 1025–1040
[32] Gabrielli G 1950 *Mech. Eng.(ASME)* **72** 775–781
[33] Mintchev S and Floreano D 2016 *IEEE Robotics & Automation Magazine* **23** 42–54
[34] Liu C, Wohlever S J, Ou M B, Padir T and Felton S M 2021 *IEEE Transactions on Robotics* **38** 491–506
[35] Doyle C E, Bird J J, Isom T A, Kallman J C, Bareiss D F, Dunlop D J, King R J, Abbott J J and Minor M A 2012 *IEEE/ASME Transactions On Mechatronics* **18** 506–517
[36] Chen S C, Huang K J, Chen W H, Shen S Y, Li C H and Lin P C 2013 *IEEE/ASME Transactions On Mechatronics* **19** 730–742
[37] Li X, McWilliams J, Li M, Sung C and Jiang C 2021 *2021 IEEE International Conference on Robotics and Automation (ICRA)* (IEEE) pp 7107–7113
[38] Liu C, Lin Q, Kim H and Yim M 2021 *arXiv preprint arXiv:2104.00800*
[39] Spinos A, Carroll D, Kientz T and Yim M 2021 *Journal of Mechanisms and Robotics* **13**
[40] Manti M, Cacucciolo V and Cianchetti M 2016 *IEEE Robotics & Automation Magazine* **23** 93–106
[41] Iida F and Laschi C 2011 *Procedia Computer Science* **7** 99–102 ISSN 1877-0509 proceedings of the 2nd European Future Technologies Conference and Exhibition 2011 (FET 11)
[42] Vanderborght B, Albu-Schaeffer A, Bicchi A, Burdet E, Caldwell D, Carloni R, Catalano M, Eiberger O, Friedl W, Ganesh G, Garabini M, Grebenstein M, Grioli G, Haddadin S, Hoppner H, Jafari A, Laffranchi M, Lefeber D, Petit F, Stramigioli S, Tsagarakis N, Van Damme M, Van Ham R, Visser L and Wolf S 2013 *Robotics and Autonomous Systems* **61** 1601–1614 ISSN 0921-8890
[43] Quinn D and Lauder G 2021 *Bioinspiration & Biomimetics* **17** 011002
[44] Nakanishi J, Radulescu A, Braun D and S V 2017 *Auton Robot 41* 273–291
[45] Nelson C A, Nouaille L and Poisson G 2020 *Mechanism and Machine Theory* **150** 103862 ISSN 0094-114X
[46] Misra S and Sung C 2022 *IEEE International Conference on Robotics and Automation (ICRA)*
[47] Sukhnandan R, Dai K and Webster–Wood V 2022 *2022 International Conference on Robotics and Automation (ICRA)* pp 11445–11451
[48] Roderick W R, Cutkosky M R and Lentink D 2021 *Science Robotics* **6** eabj7562
[49] Mengüç Y, Correll N, Kramer R and Paik J 2017 *Science Robotics* **2** eaar4527

IOP Conf. Series: Materials Science and Engineering 1292 (2023) 012004 doi:10.1088/1757-899X/1292/1/012004

Quantifying embodiment towards building better robots based on muscle-driven models

Vijay Chandiramani **Andrew T. Conn** **Helmut Hauser**

E-mail: ld19487@bristol.ac.uk a.conn@bristol.ac.uk helmut.hauser@bristol.ac.uk
Bristol Robotics Laboratory (BRL), T Block Frenchay Campus, Bristol, BS16 1QY

Abstract. In recent years, researchers have investigated different methods to quantify embodiment for a variety of robotic systems including robotic arms, grippers and legged robots. This paper will discuss some of these methods, focusing on their potential contribution to designing robotic systems based on muscle-driven models. We start with the definition of embodiment based on the relational dynamics between the system and its environments by drawing upon the idea of mutual perturbation and structural coupling between the two. We will discuss how such an understanding can provide potential approaches to quantify embodiment. These includes two information-theoretic measures which are particularly suitable for muscle-driven models. The two methods are based on (i) comparing the controller and behaviour complexity and (ii) Conditional Mutual Information, which compares the difference in distribution of the action conditional on the actuated state and purely on its morphological properties. These methods were used on muscle-driven, biologically realistic hopping models to quantify embodiment at different stages of the hopping gait. The results clearly demonstrate the contribution of morphology of the muscle fibers at different points in the hopping cycle. Furthermore, these methods have been used in latter studies to measure the contribution of embodiment across different levels in a hierarchical control system of a neuro-musculoskeletal model and also to quantify the effects of information cost during various actions in a muscle-driven robotic system. We discuss the practical implications as well as limitations and the future work in the application of these quantification methods.

1. Introduction

Traditionally, robotic systems are dominated by control laws that are based on kinematics and dynamics. However, recently, the importance of the embodiment for effective operation of intelligent machines has been recognized. Historically, the motivation for studying embodiment comes from studying a wide range of biological systems [1] which are exhibiting remarkable performances. Examples include systems such as insects [2], dogs and other mammals [3] [4]. To illustrate embodiment in these biological systems of varying complexity, we start with the simplest one, a single-cell organism, e.g., the bacterium E.Coli, see Figure 1. This remarkable organism does not possess a central nervous system and is simply propelled by its flagella to converge on nutrient sources. One could argue that only the embodiment of the organism is driving its locomotion behaviour. A much more complex example is the human body, which comprises hundreds of muscle-tendon complexes (MTC) [5], forming a complex neuro-muscular system. While neuronal structures are responsible for initiation and control of movements, it has been pointed out that the physical properties of the MTCs such as thickness, tendon stiffness, etc. contribute to the overall control as well [6]. The muscles provide low-level reflexes [7] [8]

Figure 1: A free swimming E.Coli bacterium with flagella clearly visible (photo by CDC on Unsplash)

capable of stabilising the system against external perturbations, such as sudden changes in the terrain or obstacles [9]. The above examples suggest that embodiment plays an important role in the control of muscle driven movement. To study the extent of embodiment in muscle-driven systems quantitatively, a suitable measure is required. In recent years, various information-theoretic methods have been proposed by a number of researchers, including by Ghazi-Zehadi et al. [10], Polani et al. [11], and Rückert and Neumann [12]. To the best of our knowledge, according to our literature review, the methods by Ghazi-Zehadi et al. [10] served as the main basis for a range of work on the quantification of in embodiment in muscle-driven models, e.g., [13] [14]. Due to its central role, the aim of this paper is to review the methods by Ghazi-Zehadi et al. and to provide a critical examination in their use on non-trivial, and realistic scenarios, by simulating a hopping gait on muscle-driven model[15]. We also provide an overview of subsequent research using these models on more complex models.

In the next section, we start with a conceptual understanding of embodiment, in the ontological context of robotics, primarily based on research by Quick et al. [16]. This is followed by comparing two main information-theoretic quantification methods. We discuss their applications by Haeufle et al. [14] in quantifying the extent of embodiment in various levels of hierarchy in a neuro-musculoskeletal control system, and by Polani et al. [13] to understand the impact of embodiment on informational costs.

2. Embodiment in the context of robotics

There are several studies that have defined and examined different uses and interpretations of the notion of embodiment, from its theoretical grounding in philosophy [17], to its application in computer science and robotics [18] [6]. Some of the seminal research by Brooks et al. on intelligence [19] has suggested that embodiment is almost invariably analysed in terms of the relational dynamics that exist between a system and its environment. In their research, the authors have described robots, which implement an abstract architecture, referred to as the *subsumption architecture*. This architecture embodies the fundamental ideas of decomposition into layers of task achieving behaviors, and composition through addressing the noisiness and unpredictability associated with the real world. However, no formal definition was provided to allow any pathways towards the quantification of embodiment. On the other hand, more recently, Quick et al. [16] proposed a framework to understand and exploit embodiment. It is based on the idea of *structured coupling*, which is postulated in the research by Maturana and Varela [1], who introduced it as a process that occurs when two structurally plastic systems repeatedly perturb one another in a non-destructive fashion over a period of time, leading to a structural 'fit' between the two systems. In the context of embodiment, a plastic system is one that can be affected by external events and perturbed by another system. Building on this idea of *structural coupling*, Quick et al. [16] proposed a definition of embodiment:

Definition 2.1 (Embodiment). *A system **X** is embodied in an environment E if perturbatory channels exist between the two. That is, **X** is embodied in **E** if for every time **t** at which both **X** and **E** exist, some subset of **E**'s possible states have the capacity to perturb **X**'s state, and some subset of **X**'s possible states have the capacity to perturb **E**'s state.*

This definition provides also a sound basis for articulating a system-environment relationship within robotics. Taking the example of a legged robot in a walking gait on sandy surface, the

Figure 2: A robot with a walking gait and its foot showing contact and friction forces as perturbations in the body and environment [4][20]

environment (sand) will provide a normal (contact) force and friction as perturbation to the foot (body) (see Figure 2). Conversely, the environment (sand) will be perturbed by the weight of the body with the shape of its foot, as the robot walks. A muddy surface will have different levels of perturbations due to different contact forces, friction, and compliance as compared to a sandy surface. In the next few sections, we look at methods to quantify embodiment based on this relationship between the body and the environment.

3. Quantifying Embodiment

In this section, we describe 4 properties that can potentially be used to quantify embodiment, as relevant to muscle-driven models. Quick et al. [16] introduced 2 of them, and Ghazi-Zehadi et al. [10] proposed the remaining 2. To explain these properties, we will use examples from the context of a legged robot. Specifically, we will look at the robot Solo8 (Figure 3), which is an open-source, impedance-controlled robot with 8 degrees of freedom (DoF), developed by the Open Dynamic Robot Initiative (ODRI) [21] [22]. This robot is impedance controlled. As reminder for the reader, the concept of *Impedance* $Z(s)$ is central to this robot and it is defined as the dynamic relation between the position perturbations and forces [23][24]. The underlying principle of impedance control is to define a desired end-effector dynamic behavior in response to unexpected disturbance forces. For the Solo8, an impedance controller regulates the stiffness and damping of the foot with respect to the hip, as per the control law [1] in Equation 1:

$$\tau = \mathbf{J^T}((\mathbf{D_d}(\dot{\mathbf{X}} - \dot{\mathbf{X_d}}) + \mathbf{K_d}(\mathbf{X} - \mathbf{X_d})), \tag{1}$$

where $\mathbf{X_d}$ and $\mathbf{X}$ are the desired and actual foot positions, $\mathbf{K_d}$ and $\mathbf{D_d}$ the desired stiffness and damping matrices, τ the joint torque matrix, and $\mathbf{J}$ the foot Jacobian. Having provided an

[1] The control law has been derived in the cited texts [21].

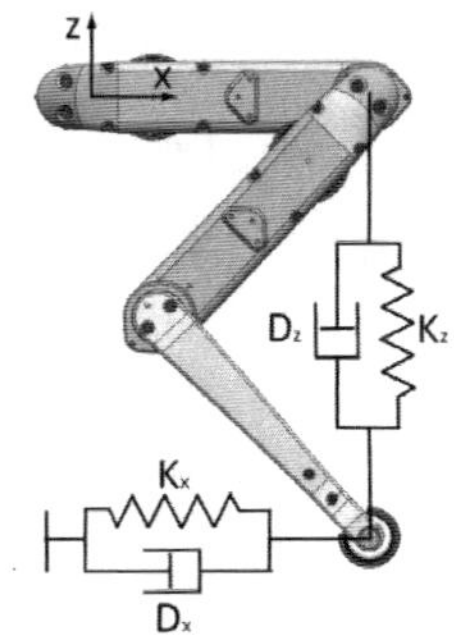

Figure 3: The Solo8 [22] 8 DoF robot based on impedance control and one leg showing the cartesian impedance

overview of the control law of the example legged robot, we will use this example to describe the 4 properties for quantifying embodiment, which are:

Size of the system's and environment's structure, where 'structure' is characterised in terms of its constituent components and its relationships. For a dynamical system like a legged robot, it would cover the size of its state space. In the example of the impedance-controlled robot above, the state space would comprise of parameters such as stiffness, damping, actuator torque, velocity and position. The size of the state space can simply be derived from the model. While it is straightforward to use to quantify embodiment, it has limited use, because it does not provide any more information other than the size of the state space. It doesn't consider the environment and only provides a discrete quantification,i.e., number of states.

Bandwidth of perturbatory channels, which essentially quantifies the effect the system and environment cause each other. If we take again our robot, it has 8 DoF (2 DoF per leg), with a force sensor on its foot. Therefore, while walking, the impact by the environment on the body, i.e. the bandwidth, will essentially comprise the forces propagating in the two joints, and friction forces. In another example, a robot designed for outdoor inspections such as the ANYMAL [4], the bandwidth will comprise forces that the body will be subjected to by the winds, in addition to the normal and friction forces from the ground. This proposed quantification can also be derived directly from the model and provides a useful measure of the number of parameters that are perturbed. However, depending on the type of perturbation it might be non-trivial to obtain an appropriated means of measurement.

Contribution of Morphological Computation (MC), where 'morphology' refers to the agent's body, explicitly including all its physical properties (shape, sensors, actuators, friction, mass distribution, etc.). In the context of embodied intelligence, MC refers to processes, which are conducted by the body (and environment) that otherwise would have to be performed by the brain [25]. MC, in this sense, captures the concept that control is partially performed by the controlled system interacting with the environment. This measure quantifies the contribution of MC as compared to active control for a specific gait or trajectory. For example, by adding compliance to the body of a given robot or adapting existing compliant elements in the body, the contribution of MC increases, as demonstrated by Mohseni et al.[5].

Level of controller complexity, which essentially defines a factor of complexity of the dynamic relationship between the system and the environment [26]. In the legged robot example, the complexity is a result of the impedance control law operating over a total of 8 DoF, along with the inner closed loop control system for the actuators. In contrast, a legged robot with an imaging system would add another layer of complexity, with the requirement of cameras and machine vision systems.

Having discussed the 4 properties, we now turn to the methods of realising quantification. This paper focuses on the last two properties, i.e. MC and the controller complexity for quantifying embodiment, since they provide more insight as compared to the other two and they were the only ones that were practically implemented so far. Interestingly, both methods are based on information theory [27]. The next sections describe these methods in more details.

3.1. Information-theoretic methods

In recent years, several attempts have been made to incorporate the measure of embodiment on information processing in artificial agents. Polani [11] minimises the controller complexity in a reinforcement learning setting. Rückert and Neumann [12] quantify MC indirectly by measuring how changes in morphology reduce or increase the required controller complexity in the context of stochastic optimal control. Ghazi-Zehadi et al. [10] proposed a total of 7 methods, based on understanding the contribution of morphology, when the system moves from one state to the next. All the above methods are based on the principle of deriving general decomposition of mutual information of key random variables relating to the interaction of the body and the environment into unique, shared, and synergistic information. However, the difference in the methods proposed by Ghazi-Zahedi et al. is that it is focused on direct quantification, whereas the Polani et al. and Rückert and Neumann quantify MC indirectly by measuring controller complexity. To the best of our knowledge based on literature review thus far, the methods of direct quantification have been found to be the main basis for quantifying MC on muscle-driven models. Furthermore, in their subsequent work Ghazi-Zehadi et al. [15] described 2 of the 7 models as being suitable for operating on the full system's state, which enable investigation on realistic muscle models without taking approximation errors into account that could potentially occur from limited sensor information. In contrast, the other 5 models are based on deriving partial information from sensors and actuators and could potentially lead to errors in quantifying MC on muscle-driven models.

Before proceeding to elaborate on the quantification methods, it is necessary to explain the concept of the *sensorimotor loop*. According to Ghazi-Zahedi et al. [10], the *sensorimotor loop* [28] is the basic control loop used in robotics, comprising a controller, which sends signals to the system's actuators, thereby affecting the system's environment. The body and environment is therefore encapsulated in a single random variable named *world*. The loop involves three

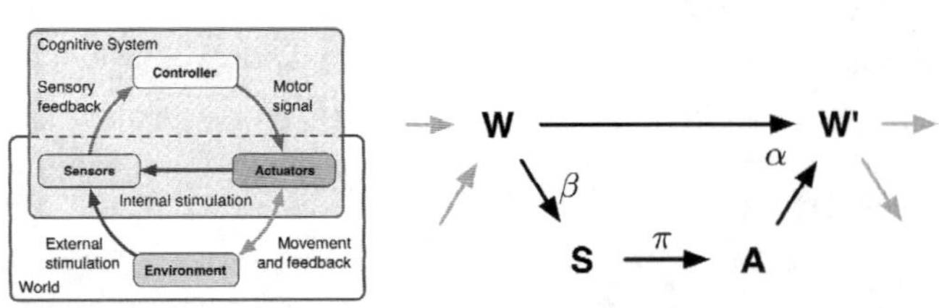

Figure 4: Sensorimotor loop showing the intrinsic and the world states, and the directed graph showing the causal nature of the sensorimotor loop [15]

(stochastic) processes $S(t), A(t), W(t)$ (see Figure 4), which take values s, a, w, in the sensor, actuator, and world state spaces. The directed edges reflect causal dependencies between these random variables, in discrete time-steps. Random variables without any time index refer to some fixed time t and primed variables to time $t + 1$, i.e., the two variables W, W' refer to $W(t), W(t + 1)$.

Starting with the initial distribution over world states, denoted by $p(w)$, the sensorimotor loop for reactive systems is given by three conditional probability distributions, β, α, π, also referred to as kernels. The sensor kernel, which determines how the agent perceives the world, is denoted by $\beta(s|w)$, the agent's controller or policy is denoted by $\pi(a|s)$, and finally, the world dynamics kernel is denoted by $\alpha(w'|w,a)$. The world dynamics kernel $\alpha(w'|w,a)$ captures the influence of the actuator signal A and the previous world state W on the next world state W'.

To measure MC, we need to measure the extent to which the system's behaviour is the result of the world dynamics (i.e., the body's internal dynamics and it's interaction with its world) and how much of the behavior is determined by the controller policy π. Using the above conditional probability distributions and kernels, Ghazi-Zahedi et al. simulated the hopping gait in a legged robot and measured MC [15] using 2 complementary methods. The first method, based on conditional mutual information captures the change in the kernel to measure MC contribution and the second, measures the difference between controller and behaviour complexity. Both are described in the next section.

3.2. Method 1 - MC as conditional mutual information (MC_W)

The principle underlying this method is based on the change to the world dynamics kernel $\alpha(w'|w,a)$, depending on the extent of the MC in the system. A complete lack of MC would mean that the behavior of the system is entirely determined by the system's controller, and, hence, by the actuator state A. In this case, the world dynamics kernel reduces to $p(w'|a)$. Every difference from this assumption means that the previous world state W had an influence, and, hence, information about W changes the distribution over the next world states W'. The discrepancy of these two distributions can be measured with the average of the Kullback-Leibler divergence $D_{KL}(\alpha(w'|w,a)||p(w'|a))$, which is also known as the conditional mutual information $I(W';W|A)$. This is computed by:

$$MC_W = \Sigma p(w',w,a) log_2 \frac{\alpha(w'|w,a)}{p(w'|a)}. \tag{2}$$

3.3. Method 2 - MC as comparison of behavior and controller complexity (MC_{MI})

This method is based on the assumption that for a given behavior, MC decreases with an increasing effect of the action A on the next world state W'. This measurement compares the complexity of the behavior with the complexity of the controller. The complexity of the behavior can be measured by the mutual information of consecutive world states, $I(W';W)$, and the complexity of the controller can be measured by the mutual information of sensor and actuator states, $I(A;S)$. The mutual information $I(W';W)$ is high if the system shows a diverse but non-random behavior, which is what we would like to see in an embodied system. On the other hand, a system with high MC should produce a complex behavior based on a controller with low complexity. Hence, we want to reduce the mutual information $I(A;S)$, because this either means that the policy has a low diversity in its output (low entropy over actuator states $H(A)$ or that there is only a very low correlation between sensor states S and actuator states A (high conditional entropy $H(A|S)$. Therefore, we define this measure as the difference of these two terms:

$$MC_{MI} = I(W';W) - I(A;S). \tag{3}$$

3.4. Experiment and results of the two information-theoretic methods.

Ghazi-Zahedi et al. [15] simulated various muscle-driven biologically realistic hopping models and quantified MC_W and MC_{MI}. The data was discretised and uni-variate random variables

IOP Conf. Series: Materials Science and Engineering 1292 (2023) 012004 doi:10.1088/1757-899X/1292/1/012004

W, W', A and S were created as per algorithms described in the paper[2]. The robot hopped periodically up to the same length of 1.07m. Figure 5 shows MC_W being measured in two

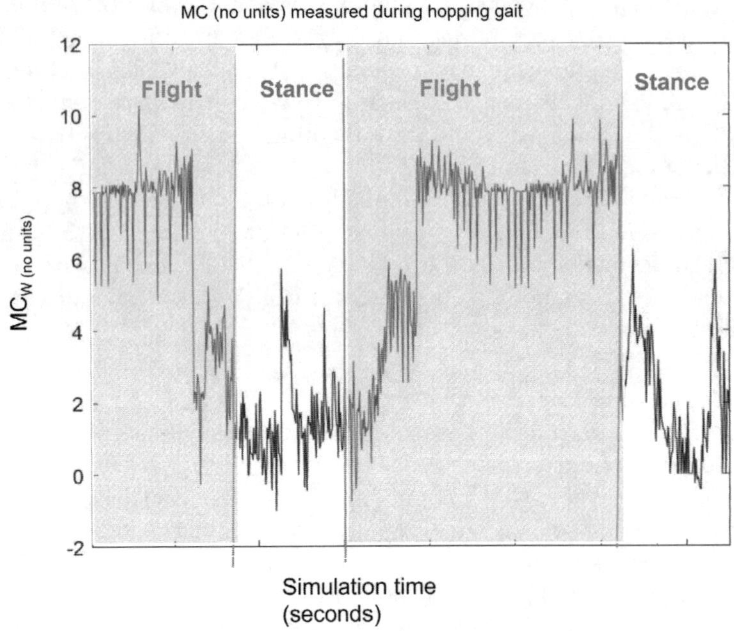

Figure 5: Extent of MC_W at various points in the hopping gait, which shows that the MC increases during flight and is maximum at the highest point. In the stance phase, the actuator control comes into play, and the MC is low.

different phases of the hopping gait, as shown by the vertical lines. First, the flight phase, during which the robot does not touch the ground, and second, the stance deceleration phase, which occurs after landing. During flight, the behavior of the system is governed only by the interaction of the body (mass,velocity) and the environment (gravity) and not by the actuator models. Also, all actuator control signals are constant during flight. As shown in Figure 5, MC drops as soon as the systems touch the ground. Shortly after touching the ground, the system shows a strong decline of MC, which is followed by a strong incline during the deceleration with the muscle. Furthermore, it was found that both MC_W and MC_{MI} have similar values, and therefore only MC_W is shown in the plot.

3.5. Discussion of the two information-theoretic methods

The numerical results of the two quantification methods, MC_W and MC_{MI} complement other findings, like [29], which showed that the minimum information required to generate hopping is reduced by the material properties of the non-linear muscle fibers. The higher MC in the muscle model can be attributed to the force–velocity characteristics of the muscle fibers [30]. These non-linear contraction dynamics reduce the influence of the controller on the actual hopping kinematics [29]. This implies that the kinematics trajectory is more predetermined by the material properties.

One of the points raised by Ghazi-Zahedi et al. [15] is whether there is a preference for MC_W or MC_{MI}. Both methods are useful, MC_W to evaluate the extent of the role that morphology

[2] The algorithms are described by the authors in their paper [15]. The data was discretised into bins and using Bayesian probability theory, the entropy and conditional mutual information measures were computed.

plays, and MC_{MI} to understand controller complexity. The latter will help in simplifying the controller design, if possible.

4. Other related work using information-theoretic approaches for the quantification of embodiment

While the provided examples are interesting, the chosen simulation setups are, in general, rather simple. It is of interest to understand how embodiment could be measured in hierarchical structure with more complexity. The first example, where this has been applied was Haeufle et al. [14], who investigated this by simulating point-to-point and oscillatory human arm movements with a neuro-musculoskeletal model. The model (Figure 6) has been developed to access all signals, i.e., low-level controller, control signal, muscles' active state (biochemistry), force, joint torques, and the resulting joint angles. MC was then quantified on the different hierarchy

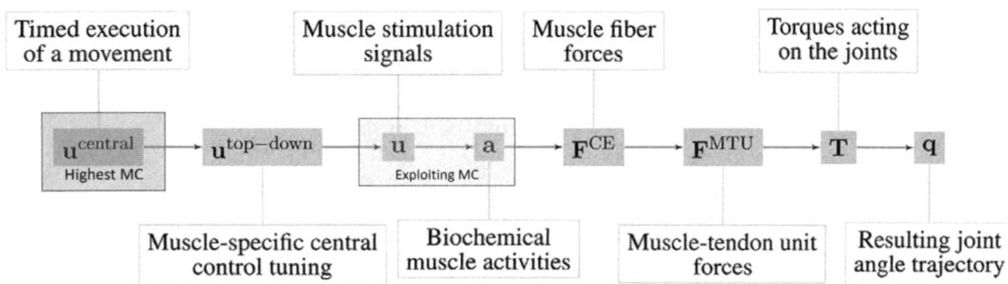

Figure 6: Different levels of hierarchy in the neuro-musculoskeletal model, from the initiation of the movement to the joint angle trajectory [14]

levels. The results show that morphological computation is highest for the most central level $u^{central}$ of the modeled control hierarchy, where the movement initiation and timing are encoded [14]. Furthermore, they show that the lowest neuronal control layer u, the muscle stimulation input (see Figure 6), exploits the morphological computation of the biochemical and biophysical muscle characteristics to generate smooth dynamic movements. This study suggests that the system's design in the mechanical as well as in the neurological structure can take over important contributions to control, which would otherwise need to be performed by the higher control levels.

Another related work has been carried out by Montúfar et al. [31], who quantified the advantages of providing a well-chosen embodiment. The authors link the the quality of an embodiment to the informational cost required to achieve it. The core message is that a good embodiment is one that makes decisions (informationally) cheaper, less complex, and more robust. It is based on the concept of *relevant information* and to identify how and when embodiment affects the decision density, in a minimalist scenario. Taking this research further, Polani et al. [13] investigated how embodiment affects information costs when, scripts, defined as predefined action sequences, are introduced instead of atomic actions. It was demonstrated that that there are cases for which the introduction of scripts allows the agent to save decisions without processing more information for an average decision, thereby reducing information costs.

5. Key considerations in quantifying embodiment

The information-theoretic methods has been widely applied for various studies and investigations, as demonstrated by a few examples in the previous section 4. However, one of the underlying considerations is that the quantification is based on a simulated neuro-musculoskeletal model that resembles the organ-level dynamics. But in reality, each muscle-tendon unit consists of many motor units that need to be controlled and we cannot rule out the effect on the overall MC. Nevertheless, this model represents the basic functional unit considering the main dynamic properties relevant for the passive contribution of muscles. Furthermore, whilst the results have been quantified on the non-trivial hopping gait using a deterministic system, the studies so far have not used these methods on high-dimensional data collected from natural systems.

Another consideration is whether MC results in the desired behaviour. Montufar et al. [32] characterised *Good MC* as body-environment interactions which contribute to the desired behavior in a way that reduces the required controller complexity, and *Bad MC* as body-environment interactions which make the desired behavior more difficult to maintain (e.g., object slipping out of the hand due to the hand's compliance). The authors provide a way of extracting information from MC_W by using dimensionality reduction methods to identify the characteristics that result in Bad MC. In that regard, quantifying the informational flows within the sensorimotor network, as described in the seminal research by Schmidt et al. [33], will also provide powerful insight into the expected behaviour.

6. Future work and potential

The above quantification methods have been tested in the hopping gait and for a robotic arm in a point to point motion trajectory and grasping action. The future work involves testing these methods using other gaits such as walking and running, and compare with the relevant bio-mechanical models. Furthermore, to the best of our knowledge, the research in this area has been conducted using different simulated muscle models and need to be tested on a real robot. Nothwithstanding the above, the above methods provide a solid basis of testing the extent of MC as part of the design process. This can be further used to test the design by augmenting the system with physical elements such as springs, dampers and other components and use the system's morphology to its advantage. This area of research shows promise with recently developed legged robot by Mohseni et al. [5] using a blended approach of virtual and physical impedance control. Another recent example is an under-actuated soft-bodied arm developed by Mahdi et al. [34] to take advantage of the morphology.

Another interesting area of research where quantification of embodiment will be useful is in the development of highly re-configurable intelligent systems with adaptive morphologies [35]. Inspired by biology, where many animals have evolved morphable bodies to extend their operational range, several terrestrial and aerial robots have been developed. Examples are *GoQBot* [36] that morphs before rolling, *RHex* [37] with legs that adapt their stiffness when morphed to run effectively over a broad range of terrains, and a few others. These systems are developed using programmable matter [35], which is concerned with the theory, design, and manufacturing of systems that reconfigure assuming different morphological configurations. This programmable matter is developed in a modular way as building blocks, starting from the simplest elements. As these elements are developed, their embodiment can be quantified to evaluate their behaviour before adding further building blocks, resulting in lower risk in the design process. Further research by Hauser et al. [38] discusses resilient machines that could enable robots to recover more effectively from damage with adaptive morphologies.

References

[1] Maturana H R and Varela F J 2012 *Autopoiesis and Cognition: The realization of the living* vol 42 (Springer Science & Business Media)
[2] Nguyen Q V and Park H C 2012 *Journal of Bionic Engineering* **9** 271–281 ISSN 1672-6529 URL `https://www.sciencedirect.com/science/article/pii/S1672652911601212`
[3] Mahapatra A, Roy S S and Pratihar D K 2020 *Multi-Legged Robots—A Review* (Singapore: Springer Singapore) pp 11–32 URL `https://doi.org/10.1007/978-981-15-2953-5%5F2`
[4] Hutter M, Gehring C, Jud D, Lauber A, Bellicoso C D, Tsounis V, Hwangbo J, Bodie K, Fankhauser P, Bloesch M, Diethelm R, Bachmann S, Melzer A and Hoepflinger M 2016 *2016 IEEE/RSJ International Conference on Intelligent Robots and Systems (IROS)* pp 38–44
[5] Mohseni O, Rashty A M N, Seyfarth A, Hosoda K and Sharbafi M A 2022 *Journal of Intelligent & Robotic Systems* **105** 1–15
[6] Pfeifer R, Lungarella M and Iida F 2007 *Science* **318** 1088–1093
[7] Brown I, Scott S and Loeb G 1995 *Programmable, High-gain, Zero-delay Intrinsic Responses of Perturbed Musculoskeletal Systems. Soc Neurosci Abstr* vol 21
[8] Proctor J and Holmes P 2010 *Biological cybernetics* **102** 513–531
[9] Müller R, Häufle D F B and Blickhan R 2015 *Journal of Experimental Biology* **218** 451–457
[10] Ghazi-Zahedi K 2019 *Morphological Intelligence* (Springer)
[11] Polani D 2011 *2011 IEEE Symposium on Artificial Life (ALIFE)* (IEEE) pp 78–85
[12] Rückert E A and Neumann G 2013 *Artificial Life* **19** 115–131
[13] Riegler B, Polani D and Steuber V 2021 *Frontiers in Robotics and AI* **8** 535158
[14] Haeufle D F B, Stollenmaier K, Heinrich I, Schmitt S and Ghazi-Zahedi K 2020 *Frontiers in Robotics and AI* **7** ISSN 2296-9144 URL `https://www.frontiersin.org/articles/10.3389/frobt.2020.511265`
[15] Ghazi-Zahedi K, Haeufle D F B, Montúfar G, Schmitt S and Ay N 2016 *Frontiers in Robotics and AI* **3** ISSN 2296-9144 URL `https://www.frontiersin.org/article/10.3389/frobt.2016.00042`
[16] Quick T, Dautenhahn K, Nehaniv C L and Roberts G 2000 *AIP Conference Proceedings* **517** 649–660 (*Preprint* `https://aip.scitation.org/doi/pdf/10.1063/1.1291299`) URL `https://aip.scitation.org/doi/abs/10.1063/1.1291299`
[17] Dreyfus H L 1996 *The Electronic Journal of Analytic Philosophy* **4** 1–16
[18] Mautner C and Belew R K 1999 *Proceedings of the GECCO-99 ICGA and GP joint conference*
[19] Brooks R A 1991 *Artificial Intelligence* **47** 139–159
[20] Mathworks 2021 Contact Forces `https://blogs.mathworks.com/student-lounge/2017/09/12/introduction-to-contact-modeling/` last accessed 27 May 2021
[21] Grimminger F, Meduri A, Khadiv M, Viereck J, Wuthrich M, Naveau M, Berenz V, Heim S, Widmaier F, Flayols T and et al 2020 *IEEE Robotics and Automation Letters* **5** 3650–3657 ISSN 2377-3774
[22] ODRI 2020 ODRI Github Repository `https://open-dynamic-robot-initiative.github.io/` last accessed 25 March 2021
[23] Semini C, Barasuol V, Boaventura T, Frigerio M, Focchi M, Caldwell D G and Buchli J 2015 *The International Journal of Robotics Research* **34** 1003–1020 (*Preprint* `https://doi.org/10.1177/0278364915578839`) URL `https://doi.org/10.1177/0278364915578839`
[24] Song P, Yu Y and Zhang X 2017 *2017 2nd International Conference on Cybernetics, Robotics and Control (CRC)* pp 51–55
[25] Hauser H, Ijspeert A J, Füchslin R M, Pfeifer R and Maass W 2011 *Biological Cybernetics* URL `https://doi.org/10.1007/s00422-012-0471-0`
[26] Nehaniv C L 1997 *Mathematical Hierarchies and Biology. Amer. Math. Soc* 29–42
[27] Shannon C E 1948 *The Bell system technical journal* **27** 379–423
[28] Ay N and Zahedi K 2014 *On the Causal Structure of the Sensorimotor Loop* (Berlin, Heidelberg: Springer Berlin Heidelberg) pp 261–294 ISBN 978-3-642-53734-9
[29] Haeufle D, Grimmer S, Kalveram K T and Seyfarth A 2012 *Journal of The Royal Society Interface* **9** 1458–1469
[30] Walker K and Hauser H 2019 *Journal of Biomechanics* **87** 28–36 ISSN 0021-9290 URL `https://www.sciencedirect.com/science/article/pii/S0021929019301393`
[31] Montúfar G, Ghazi-Zahedi K and Ay N 2015 *PLoS Computational Biology* **11** e1004427
[32] Ghazi-Zahedi K, Deimel R, Montúfar G, Wall V and Brock O 2017 *2017 IEEE/RSJ International Conference on Intelligent Robots and Systems (IROS)* (IEEE) pp 464–469
[33] Schmidt N M, Hoffmann M, Nakajima K and Pfeifer R 2013 *Advances in Complex Systems* **16** 1250078

[34] Haghshenas-Jaryani M 2022 *Journal of Intelligent & Robotic Systems* **105** 1–22
[35] Mintchev S and Floreano D 2016 *IEEE Robotics & Automation Magazine* **23** 42–54
[36] Lin H T, Leisk G G and Trimmer B 2011 *Bioinspiration & biomimetics* **6** 026007
[37] Saranli U, Buehler M and Koditschek D E 2001 *The International Journal of Robotics Research* **20** 616–631
[38] Hauser H 2019 *Nature Machine Intelligence* **1** 338–339

IOP Conf. Series: Materials Science and Engineering 1292 (2023) 012005 doi:10.1088/1757-899X/1292/1/012005

Energy, Passivity, and Morphological Capacities

Pablo Borja[1,2], **Cosimo Della Santina**[2,4], **and Azita Dabiri**[3]

[1] School of Engineering, Computing and Mathematics, University of Plymouth, Plymouth, U.K.
[2]Department of Cognitive Robotics, Delft University of Technology, Delft, The Netherlands.
[3]Delft Center for Systems and Control, Delft University of Technology, Delft, The Netherlands.
[4]Institute of Robotics and Mechatronics, DLR, Oberpfaffenhofen, Germany.

E-mail: pablo.borjarosales@plymouth.ac.uk;{c.dellasantina,a.dabiri}@tudelft.nl

Abstract. The morphology of a system may be exploited to reduce the controller workload or to perform specific tasks, e.g., locomotion processes. However, understanding and characterizing these capacities in terms of physical quantities remains an open problem. In this paper, we pose the question: *are energy and dissipation useful physical quantities to characterize morphological capacities?* First, to illustrate the relevance of this question, we present some basic concepts of passivity theory that are useful for carrying out an energy-based analysis of the behavior of dynamical systems. Then, we analyze the behavior of a mechanical system, and we show how its morphology affects its energy. Finally, we present a detailed analysis of a passive walker and explain how its locomotion capacities can be interpreted in terms of energy.

1. Introduction

The relevance of the morphology in biological systems has been studied for a long time (see, for instance, the discussion provided in [1, 2]). Similarly, there is a vast range of applications where the morphology of systems has been engineered for diverse purposes. For instance, the design of soft robots with different shapes and made of different materials is a current hot-topic in robotics [3, 4]. Regarding art, the *Strandbeests* created by Theo Jansen (see [5]) illustrate how to exploit the morphology for locomotion purposes. In particular, the morphology of an embodied system may exhibit capacities to perform tasks often associated with the system's brain (control). Depending on the extent of such capacities, this phenomenon can be understood as a degree of *embodied intelligence.* Examples of bodies performing tasks with minimal—or without—intervention of the brain are abundant in nature, e.g., an octopus arm, the whiskers of a cat, and the muscles of our legs, while passive walkers are a celebrated example in robotics [6]. More examples of tasks performed by the morphology of a system are mentioned in detail in [7, 8].

Several terms have been coined to describe how and when the morphology of a system reduces its brain's workload. Among the proposed terminology, the concept of *morphological computation* has been thoroughly described and explained, e.g., [8, 9]. Notably, a great effort has been made to unify concepts and results regarding the relationship between the morphology and intelligence of physical systems. Nevertheless, several questions remain to be addressed in this field. In particular, how to intuitively characterize the morphological capacities of an embodied system in terms of physical quantities. This question is the cornerstone for the design, analysis, and control of embodied systems in an efficient way.

A thorough exposition on morphological capacities is provided in [7], where the authors recurrently underscore how energy storage and dissipation affect an embodied system. Interestingly, the concepts of energy and dissipation have been used in fields such as mechanics, dynamical systems, and control theory to study the behavior of complex systems. Following this rationale, in this paper we pose the question: *are energy and dissipation useful physical quantities to characterize morphological capacities?* To study this question formally, we use passivity theory, which has roots in electrical circuits theory but has additionally proven extremely useful in control systems [10, 11] and robotics [12, 13].

While there exists literature where morphological capacities are quantified and mathematically formalized, e.g., [14, 15, 16], here we propose for the first time—to the best of our knowledge—passivity, energy, and dissipation as concepts suitable to understand, characterize, and formalize the morphological capacities of an embodied system. In particular, our goal is to pave the way to develop an alternative framework to analyze such capacities intuitively. However, we stress that such an alternative is not in conflict with the existing results in this field. On the contrary, it may represent an excellent complement to them and, in this way, contribute towards a unified framework for the study of morphological capacities and their relationship with embodied intelligence.

2. Preliminaries

2.1. Notation

The symbol $\mathbf{0}$ represents a vector or matrix whose entries are zeros. The ith element of the vector $x \in \mathbb{R}^n$ is denoted by x_i. Consider a scalar multivariable function $f : \mathbb{R}^n \to \mathbb{R}^n$. We adopt the convention

$$\frac{\partial f(x)}{\partial x} = \begin{bmatrix} \frac{\partial f(x)}{\partial x_1} \\ \vdots \\ \frac{\partial f(x)}{\partial x_n} \end{bmatrix}.$$

We use Newton's notation for differentiation with respect to time, i.e.,

$$\dot{x} = \frac{dx(t)}{dt}.$$

The symbols e and sgn represent the exponential and signum functions, respectively.

2.2. Morphological Capacities

The term morphological computation is commonly used to refer to the capacity of the body to perform tasks without the intervention of the brain in an embodied system. However, we remark that the definition of the mentioned term can vary from one reference to another, where a fundamental difference lies in defining when systems compute. For instance, in [17], the authors understand morphological computation as the phenomenon of morphology and material taking over some of the functions often attributed to the brain of a system. However, the mentioned definition does not meet the requirements established in [8], where the authors define three roles of morphology: morphology facilitating control, morphology facilitating perception, and morphological computation. We refer the reader interested in this discussion to [7, 18, 19, 9] and the previous references.

To avoid controversy, we use the term morphological capacities to indicate that: (a) the morphology permits performing specific tasks without requiring a control input (brain action), or (b) the morphology simplifies the control signal (brain's action) required to perform a task. In particular, we focus on tasks related to locomotion. However, the ideas exposed in this paper can be adapted to perception tasks. For example, in [20], the authors use a similar analysis to detect collision in a robotic arm. Moreover, the characterization of morphological capacities presented in this paper can be adapted to the different definitions of morphological computation and other roles of morphology.

2.3. Passivity

In this section, we revisit some passivity theory concepts. In particular, we restrict our attention to input-affine nonlinear systems in state-space representation for two reasons: (i) these systems are convenient to illustrate the main idea of this manuscript, and (ii) we avoid unnecessarily dense mathematical content. However, the reader interested in general dissipativity and passivity theory is referred to the seminal works [21, 22] and the more recent references [23, 10].

The concept of passivity has its roots in circuit theory. However, it has proven extremely useful for analyzing and controlling dynamical systems, particularly physical ones. Passivity is an input-output property, where, informally, we consider that a system is passive if it cannot generate energy. To formalize this idea, we consider a system, whose dynamics are described by

$$\dot{x} = f(x(t)) + g(x(t))u(t), \tag{1}$$

where $t \in \mathbb{R}_{\geq 0}$ denotes time, $x : \mathbb{R}_{\geq 0} \to \mathbb{R}^n$ represents the states of the system; $u : \mathbb{R}_{\geq 0} \to \mathbb{R}^m$ denotes the input vector, with $m \leq n$; $f : \mathbb{R}^n \to \mathbb{R}^n$; and $g : \mathbb{R}^n \to \mathbb{R}^{n \times m}$. Additionally, we consider that the output of (1) is a signal of dimension m, given by

$$y(t) = h(x(t)), \tag{2}$$

where $h : \mathbb{R}^n \to \mathbb{R}^m$.

Throughout this paper, we assume that the solution $x(t)$ of (1) is unique; the signals $u(t)$ and $y(t)$ are bounded, and the integral $\int_0^t u^\top(\tau)y(\tau)d\tau$ is well-defined.

Definition 1 (Passive system) *A system of the form* (1) *is said to be passive if there exists a function* $S : \mathbb{R}^n \to \mathbb{R}_{\geq 0}$ *such that*

$$S(x(t)) \leq S(x(0)) + \int_0^t u^\top(\tau)y(\tau)d\tau, \qquad \forall\, t \in \mathbb{R}_{\geq 0}. \tag{3}$$

In passivity theory jargon, $S(x(t))$ is referred to as the *storage function* and the inequality (3) is known as the *dissipation inequality*. Moreover, if (3) holds, $y(t)$ is referred to as the *passive output* of (1).

For our purposes, we consider that the storage function of a physical system denotes its energy. Furthermore, the units of $u(t)$ and $y(t)$ are such that their inner product has power units, i.e., $u^\top(t)y(t)$ represents the power injected into the system. Thus, (3) implies that the energy of the system (1) at t cannot be greater than the energy previously stored by the system, plus the energy injected during the interval $[0, t]$. Henceforth, to ease the readability, we omit the argument t from the state x and the signals u and y.

Note that, under some differentiability assumptions, we can rewrite (3) as

$$\dot{S} \leq u^\top y, \tag{4}$$

which is referred to as the *differential dissipation inequality*. In this case, (4) implies that the power that can be extracted from the system at t should be less than or equal to the power injected into the system.

Definition 2 (Lossless system) *A system of the form* (1) *is said to be lossless if* (3)*—alternatively,* (4)*—holds with equality.*

Note that the energy of a lossless system, at t, equals the sum of the stored energy and the energy injected into the system during the interval $[0, t]$. In this regard, every lossless system is passive, but the converse is not true.

Passivity theory is not limited to linear systems. Moreover, a broad range of physical systems is passive. Therefore, the arguments exposed above are suitable to analyze the behavior of a wide variety of systems even if their dynamics are highly complex, e.g., nonlinear systems.

3. Energy as the Orchestra's Director

Intuitively, the behavior of a physical system is ruled by its energy. Consequently, energy-based modeling frameworks have been extensively used to describe the behavior of physical system, e.g., the Euler-Lagrange formalism [24] and the port-Hamiltonian framework [25, 26].

To illustrate the role of energy in the behavior of a physical system, let us consider the system depicted in Fig. 1, which consists of two pendulums moving along the horizontal plane, where the first pendulum is attached to the reference through an ideal torsion spring and the second pendulum can rotate freely. Additionally, m_1 and m_2 represent the mass of the first and second pendulum, respectively; ℓ_1 and ℓ_2 denote the length of the first and second pendulum, respectively; and k represents the spring constant.

A suitable choice for the state vector is

$$x = \begin{bmatrix} q \\ p \end{bmatrix}; \qquad \begin{matrix} q \in \mathbb{R}^2 \\ p \in \mathbb{R}^2, \end{matrix} \tag{5}$$

where q represents the angular position of the pendulums and p the corresponding momenta. For simplicity, we consider that the rods are rigid, massless, and inextensible; moreover, we assume that the masses at the tip of each pendulum cannot rotate on their own axis. Under these assumptions, the expression that characterizes the potential energy is

$$V(q) = \frac{1}{2} k q_1^2, \tag{6}$$

whereas the kinetic energy is given by

$$T(q,p) = \frac{1}{2} p^\top M^{-1}(q) p; \qquad M(q) := \begin{bmatrix} \ell_1^2(m_1 + m_2) & \ell_1 \ell_2 m_2 \cos(q_1 - q_2) \\ \ell_1 \ell_2 m_2 \cos(q_1 - q_2) & \ell_2^2 m_2 \end{bmatrix}. \tag{7}$$

Hence, the total energy of the system—i.e., the system's Hamitonian—is

$$H(q,p) = T(q,p) + V(q). \tag{8}$$

Moreover, the dynamics of the system are described by

$$\begin{aligned} \dot{q} &= \frac{\partial H(q,p)}{\partial p}, \\ \dot{p} &= -\frac{\partial H(q,p)}{\partial q}. \end{aligned} \tag{9}$$

Note that, given the choice of state variables q and p, the dynamics of the system are completely determined by its energy. At this point, we make the following remarks:

- This system only stores *elastic* potential energy because of the spring. Consequently, the spring exerts a force on the first link when this is not aligned with the Y-axis.
- Because we are not considering friction or dampers, there is no dissipation, i.e., this system is lossless.

Observing the schematic provided in Fig. 1, we can obtain the configuration of the system in the cartesian plane via the following expressions:

$$\begin{aligned} X_{\mathtt{m}_1} &= \ell_1 \sin(q_1), & X_{\mathtt{m}_2} &= \ell_1 \sin(q_1) + \ell_2 \sin(q_2), \\ Y_{\mathtt{m}_1} &= \ell_1 \cos(q_1), & Y_{\mathtt{m}_2} &= \ell_1 \cos(q_1) + \ell_2 \cos(q_2). \end{aligned} \tag{10}$$

Fig. 1 depicts the behavior of this system in cartesian coordinates for different initial conditions of the second link, where we consider $\ell_1 = 2$ [m], $\ell_2 = 0.5$ [m], $m_1 = 1$ [kg], $m_2 = 2.5$ [kg], and $k = 1$ [N/m] for simulations purposes. In the three cases, we consider that the initial conditions of the momenta are zero, i.e., $p_0 = \mathbf{0}$. Figs. 1 and 2 show that different initial conditions of the second link yield completely different trajectories, while the first link describes the same semi-circular trajectory for the three cases. Such behavior could seem random. Nevertheless, it is entirely characterized by the energy of the system. To note this, we recall that the system is lossless. Thus,

$$\dot{H} = 0 \implies \int_0^t \dot{H} dt = 0 \implies H(q(t), p(t)) = H(q_0, p_0); \; \forall \, t \in \mathbb{R}_{\geq 0}, \tag{11}$$

where $q_0 = q(0)$ and $p_0 = p(0)$. Therefore, because we consider $p_0 = \mathbf{0}$, the expression that describes the system's energy reduces to

$$H(q(t), p(t)) = \frac{1}{2} k q_1^2(0), \; \forall \, t \in \mathbb{R}_{\geq 0}. \tag{12}$$

Consequently, (12) implies that the trajectories of the system must guarantee that the energy remains constant all the time.

We consider the same initial condition $q_1(0)$ in the three proposed simulations. Hence, the total energy should be the same even if the kinetic and potential energies evolve differently for each set of initial conditions, which is corroborated in Fig. 2.

Suppose that the system consists only of the links, and the spring is an external body. Under this hypothesis, the energy of the system is given only by $T(q, p)$. Moreover, the dynamics of the system are

$$\begin{aligned} \dot{q} &= \frac{\partial T(q,p)}{\partial p}, \\ \dot{p} &= -\frac{\partial T(q,p)}{\partial q} - A F_{\mathtt{k}}(q_1), \end{aligned} \tag{13}$$

where

$$F_{\mathtt{k}}(q_1) = k q_1, \qquad A := \begin{bmatrix} 1 \\ 0 \end{bmatrix}. \tag{14}$$

Hence, some simple computations, omitted for brevity, yield

$$\dot{T} = F_{\mathtt{k}}(q_1) \dot{q}_1, \tag{15}$$

where the passive output is given by $\dot{q}_1$. Note that the system is still lossless, but we consider the force $F_{\mathtt{k}}(q_1)$ as an external input. Therefore,

$$p_0 = \mathbf{0} \implies T(q_0, p_0) = 0. \tag{16}$$

Accordingly, the (kinetic) energy varies only if the supply rate $F_{\mathtt{k}}(q_1)\dot{q}_1$ is different from zero. The physical interpretation of this analysis is that the system moves only if the spring exerts a force on the first link, i.e., some power is injected into the system.

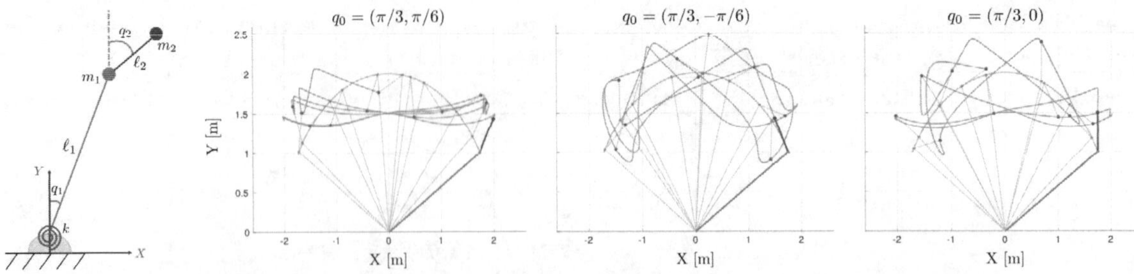

Figure 1. **(Left)** Schematic of the two-links system moving along the horizontal plane. **(Right)** Trajectories of the system under different initial conditions, denoted as q_0. The initial configuration is indicated with thicker lines.

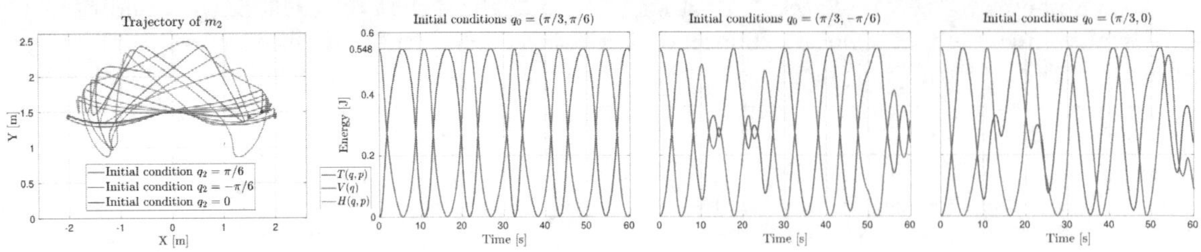

Figure 2. **(Left)** Evolution of m_2 under different initial conditions for q_2. **(Right)** Evolution of the potential, kinetic, and total energy of the system for different initial conditions.

3.1. Dissipation and equilibria

Dissipation plays an essential role in the stabilization of mechanical systems. Loosely speaking, dissipative elements (e.g., dampers and friction) make the system lose energy until it eventually reaches *a* minimum energy value. Such a phenomenon is related to the concept of equilibrium points, namely, points (configurations) where the system stays if there are no external inputs. Note that a configuration is an equilibrium point only the dynamics of the system are zero at that point. Thus, intuitively, equilibrium points are characterized by the critical points of the system's Hamiltonian (energy). Moreover, because the kinetic energy is quadratic in p, the equilibrium points of a mechanical system are given by points of the form $(\bar{q}, \mathbf{0})$, where $\bar{q} \in \mathbb{R}^n$ is a critical point of the potential energy.

Consider the two-links system, depicted in Fig. 1, and suppose there exists friction between the second link and a surface. The dynamics of the system become[1]

$$\begin{aligned} \dot{q} &= \frac{\partial H(q,p)}{\partial p}, \\ \dot{p} &= -\frac{\partial H(q,p)}{\partial q} - AF_{\mathfrak{f}}(\dot{q}_1), \end{aligned} \tag{17}$$

where $H(q,p)$ and A are defined in (8) and (14), respectively, and

$$F_{\mathfrak{f}}(\dot{q}_1)\dot{q}_1 > 0, \ \forall \ \dot{q}_1 \neq 0. \tag{18}$$

The critical points of the energy of this system are characterized by $q_1 = 0$ and $p = \mathbf{0}$. Therefore, this system has an infinite number of equilibrium points. In particular, any point of the form

[1] Note that $\dot{q}$ is not a state of the system in our analysis. However, we introduce the abuse of notation $F_{\mathfrak{f}}(\dot{q}_1)$ to preserve the physical intuition of this force.

IOP Conf. Series: Materials Science and Engineering 1292 (2023) 012005 doi:10.1088/1757-899X/1292/1/012005

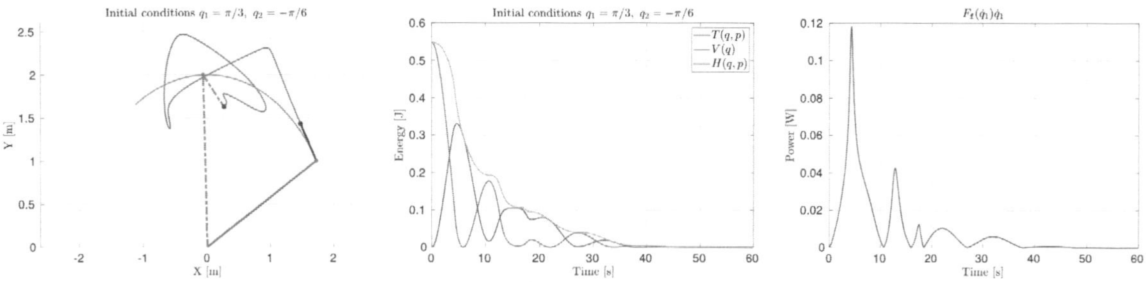

Figure 3. **(Left)** Trajectories of the system. The initial configuration is depicted with solid lines, while the configuration after 60 seconds is plotted with dashed lines. **(Middle)** Evolution of the potential, kinetic, and total energy of the system. **(Right)** Supply rate associated with friction.

$(0, q_2, 0, 0)$ is an equilibrium for the system.

If we consider the force $F_{\mathfrak{f}}(\dot{q}_1)$ as an external input, then the system (17) is passive, where the passive output is given by $\dot{q}_1$. In particular, we have that

$$\dot{H} = -F_{\mathfrak{f}}(\dot{q}_1)\dot{q}_1. \tag{19}$$

Therefore, from (18), the supply rate is negative for $\dot{q}_1$ different from zero. In other words, friction implies that power is extracted from the system until the second link stops moving. Fig. 3 shows the simulation of the behavior of (17), where, inspired by [27], we consider the friction model

$$F_{\mathfrak{f}}(\dot{q}_1) = \left[b_1 \mathrm{e}^{-(\dot{q}_1/b_2)^2} + b_3\right] \operatorname{sgn}(\dot{q}_1) + b_4\dot{q}_1, \tag{20}$$

with $b_1 = 0.05$, $b_2 = 0.5$, $b_3 = 0.01$, and $b_4 = 0.8$. In the mentioned figure, we notice that the first link converges to a configuration close to $q_1 = 0$. Note that the first link does not stabilize exactly at zero because of the signum function in (20), which is related to dry friction. Moreover, the trajectories of the second link are different than the ones shown in Fig. 1 under the same initial conditions. Thus, we conclude that the presence of friction in the first link also affects the behavior of the second link. Also in Fig. 3, we observe that the stored energy converges to zero and that the power related to friction decreases as the system's energy decreases. Note that this supply rate represents the amount of energy extracted from the system. Hence, it is expected to become zero when the system has no stored energy. Moreover, because friction dissipates energy, the supply rate should not be negative,[2] implying that some energy is injected into the system. Both properties can be corroborated in Fig. 3.

3.2. Other external inputs

As explained in Subsection 3.1, the effect of dissipation can be understood as a supply rate that extracts power from the system. Note that this supply rate results from the interaction of the system with its environment. However, supply rates may originate from signals that could exist independently of the body of the system, i.e., signals that are independent of the system's dynamics. Examples of these signals are control inputs. These supply rates can drastically affect the energy and, thus, the natural behavior of the system. To illustrate this, we consider the two-links system neglecting any dissipation but supposing there is an external force $u : \mathbb{R}_{\geq 0} \to \mathbb{R}$ acting on the first link. Hence, the dynamics of the system become

[2] Notice the minus sign in (19).

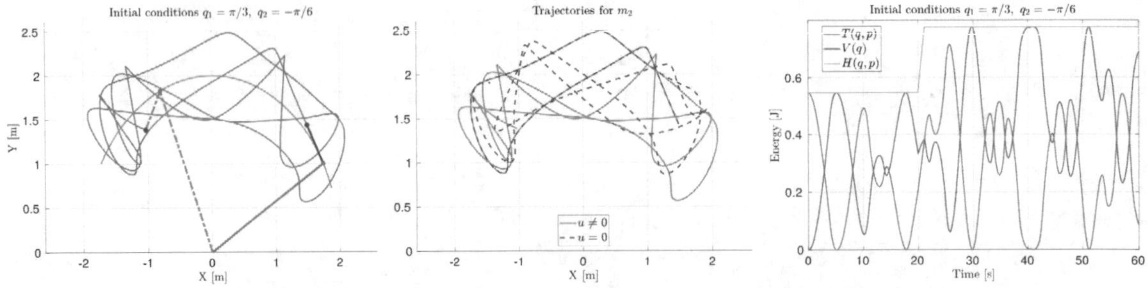

Figure 4. **(Left)** Trajectories of the system. The initial configuration is depicted with solid lines, while the configuration after 60 seconds is plotted with dashed lines. **(Middle)** Evolution of m_2 under the same initial conditions with and without external input. **(Right)** Evolution of the potential, kinetic, and total energy of the system.

$$\begin{aligned} \dot{q} &= \frac{\partial H(q,p)}{\partial p}, \\ \dot{p} &= -\frac{\partial H(q,p)}{\partial q} + Au(t). \end{aligned} \tag{21}$$

We consider that $u(t)$ is a unitary pulse starting at $t = 20$ and lasting 1.2 seconds for illustration purposes. Fig. 4 shows the simulations of this system, where we observe the following:

(Left plot). The oscillations of the first link exceed the initial conditions. In particular, the light red curve that describe the trajectory of the first mass overcomes the initial conditions depicted with a solid red dot.

(Middle plot). The external input also affects the trajectory of the second mass. In this plot, both trajectories are identical until the external input modifies the behavior of the system.

(Right plot). At $t = 20$ [s], the energy stored in the system increases because of the power injected by the external input.

Remark 1 *External inputs play a crucial role in control systems, where these signals are designed to guarantee that the system stabilizes at a desired point or follows a desired trajectory. Customarily, these problems are referred to as set-point regulation and trajectory tracking, respectively. Regarding nonlinear control techniques, some passivity-based control strategies consist in designing a control signal that modifies the energy such that the desired equilibrium point becomes a minimum of the energy function. Then, another control input injects damping into the system to ensure its convergence to the desired equilibrium, i.e., the minimum of the closed-loop energy.*

4. Understanding Morphological Capacities in Terms of Energy

The morphology of a system is closely related to its energy, e.g., the potential energy captures any compliant behavior in a mechanical system in the form of elastic potential energy. Also the forces resulting from gravity—interaction with the environment—are characterized by the potential energy. Two common strategies to model mechanical systems are the Euler-Lagrange formalism (see [24]), where the equations of motion are given by

$$M(q)\ddot{q} + C(q,\dot{q})\dot{q} + D(\dot{q})\dot{q} + \frac{\partial V(q)}{\partial q} = A(q)u(t)$$

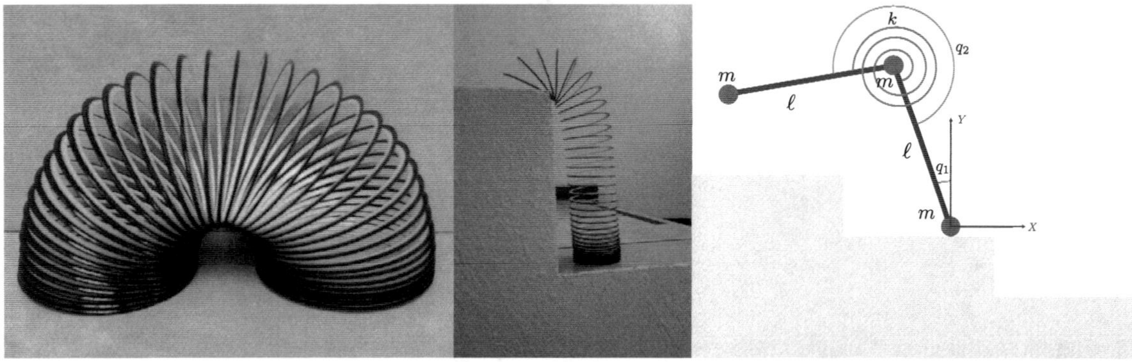

Figure 5. **(Left)** Slinky toy. **(Middle)** Slinky toy walking down some stairs. **(Right)** Simplified schematic of the slinky toy walking down stairs.

and the port-Hamiltonian approach (see [25, 26]), where the dynamics of the system are given by

$$\begin{aligned}\dot{q} &= \frac{\partial H(q,p)}{\partial p}, \\ \dot{p} &= -\frac{\partial H(q,p)}{\partial q} - D(q,p)\frac{\partial H(q,p)}{\partial p} + A(q)u(t).\end{aligned}$$

Note that in both cases some simple computations lead to the conclusion that the equilibrium of a mechanical system is given by $(q_\star, \mathbf{0})$, where $q_\star \in \mathbb{R}^n$ is determined by the values of the configuration variables q such that

$$\frac{\partial V(q)}{\partial q} = \mathbf{0}.$$

Consequently, if there are no external inputs—i.e., $u(t) = \mathbf{0}$—any mechanical system with dissipation converges to a critical point of its potential energy. Hence, modifying the morphology of a system determines where it naturally stabilizes. An illustrative example is the Gömböc, a mono-static body whose potential energy has only two equilibria: one stable and the other unstable. Accordingly, whenever the system is placed in an initial condition different from the unstable equilibrium, it self-stabilizes at the stable equilibrium. A similar principle is exhibited by the shape of some turtles' shells that can self-right themselves with minimum effort. For more discussion on this topic, we refer the reader to [28, 29].

An example of how morphology can facilitate locomotion is given by slinky toys, such as the one depicted in Fig. 5. These systems have some interesting properties due to their morphology, e.g., [30]. In particular, their compliant nature lets them move downstairs without any external (control) input involved. In this section, inspired by a slinky toy, we propose the mechanical system presented in the right image of Fig. 5, which consists of three identical masses, two identical rods, and a torsion spring. Then, we model how the system descends each step. To this end, we modify the model provided in [31] by changing the references to illustrate more transparently the role of energy in locomotion. The resulting model is of the form provided in (9), with

$$M = \ell^2 \begin{bmatrix} 2 & -\cos(q_1 - q_2) \\ -\cos(q_1 - q_2) & 1 \end{bmatrix}, \quad V(q) = \frac{1}{2}k(q_1 + q_2)^2 + mg_{\mathrm{r}}\ell\left(\cos(q_1) - \cos(q_2)\right). \tag{22}$$

Note that q_1 corresponds to the angular position of middle mass, while q_2 denotes the angular position of the mass at the tip of the rear leg (rod). However, once the system walks down a

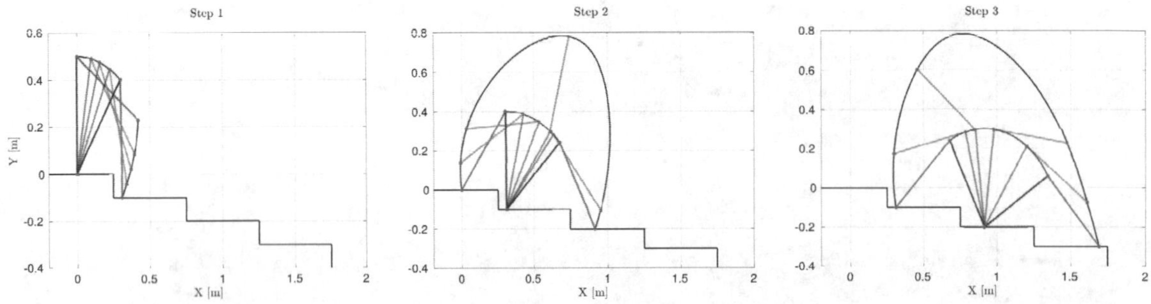

Figure 6. Motion of the slinky-like system while walking down the steps. The rear and front legs are depicted in red and blue, respectively. Moreover, the initial and final configuration of each step are presented in solid lines. Additionally, we show the trajectories of q_1 and q_2 in green and purple, respectively.

step, the front leg becomes the rear one and vice-versa. For simulation purposes, we consider three steps 0.1 [m] height and 0.5 [m] of width. Moreover, the values of the parameters used for simulation are: $\ell = 0.5$ [m], $m = 0.1$ [kg], and $k = 0.5$ [N/m]. Fig. 6 depicts the evolution of the system while walking down the three steps. Moreover, the left image of Fig. 7 shows the initial condition of the system and its position when making contact with each step. We observe that the angle between both legs increases with each step. Hence, its (standalone) locomotion capacity is limited. However, it can perform a task relying only on its morphology—particularly the torsion spring—and its interaction with the environment—the effect of gravity. We stress that we neglect any dissipation phenomenon. Therefore, the process is lossless, as is shown in the middle plot of Fig. 7. Similarly to the two-links example in Section 3, we can consider the force of exerted by the spring as an external force to analyze the locomotion process. Thus, the dynamics take the form (13), with

$$F_{\mathrm{k}}(q) = k(q_1 + q_2), \qquad A = \begin{bmatrix} 1 \\ 1 \end{bmatrix}. \tag{23}$$

Moreover, the differential dissipation inequality takes the form

$$\dot{H} = -\dot{q}^{\top} A F_{\mathrm{k}}(q) = -k(q_1 + q_2)(\dot{q}_1 + \dot{q}_2) \tag{24}$$

The right plot of Fig. 7 shows the profile of the power (supply rate) associated with the spring. This supply rate can be interpreted as a control signal commanded by the body instead of the brain, i.e., the morphology has the capacity to generate control inputs to perform a specific task. While the system's ability to learn or adapt is limited in this example, the proposed analysis can be extended to more complex morphologies.

In [15], the authors highlight the overlapping of morphological computation with control theory, and in [32], the authors borrow concepts from control systems to analyze the importance of feedback in morphological computation. Notably, both control systems and morphological computation are closely related to the dynamics of the system to be studied. As mentioned in Section 3, such dynamics are mainly determined by energy, dissipation, and external inputs. Moreover, even if the body of a system cannot perform a task by itself, it may reduce the control effort. Examples of this phenomenon are the bipedal robots designed in [33], and the locomotion approach proposed in [34].

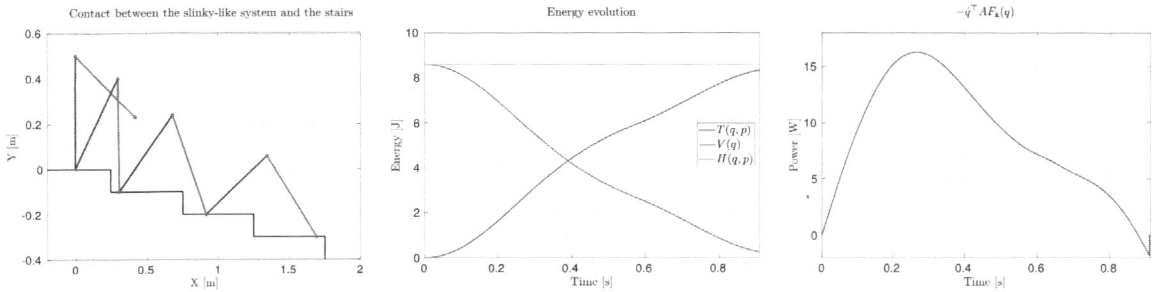

Figure 7. (**Left**) Initial configuration and contact postures of the slinky-like system. The rear and front legs are plotted in red and blue, respectively. (**Middle**) Evolution of the potential, kinetic, and total energy of the system corresponding to the locomotion process from step 1 to step 2. (**Right**) Power related to the torsion spring during the locomotion process from step 1 to step 2.

5. Concluding Remarks and Future Work

Analyzing the energy of a system provides relevant information about its behavior. The morphology of a system determines how energy is dissipated and internally stored in a system. Hence, through an energy-based analysis, we can determine how morphology affects some capacities of the system—e.g., locomotion capacities—while providing a physically-intuitive interpretation of the process. In this paper, we have illustrated how to use some basic concepts of passivity theory to interpret locomotion processes originated by the morphology of a system. However, this is just a humble example of what could be achieved by using energy-based analyses to understand and characterize morphological capacities.

As future work, we envisioned the use of energy-based methods to understand, quantify, and characterize the morphological computation of more complex systems, e.g., systems that exhibit physical reservoir computing (see [8]). We stress that passivity theory only requires information about the system's dynamics and energy to provide a physically intuitive analysis of its behavior. Hence, this theoretical framework may be the cornerstone that complements, unifies, and enlarges current methodologies that quantify and study morphological capacities, such as locomotion.

6. Acknowledgement

Pablo Borja thanks Eduardo Duarte for the exciting discussions on this topic, especially for pointing out the Gömböc system.

References

[1] Allen G E 1981 *Journal of the History of Biology* 159–176
[2] Tamborini M 2020 The twentieth-century desire for morphology
[3] Rus D and Tolley M T 2015 *Nature* **521** 467–475
[4] Schmitt F, Piccin O, Barbé L and Bayle B 2018 *Frontiers in Robotics and AI* **5** 84
[5] Jansen T 2008 *Architectural Design* **78** 22–27
[6] McGeer T 1993 *Journal of theoretical biology* **163** 277–314
[7] Pfeifer R and Bongard J 2006 *How the body shapes the way we think: a new view of intelligence* (MIT press)
[8] Müller V C and Hoffmann M 2017 *Artificial life* **23** 1–24
[9] Hauser H, Fuechslin R M and Nakajima K 2014 *Opinions and Outlooks on Morphological Computation* (Self-published) pp 226–244
[10] van der Schaft A J 2016 *L_2-Gain and Passivity techniques in nonlinear control.* 3rd ed (Berlin: Springer)
[11] Ortega R, Romero J G, Borja P and Donaire A 2021 *PID Passivity-Based Control of Nonlinear Systems with Applications* (John Wiley & Sons)
[12] Spong M W and Vidyasagar M 2008 *Robot dynamics and control* (John Wiley & Sons)
[13] Albu-Schäffer A, Ott C and Hirzinger G 2007 *The international journal of robotics research* **26** 23–39

[14] Hauser H, Ijspeert A J, Füchslin R M, Pfeifer R and Maass W 2011 *Biological cybernetics* **105** 355–370
[15] Füchslin R M, Dzyakanchuk A, Flumini D, Hauser H, Hunt K J, Luchsinger R H, Reller B, Scheidegger S and Walker R 2013 *Artificial life* **19** 9–34
[16] Zahedi K and Ay N 2013 *Entropy* **15** 1887–1915
[17] Pfeifer R and Gómez G 2009 *Morphological Computation – Connecting Brain, Body, and Environment* (Berlin, Heidelberg: Springer Berlin Heidelberg) pp 66–83
[18] Pfeifer R, Iida F and Gómez G 2006 *International Congress Series* vol 1291 (Elsevier) pp 22–29
[19] Horsman C, Stepney S, Wagner R C and Kendon V 2014 *Proceedings of the Royal Society A: Mathematical, Physical and Engineering Sciences* **470** 20140182
[20] Califano F, van Dijk D and Roozing W 2022 *IEEE Robotics and Automation Letters* **7** 8791–8798
[21] Willems J C 1972 *Arch Rational Mechanics and Analysis* **45** 321–351
[22] Hill D and Moylan P 1975 Cyclo-dissipativeness, dissipativeness, and losslessness for nonlinear dynamical systems Tech. Rep. EE7526 University of Newcastle
[23] Moylan P 2014 *Dissipative systems and stability.*
[24] Goldstein H, Poole C P and Safko J L 2001 *Classical Mechanics* 3rd ed (Addison-Wesley)
[25] Duindam V, Macchelli A, Stramigioli S and Bruyninckx H 2009 *Modeling and control of complex physical systems: the port-Hamiltonian approach* (Springer Science & Business Media)
[26] van der Schaft A J and Jeltsema D 2014 *Foundations and Trends in Systems and Control* **1** 173–378
[27] Armstrong B and Canudas de Wit C 1996 *The control handbook* **77** 1369–1382
[28] Várkonyi P L and Domokos G 2006 *Math. Intelligencer* **28** 34–38
[29] Domokos G and Várkonyi P L 2008 *Proceedings of the Royal Society B: Biological Sciences* **275** 11–17
[30] Cross R and Wheatland M 2012 *American journal of physics* **80** 1051–1060
[31] Hu A P 2010 *American Journal of Physics* **78** 35–39
[32] Hauser H, Ijspeert A J, Füchslin R M, Pfeifer R and Maass W 2012 *Biological cybernetics* **106** 595–613
[33] Collins S, Ruina A, Tedrake R and Wisse M 2005 *Science* **307** 1082–1085
[34] Calzolari D, Della Santina C, Giordano A M and Albu-Schäffer A 2022 *2022 American Control Conference (ACC)* (IEEE) pp 506–513

IOP Conf. Series: Materials Science and Engineering 1292 (2023) 012006 doi:10.1088/1757-899X/1292/1/012006

Mechanisms boosting material embodied intelligence to realize self-healing soft robots

Seyedreza Kashef Tabrizian[1*], Seppe Terryn[1,2], Joost Brancart[1,2], Guy Van Assche[2], Bram Vanderborght[1]

[1]Brubotics, Vrije Universiteit Brussel (VUB) and Imec; Pleinlaan 2, 1050 Brussels, Belgium.
[2]Physical Chemistry and Polymer Science (FYSC), Vrije Universiteit Brussel (VUB); Pleinlaan 2, 1050 Brussels, Belgium.

[*]Seyedreza.kashef.tabrizian@vub.be

Abstract. The recent introduction of self-healing soft materials in robotics is a major step towards sustainable next generation robots. By manufacturing soft robots out of these smart materials, we integrate a self-healing ability and increase the physical intelligence of these systems. However, the embodied intelligence in the material level needs to be augmented by incorporating assistive mechanisms in the system level with minimized control, enabling healing of damage in different sizes and in diverse working conditions. These assistive mechanisms can provide damage detection, damage closure, healing stimuli providing, health monitoring, or a combination of the previous. In this paper, we present two different mechanisms for an autonomous healing of damages; (i) Embedding a healable heater in a self-healing soft actuator to increase the temperature required for an efficient healing, while it allows detecting the damage and monitoring the health of the system. (ii) Incorporating shape memory alloy wires in a self-heling soft bending actuator, with simultaneous sealing through contraction and heating abilities. Apart from assisting in the healing action, both mechanisms play a part in the actuation of the bending robots as strain limiting elements. These assistive mechanisms will overcome the limitation on the material level, leading to robots that can self-heal in applications outside of laboratories and factories.

1. Introduction

Despite the remarkable advancement of soft robotics systems, they have not yet been integrated into a large variety of industrial and commercial applications to their full potential [1]. One reason that hinders their wide adoption is the inherent vulnerability of their soft matter that reduces their lifetime. In addition, this limited lifetime and the non-recyclability of the polymeric materials have negative environmental effects [2]. To tackle these issues, soft robots can be made out of recyclable and healable materials to recover their functional performance after an incurred damage [3,4].

Self-healing in soft robots is in general highly beneficial, whether the process of healing is carried out completely autonomously or by human intervention. Considering that many applications of soft robotics systems are where humans have limited access to, e.g., robots for earth or space exploration, the importance of having a fully autonomous healing system becomes more significant. This autonomous healing can originate from the selected self-healing material, e.g. materials that can be healed at room temperature [5–7], or provided by integration of some assistive mechanisms [5]. However, to increase the robustness of autonomous self-healing processes, assistive mechanisms will

IOP Conf. Series: Materials Science and Engineering 1292 (2023) 012006 doi:10.1088/1757-899X/1292/1/012006

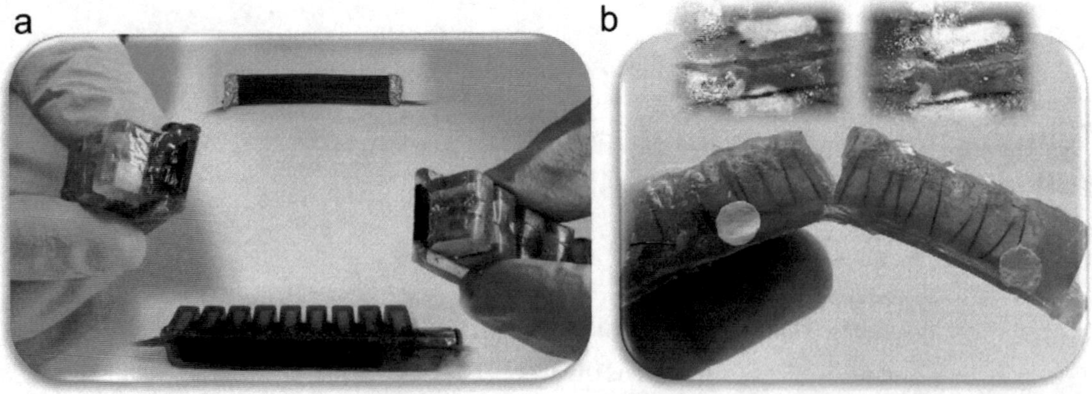

Fig. 1. Integrated assistive mechanisms for autonomous healing. **(a)** An embedded healable heater in a PneuNet bending actuator that provides stimulus for healing, detects the damage and monitor the health of the robot [15]. **(b)** Embedded shape memory alloy wires to make an SMA wire-reinforced bending actuator with the ability of damage closure and providing stimulus for healing [16].

be unavoidable. This robustness refers to the ability to perform multiple damage-healing cycles, to heal different types and sizes of damages, to heal at different environmental conditions and to heal without any human intervention.

A fully autonomous healing procedure, starts with an autonomous detection of damage[8–10]. Secondly, as synthetic material cannot be grown, damage closure (sealing) is essential for healing and should be performed autonomously as well [11,12]. Thirdly, depending on the material, providing a stimulus, in the form of heat or light, is needed for triggering the healing or for accelerating it [13,14]. Finally, the system should monitor and re-evaluate its health after healing, e.g. defining the healing efficiency, to make sure that the healing was successful enough to continue operation. Here, we provide two different mechanisms, integrated in self-healing soft bending actuators, to facilitate and assist the autonomous healing in the robots.

2. Embedded healable heater

A self-healing heater, made of a hybrid composite of self-healing Diels-Alder based material, carbon black and nanoclay, has been developed [15]. The addition of the two fillers makes the material electrically conductive, while maintaining its self-healing ability. These two unique properties, healing and conductivity, greatly enhance the embodied intelligence of the material. Due to the Joule-effect, the material heats upon applying a current, generating the required temperature increase to healing autonomously. Furthermore, the heater is intelligent enough to self-target the heat towards the damaged region and provides localized heating/healing. Via resistance tracking in normal operation, both deformation as well as damage can be detected, while monitoring the resistance during and after healing allows to monitor its health and recovery. Moreover, we showed that the material is also able to self-sense the change in temperature.

As seen in Figure 1a, the heater is embedded in a soft self-healing PneuNet bending actuator to assist in the healing procedure of the robot. Illustratively, by passing electrical current through the heater, it heats up, and can provide enough stimulus for the healing of the robot [15]. Note that some self-healing materials need to be thermally triggered for either enabling or accelerating the healing action. Self-sensing of the temperature is very helpful in this regard. Overheating in some self-healing materials, may cause loss of the structure of the robot due to crossing of a solid-liquid transition (degelation). As a result, controlling the temperature is important and will enhance the robustness of the system, successfully healing without human intervention in diverse working conditions. As a damage sensing element, tracking the resistance of the heater which is changed upon damage provides information about its occurrence, however not its location. However, damage leads to heat concentration at the corresponding location. This can be monitored by a thermal camera and used to localize the damage [15]. As a health monitoring system, comparing the resistance of the heater after a damage-healing cycle with the initial amount before the damage can be a criterion for health monitoring. Additionally, a successful healing procedure will distribute the heat generation along the heater, not just concentration at the damage location [15]. Consequently, generating the heatmap after healing provides information about the performance of the healing process. All these contributions level up the intelligence of the system and remove human assistance from the healing practice in the robot. Being stiffer than the material of the actuator, the heater limits the strain of the robot locally, which improves the bending performance of the actuator. It is worth mentioning that there is a strong interfacial bonding between the heater and the actuator as they are made from a self-healing polymer and composite that benefit from the same chemistry in their polymer network structure. This expands the application of self-healing material for soft robotics and can provide a solution for multi-material robotics design where delamination and interfacial debonding is a serious problem.

3. Embedded shape memory alloy wires

Like in most biological systems, damage can only be healed properly if it is first closed (assuming that the damaged area is clean). In the human body, large wounds must be closed by stitching and bone fractures should be realigned and fixed. In self-healing soft robots, in case of a puncture or a small damage, closure may happen by the elastic response of the material, whereas large damage need to be closed by external forces. In figure 1.b, an SMA wire-reinforced self-healing soft bending actuator is shown [16]. The embedded SMA wires are pre-strained and able to recover their permanent shape if heated above their transition temperature [17]. They can be quickly heated to that temperature by passing electrical current through them. Being integrated into the actuator system, the contraction of the wires will contract the chamber and close the damage. Furthermore, the produced Joule heating that activates the wires will also transfer to the chamber and heat it up simultaneously. As such, the wires can not only close the damage, but also provide the stimulus for heating and healing of the actuator [16]. Upon cooling, the wires can be re-elongated by the elastic forces of the chamber or upon actuation of the robot, making the damage closure possible for multiple cycles. Based on the length of the wires and the amount of pre-strain, they can close gaps as large as 2 mm [16]. For healing, usually the material is heated for a few minutes. Continuation of the contraction forces of the embedded wires during heating ensures that the damage remains closed throughout the healing procedure, which is in contrast to when the robot is manually closed and heated up via an oven. In many cases, the damage opens while the specimens or robotic systems are heated up in the oven, due to different thermal expansion of the material or object in different directions. Another advantage of this system is the ability to perform in-situ healing, meaning that there is no need to disassemble the actuator from the working station and take it to a rest position. Thanks to the forces and heat of the SMA wires, the system has a higher level of autonomy in the healing procedure. Finally, the strong SMA wires reinforce the chamber radially and longitudinally which is crucial for the bending motion of the chamber when pneumatically inflated, preventing it from undesired expansions and ballooning. At the same time, they support the robot against very serious injuries, e.g. being cut it two.

Acknowledgment

This work has received funding from the European Union’s Horizon 2020 research and innovation programme under the Marie Sklodowska-Curie grant agreement No 860108 (SMART) as well as the personal FWO grants of Terryn (FWOTM784) and Brancart (12E1123N).

IOP Conf. Series: Materials Science and Engineering 1292 (2023) 012006 doi:10.1088/1757-899X/1292/1/012006

References

[1] Rothemund P, Kim Y, Heisser R H, Zhao X, Shepherd R F and Keplinger C 2021 Shaping the future of robotics through materials innovation *Nature Materials* **20** 1582–7

[2] Terryn S, Langenbach J, Roels E, Brancart J, Bakkali-Hassani C, Poutrel Q A, Georgopoulou A, George Thuruthel T, Safaei A, Ferrentino P, Sebastian T, Norvez S, Iida F, Bosman A W, Tournilhac F, Clemens F, Van Assche G and Vanderborght B 2021 A review on self-healing polymers for soft robotics *Mater. Today* **47** 187–205

[3] He Q, Wang Z, Wang Y, Song Z and Cai S 2020 Recyclable and Self-Repairable Fluid-Driven Liquid Crystal Elastomer Actuator *ACS Applied Materials and Interfaces* **12** 35464–74

[4] Terryn S, Brancart J, Lefeber D, van Assche G and Vanderborght B 2017 Self-healing soft pneumatic robots *Sci. Robot.* **2** 1–13

[5] Tan Y J, Susanto G J, Anwar Ali H P and Tee B C K 2020 Progress and Roadmap for Intelligent Self-Healing Materials in Autonomous Robotics *Adv. Mater.* **33** 2002800

[6] Hardman D, George Thuruthel T and Iida F 2022 Self-healing ionic gelatin/glycerol hydrogels for strain sensing applications *NPG Asia Mater.* **14** 1–13

[7] Terryn S, Brancart J, Roels E, Assche G van and Vanderborght B 2020 Room Temperature Self-Healing in Soft Pneumatic Robotics *IEEE Robot Autom Mag* **27** 44–55

[8] Markvicka E J, Tutika R, Bartlett M D and Majidi C 2019 Soft Electronic Skin for Multi-Site Damage Detection and Localization *Adv Funct Mater* **29** 1–10

[9] Khatib M, Zohar O, Saliba W and Haick H 2020 A Multifunctional Electronic Skin Empowered with Damage Mapping and Autonomic Acceleration of Self-Healing in Designated Locations *Adv Mater* **32** 1–7

[10] Thuruthel T G, Bosman A W, Hughes J and Iida F 2021 Soft Self-Healing Fluidic Tactile Sensors with Damage Detection and Localization Abilities *MDPI Sensors* **21** 8284

[11] Cerdan K, Van Assche G, van Puyvelde P and Brancart J 2020 A novel approach for the closure of large damage in self-healing elastomers using magnetic particles *Polymer* **204** 122819

[12] Alabiso W, Manuel Hron T, David R, Daniel B-A and Schlögl S 2021 Shape memory-assisted self-healing of dynamic thiol-acrylate networks *Polym. Chem.* **12** 5704–14

[13] Willocq B, Bose R K, Khelifa F, Garcia S J, Dubois P and Raquez J M 2016 Healing by the Joule effect of electrically conductive poly(ester-urethane)/carbon nanotube nanocomposites *Journal of Materials Chemistry A* **4** 4089–97

[14] Orellana J, Moreno-villoslada I, Bose R K, Picchioni F, Flores M E and Araya-hermosilla R 2021 Self-Healing Polymer Nanocomposite Materials by Joule Effect

[15] Kashef Tabrizian S, Sahraeeazartamar F, Brancart J, Roels E, Ferrentino P, Legrand J, van Assche G, Vanderborght B and Terryn S 2022 A Healable Resistive Heater as a Stimuli-Providing System in Self-Healing Soft Robots *IEEE Robot. Autom. Lett.* **7** 4574–81

[16] Kashef Tabrizian S, Terryn S, Costa Cornellà A, Brancart J, Legrand J, Van Assche G and Vanderborght B 2022 Close-then-heal in soft sustainable robots by shape memory alloy wires *Sci Rep.* Under revision

[17] Kirkby E L, Michaud V J, Månson J E, Sottos N R and White S R 2009 Performance of self-healing epoxy with microencapsulated healing agent and shape memory alloy wires *Polymer* **50** 5533–8

Embodied Active Tactile Perception

Liang He and Perla Maiolino

Oxford Robotic Institute, Department of Engineering Science, University of Oxford, United Kingdom

E-mail: Liang.he@eng.ox.ac.uk;Perla.maiolino@eng.ox.ac.uk

Abstract. Tactile perception plays an important role in an agent safely interacting with the environment while acquiring information about it. Bio-inspired robotics opens up possibilities for a new paradigm leveraging the morphology of the body, which filters the tactile information in physical interactions and enables investigations of new designs for embodied active tactile perception. The subjects of morphology embodied active perception and motor embodied active perception is defined and discussed in this chapter. In the scope of morphology embodied active perception, sensor optimization and sensor adaptation are further defined to describe the change of sensor morphology in the design phase and the interacting phase, respectively. More specifically, the concept of online and offline sensor adjustment is presented. Sensor optimization is solely considered in the offline process for optimization and evolution design of the sensor structure and characteristics. Sensor adaptation and motor embodied active perception are considered in the online process to actively shape the sensing process with the morphology change of the sensors themselves and the action of the body where the sensors are placed, respectively. "Design as a whole" is proposed as an inverse problem to address the sensing tasks. The design of new tactile sensors should not focus on the sensor per se but should also include design parameters for sensor optimization, sensor adaptation, and motor actions.

1. Introduction

Developing sensor technologies to allow an artificial agent to perceive and understand the environment has always been a frontier of research in robotics [1, 2]. This capability allows robots to perform in the real world and to cope with uncertainty, reacting quickly to changes in the environment. Artificial vision significantly increased the impact of robotic applications in unstructured environments owing to the advance in camera-based hardware technologies, computer vision for data processing, and learning-based solutions for recognition [3]. In contrast, the development of systems to endow robots with the sense of touch is less explored due to the nature of touch elaborated on more complex sensory modalities and contact models, requiring to solve system level challenges and to implement complex control strategies to deal with physical interaction with the environment [2, 4].

When we touch an object, we receive and process information from two sensory modalities: kinaesthetics and cutaneous [5]. Kinaesthetics sense refers to large-scale force, motion, and configuration information, emphasizing spatial and loading realization, equivalent to what humans feel at the tendons and spindles level. Cutaneous sense reflects fine features like temperature, texture, slippage, vibration, and low-level pressure, emphasizing the resolution and sensitivity of touch. Humans rely on the mechanoreceptors underneath the skin (around 17, 000 are distributed only in the human hand) to detect and filter these various tactile features,

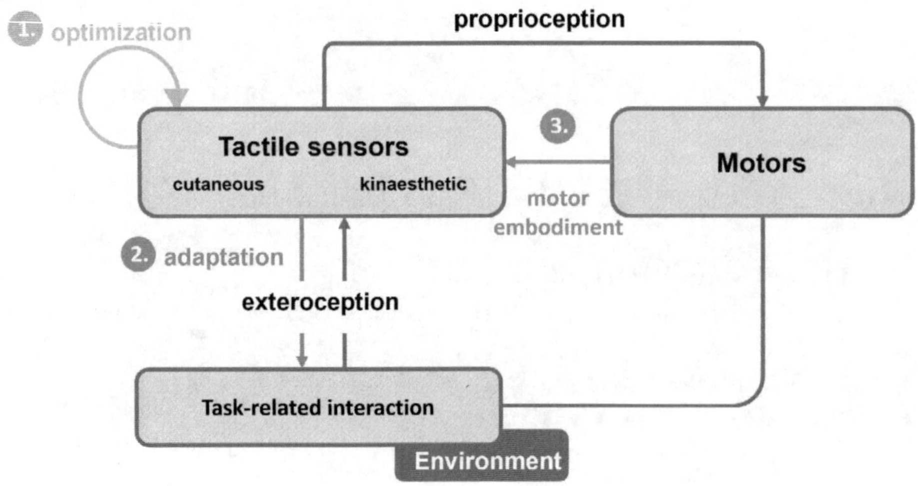

Figure 1. The schematic of a tactile-based interaction between an agent and the external environment. A conceptual classification of tactile sensors and motors is modelled in this diagram. Sensors are considered solely the sensory system itself, including receptors for detecting both cutaneous and kinaesthetic features. Motors are considered as the mechanical system where the sensors are attached. In the case of a human recognizing an object through touch, the mechanoreceptors, tissues, and skin on the hand are considered the sensory system, while the motion generated by fingers, wrist and arms control are considered motor actions. Within the framework, this chapter focuses on three subjects for embodied active tactile perception 1. sensor optimization (section.2.1), 2. sensor adaptation (section.2.2), and 3. motor embodiment (section.3). 1. sensor optimization and 2. sensor adaptation represented the active morphology change of the sensory system itself, including the modalities, materials, shape, spatial resolutions, the number of receptors, etc. In contrast, 3. motor embodiment represented the process of active perception, where the motor actions affect the perceived information of the sensory system.

enabling the execution of dexterous tasks, and affective touch-based communication [6]. The effective processing and interpretation of tactile information rely on the multimodal integration of both kinaesthetics and cutaneous feedback, reflecting on the challenges of developing robotic tactile sensors [7].

In addition, humans utilize not only the physical sensors but also the actions to explore and perceive via the coordination of “exteroceptive” and “proprioceptive” procedures (sensorimotor coordination process) [8, 9]. This adaptive process ultimately establishes various contact models between the agent and the environment, filtering the perceived touch-based information.

Touch is complex, but despite its complexity, it remains an important sensing modality which is at the base of developing robots cognitive behaviours [10, 11]. In the last 40+ years extensive research has been done in robotics to reproduce similar tactile function of mechanoreceptors by focusing on design of new tactile sensing technologies via capacitive [12], piezoelectric [13], piezo-resistive[14], and optical [15] sensors. However, none of these state-of-the-art solutions can reach or is even comparable to the level of tactile sensation we perceive in its biological/natural counterpart [16].

The design of these sensing technologies was aimed at the detection of different types of contact events and, in particular, focused on achieving high spatial resolutions, large static and dynamic sensing range, linearity, high sensitivity, repeatability, the durability of the sensors and, in the case of artificial skins, conformability, modularity, multimodality, networking, efficient data processing, low latency, etc. [4].

Besides the development of appropriate sensing technologies which can detect the desired contact event, a big challenge is related to the interpretation of the data coming from these sensors. As we said, the complexity of tactile perception in robotics is also related to the fact that the acquisition of tactile information is an active process which involves interaction with the environment and appropriate control strategies to deal with unexpected physical interactions. However, this interaction deeply affects the structure of the acquired data and the subsequent inference process, and it is mediated through the body. In particular, the mechanical properties of the skin, the stiffness of the joints, the morphology and degree of freedoms of the body (and its physical constraints) play an active role in shaping what is perceived by filtering the physical contact event felt by the mechanoreceptors and by allowing for particular motor actions which define the space of the interaction [17].

Active tactile perception research has investigated how humans use and select appropriate stereotyped movements (exploratory procedures taxonomy) to elicit information related to the properties of objects and has implemented the same process for improving robot's tactile perception [18], but this approach was based on a fixed taxonomy of exploration actions under the classical sense-plan-act architecture. However, as proposed in [19], the behaviour of a system comes from the dynamic coupling of the brain (control) body and environment, and if this is considered in the design process, it can lead to simplifying the inference process in the brain/controller because some control and processing aspects are distributed to the body and behaviours emerging from the interaction with the environment.

This chapter focuses on this new paradigm, i.e., the implication of embodiment in tactile perception. The schematic of a tactile-based interaction between an agent and the external environment is illustrated in Fig. 1. Three subjects of embodied active tactile perception are defined. Sensor optimization and sensor adaptation are discussed as the two subbranches of the morphology embodied active perception to describe the offline and online changes in the sensor morphology. Motor embodied active perception is discussed as an online process to exploit the role of motion in active perception. This paradigm has not yet been sufficiently explored, and this chapter wants to highlight the work that has been done so far in this domain. In particular, we will focus on the role of the morphology of the sensor, the active change of it and the mediated interaction between the body and the environment in structuring the information received by the sensory system. This not only contributes to reducing the complexity of the inference task at the brain level but can also be exploited to achieve adaptable and resilient sensors.

2. Morphology Embodied Active Perception

In biological systems, the variations of the sensor morphology represent the ecological performance of the organism, reflecting its unique living environment, behavioural engagement, and evolution [20]. The role of sensor morphology in embodied tactile perception reflects the advantages of processing touch-related information while interacting with external stimuli. In this chapter, sensor morphology is defined as the characteristics of a sensor, including its number of receptors, shape, spatial resolutions, materials, modalities, etc. The change of these characteristics can be adopted in both sensor optimization and sensor adaptation.

The variation of the sensor morphology of an agent is selected in its evolution process as a biological agent or iterative design process as an engineering system. The 'suboptimal' choices of the sensor morphology, determined in an "offline" selecting process, ensured the survival of the agent in the suitable environment/scenario. This process is defined as sensor optimization in this chapter, reflecting recent design choices of morphology in sensor development [21].

In contrast, the term sensor adaption is defined in this chapter as the ability of an agent to actively tune its morphology to favour specific tactile signals in perception. Unlike artificial sensors, which are usually designed with a fixed characteristic during operation, the morphology

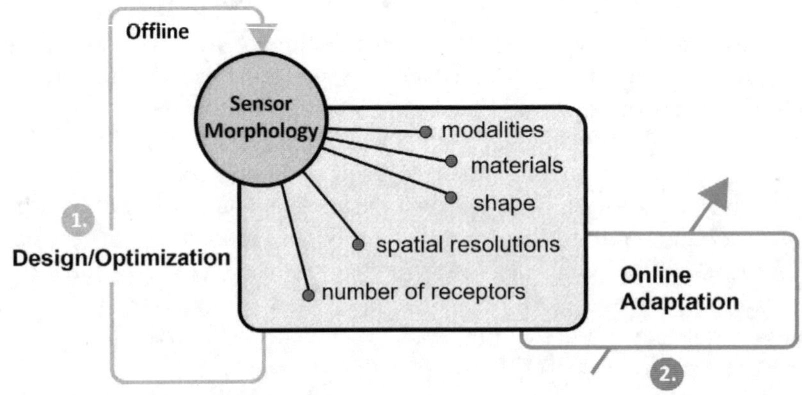

Figure 2. The schematic of morphology embodied active perception. Sensor morphology determines the primary contact and interaction of the agent with the environment. Both physical characteristics and sensing characteristics are considered sensor morphology in this chapter. Physical characteristics include the materials (stiffness, stretchability, etc.), shape, size, etc. Example sensing characteristics are the sensing modalities (the response of different types of receptors), the spatial resolution (distribution of the receptors), and the number of receptors. Sensor optimization refers to the iterative changes of these characteristics in the design phase where the designed morphology of the sensor remains unchanged regardless of the presence of the task-related interaction. Examples of this offline process are the iterative design-fabrication-testing engineering production or biological evolution of an animal. The design/biological structure remains unchanged for the morphology of the individual agent. In contrast, sensor adaptation refers to the online process where the sensing characteristics are tuned in association with the task-related interaction.

of a biological sensor is not defined with a constant geometry, sensitivity, and physical properties. The morphology of the sensors is continuously adapted to the change in the environment to ensure the agent effectively detects and evaluates the stimuli regardless of the changes of the ambient conditions. This "online" process ensured the adaptation of the agent to environmental changes without the need to learn complicated motor behaviours. This process is highlighted in this chapter, reflecting recent solutions in tuning sensor morphology to adapt to the environment and adjust the perceived information [19].

2.1. Sensor Optimization

The "Offline" optimization process to determine and select sensor characteristics based on tasks or environment is essential for sensor development. The challenges of the process are exhibited with the increase in design space dimensions and complexity of interactions. The design of a single-taxel tactile sensor would only need to consider its sensory structure and its interactive cover. In contrast, an array-based tactile sensor skin would face additional challenges with spatial resolution, deformability, adaptability, and sensor coverage.

The sensor morphology design developed naturally from analysis (e.g., theoretical modelling, Finite Element Modeling) to optimization (with known design specifications). An emerging interest is also observed for sensor development based on Sim2Real approaches to iterative evaluate sensor performance with a physical simulator [22, 23]. Preliminary work on sensory morphology and controller has been done with a genetic algorithm to select simple sensor morphologies in the scope of a line-following robot [24]. This section reviews several approaches that address the role of morphology in tactile sensors, showing how the embodiment can act as a physical filter to alter the inference process of the received tactile information.

The elastic cover of the tactile sensors is considered as a key parameter for "Offline" tuning the sensor characteristics due to its mechanical filtering effect (also known as mechanical information-coding). An early study with Shimojo [25] analyzed the low-pass spatial filtering effect of the cover, modelled the underline stress distribution as the convolution of the surface stress distribution. The finite-element method was used to calculate the filtering gain for different elastic covers, studying the effect of cover thickness, Young's Modulus, and Poisson's Ratio [25]. Work in [26] studied the effects of sheltering rubber layer, applied on single-crystalline silicon 3D-force sensors. With a theoretical model to forecast the sensor response, sensor optimization can be achieved by tuning the cover to have a better receptive field size at the expense of sensitivity. The investigation carried out by Cabibihan et al. [27] shows the effects of skin thickness on discriminating the object shape (subsurface pressure profiles of flat, curved, and braille) that is in contact at the surface of the sensor. The performance of the mechanical filter ultimately depends on the environment characteristics [28].

Sensor placement and deployment are another key "Offline" parameters for altering the inference process of perceived information. This subject is particularly critical when embedding sensors with soft composite and soft robotics. The infinite degrees of freedom of the soft body act as an inherited mechanical filter when interacting with the environment. The sensor placement and deployment need to comply with the mechanical response of the soft body and utilize it for better proprioception or exteroception. Culha et al. programmed the fibre morphology of conductive thermoplastic elastomer-based strain gauge sensors to embed into soft elastic bodies [29]. By analyzing the soft elastic deformations and extracting strain vectors, the programmed sensor morphology can effectively discriminate different motion patterns of the soft body [29].

2.2. Sensor Adaptation

Sensor adaptation, indicating an "Online" tuning capability of the sensor morphology, supports the process of active sensing with embodied intelligence for inference shaping. The adaptation has been well adopted in biological sensors to mitigate environmental variations. For instance, the rats control the muscle stiffness of the follicle where the whisker is connected to bias the optimal range of sensing from low to high frequencies [30]. Humans change their finger stiffness and joint impedance to maximize the gained tactile information in the context of haptic explorations [31]. In theory, all physical parameters for "offline" sensor optimization can be explored with "online" sensor adaptation. However, some parameters are directly associated with the construction of the sensor, resulting in a limited tuning range. He et al. proposed a tunable stiffness soft sensor that the sensing characteristics (such as sensitivity and sensing range) can be adjusted online via positive pressure [32]. A limited tunable range of contact stiffness can be achieved for sensor adaptation, while the baseline sensor characteristics are provided via 3D printing elastomeric material with different stiffness for sensor optimization. A wrinkled soft tactile sensor with tunable morphology and stiffness was developed, demonstrating the potential of maximizing sensor performance by adapting morphology in a specific task, such as force sensing, shape discrimination, and texture detection [33, 34]. Costi et al. proposed a magneto-active elastomer filter for online stiffness tuning by placing a magnet on the end-effector, indicating an increased classification accuracy with adaptive stiffening [35]. Phase-changing material such as granular jamming can also be used as an online tunable mechanical filter when placed over a tactile sensor [36].

Morphology sensor adaptation can also provide additional sensor resilience. Conventional, the reflection of sensitivity changes of polymer-based material in tactile sensing due to ageing, temperature dependence and nonlinearity is considered a critical limitation. For instance, Maiolino et al. employed temperature compensation in the tactile sensor design to address the change of capacitance of the sensor material at different temperatures [12]. With an actuated adaptive sensing system, compensation can be achieved via altering the sensor morphology. An example is the soft whisker sensor system proposed by Nguyen and Nhan Huu [37], which compensates for the tactile deficiency upon the whisker being broken, torn or trimmed by changing the stiffness. Resilient sensing can also be achieved by altering the sensor spatial resolution at the software level according to the tasks. For instance, selectively reading and processing a reduced number of sensors (receptors) can mitigate sensor impairment and redundancy while reducing the computation cost.

3. Motor Embodied Active perception

Illustrated in the schematic shown in Fig. 1, tactile perception is an integrated process of "exteroception" and "proprioception" based on motor actions. The measured tactile information ultimately depends on the sensor's physical properties and the associated motor action when the agent interacts with the environment. The adaptation of the motor actions creates a dynamic physical reservoir that filters the tactile information for active sensing. Examples of these motor embodied perception are widely seen in nature. For instance, humans modulate finger behaviour and explore patterns better perceive touch-related information for recognition and discrimination [38]. Humans also actively control the leg joint impedance and gait pattern to maintain stability while sensing in unstructured terrain [39].

Motor embodied active perception plays a significant role in the adaptation process of an agent, understanding the environment and evolving cognitive capabilities through both passive and active interaction with the environment [40]. Passive interaction is considered a process for the agent to respond to active stimuli and better perceive and understand the intention of others. For instance, in the process of "being touched" with aroused "affective emotions", the agent is conditioning the motor actions based on the interpreted emotion of the agent who is performing the touch [41]. The realization relies not only on the sensor spatial resolution and the type of mechanoreceptors but also on the interaction determined by the active motor actions. The subject is less explored in the field of robotics, considering the difficulties in creating a large-area tactile skin with high spatial resolution and the challenges in collecting high-quality/robust training data [42]. In contrast, active tactile perception is more developed in the field, demonstrating applications in environment exploration, dexterous in-hand manipulation, tactile feature recognition, and human-robot interaction [43].

Robotic whiskers are classic examples of motor embodied active perception in environment exploration [44, 45]. By attaching artificial whiskers to a mobile robot platform and an actuated moving mechanism, the robot can effectively increase the whisker sensing performance. For instance, each whisker module on the Shrewbot consists of a motor and shaft encoder, a 3-axis Hall effect sensor, and an embedded microprocessor [44]. Adopting low latency motor control of the rhythmic motion of the whiskers in response to contact-induced stimuli, the Shrewbot constrains the sensory range while maximizing the number of whisker contacts [44]. Another whisker design that utilizes material compliance instead of individual whisker motion is the TacWhisker design. By modifying a vision-based tactile sensor, TacWhisker used one motor to actuate 10 whiskers to create an open/close whisking motion for object localization [46].

In the field of object recognition with tactile information, motor embodied active perception allows information-rich data to be extracted with related actions from raw tactile signals. For instance, Kaboli et al. used a robotic hand with fingertip tactile sensors to distinguish different in-hand objects through texture properties detected via small sliding movements of the fingertips of the robot over the object surface [47]. The motor action is similar to what a human does to identify an in-hand object.

The motor embodied active perception has also been demonstrated at the wrist and arm level when performing active tactile exploration with a robotic arm. Inspired by how practitioners regulate their hand and wrist motion in medical palpation to diagnose the abnormalities

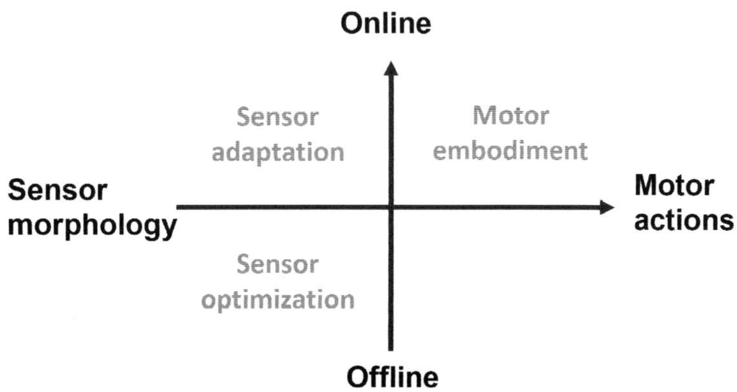

Figure 3. The distinction of sensor optimization, sensor adaptation, and motor embodiment in active embodied tactile perception.

underneath the human skin, Scimeca et al. investigated how robot actions can augment the tactile perception in soft-body palpation [48]. Validated on a robotic patient, the study shows that small changes in the motor action would significantly affect the sensor performance [49].

4. Designing as a whole for active perception

High-performance tactile sensing required a joint effort of both sensor morphology optimization/adaptation and strategic control of the motor actions. In the previous sections, we discussed morphology embodied active perception (including sensor optimization and sensor adaptation) and motor embodied active perception. The baseline we want to emphasize here is "designing as a whole" with sensor optimization, sensor adaptation, and motor embodiment. Considering embodied active tactile perception as the process of shaping the inference process based on the environment, both offline and online adaptation is essential for the optimization of sensing (Fig. 3). Sensor optimization is defined as the offline selection and optimization process to modify the sensor characteristics. In contrast, both sensor adaptation and motor embodiment are considered online processes, where sensor adaptation is the change of characteristics of the sensor itself, and motor embodiment is the change of motor actions of an actuator where the sensor is attached. The concept of "designing as a whole" describes sensor development as an inverse problem that goes beyond the sensor itself. Designers should understand the niche of the problem, identify the broad scope of the tasks, and then design body-sensing-action together. When considering the sensory system and the motor system as a whole, the boundary between sensor adaptation and motor embodied active perception is blurred. The difference between these two subjects is defined in this chapter for the purpose of modelling. However, in reality, the classification of sensors and motors works in a continuum domain. Both sensor adaptation with tunable sensor morphology and motor embodied active perception create an online adjustment on the mediated tactile interaction. By extending the scope and definition of the sensory system, considering the motor system as part of the active change of the sensor morphology, designers can determine the contribution of each subject to achieve the same objective of online adaptation.

It also needs to be noted that, within the framework of "designing as a whole", the designer should evaluate the performance and capability of both the sensor morphology and the motor actions. In many cases, the motor actions determined by the body of the agent provided the constraints on the level of adaptation. For instance, a single degrees of freedom (DoF) gripper can only create a single-DoF motor action to alter the embodied active sensing. With increased degrees of freedom in the motor system, the complexity of creating optimal motor embodied active perception is increased. Similar challenges are also considered in the field of sensorimotor coordination and active robot control, where the effect of learning is considered. In the case of sensing in soft robotics with infinite degrees of freedom, the role of the environment is more significant. By implementing "designing as a whole" for the sensors and motors development, the adaptation to the environment should be considered in both the online and offline phases.

5. References

[1] J. Troccaz, G. Dagnino, and G.-Z. Yang, "Frontiers of medical robotics: from concept to systems to clinical translation," *Annual review of biomedical engineering*, vol. 21, pp. 193–218, 2019.
[2] R. S. Dahiya and M. Valle, *Robotic tactile sensing: technologies and system.* Dordrecht: Springer, 2013.
[3] G. N. DeSouza and A. C. Kak, "Vision for mobile robot navigation: A survey," *IEEE transactions on pattern analysis and machine intelligence*, vol. 24, no. 2, pp. 237–267, 2002.
[4] M. H. Lee and H. R. Nicholls, "Review article tactile sensing for mechatronics—a state of the art survey," *Mechatronics*, vol. 9, no. 1, pp. 1–31, 1999.
[5] M. A. Otaduy, A. Okamura, and S. Subramanian, "Haptic technologies for direct touch in virtual reality," in *ACM SIGGRAPH 2016 Courses*, SIGGRAPH '16, (New York, NY, USA), Association for Computing Machinery, 2016.
[6] R. S. Johansson and Å. B. Vallbo, "Tactile sensory coding in the glabrous skin of the human hand," *Trends in neurosciences*, vol. 6, pp. 27–32, 1983.
[7] L. Zou, C. Ge, Z. J. Wang, E. Cretu, and X. Li, "Novel tactile sensor technology and smart tactile sensing systems: A review," *Sensors*, vol. 17, no. 11, p. 2653, 2017.
[8] S. Balasubramanian, A. Melendez-Calderon, A. Roby-Brami, and E. Burdet, "On the analysis of movement smoothness," *Journal of neuroengineering and rehabilitation*, vol. 12, no. 1, pp. 1–11, 2015.
[9] R. Pfeifer and C. Scheier, "Sensory—motor coordination: The metaphor and beyond," *Robotics and autonomous systems*, vol. 20, no. 2-4, pp. 157–178, 1997.
[10] A. Sheya and L. B. Smith, "Development through sensorimotor coordination," *Enaction: Toward a new paradigm for cognitive science*, pp. 123–144, 2010.
[11] C. J. Cascio, D. Moore, and F. McGlone, "Social touch and human development," *Developmental cognitive neuroscience*, vol. 35, pp. 5–11, 2019.
[12] P. Maiolino, M. Maggiali, G. Cannata, G. Metta, and L. Natale, "A flexible and robust large scale capacitive tactile system for robots," *IEEE Sensors Journal*, vol. 13, no. 10, pp. 3910–3917, 2013.
[13] L. Seminara, L. Pinna, M. Valle, L. Basiricò, A. Loi, P. Cosseddu, A. Bonfiglio, A. Ascia, M. Biso, A. Ansaldo, *et al.*, "Piezoelectric polymer transducer arrays for flexible tactile sensors," *IEEE Sensors Journal*, vol. 13, no. 10, pp. 4022–4029, 2013.
[14] S. Stassi, V. Cauda, G. Canavese, and C. F. Pirri, "Flexible tactile sensing based on piezoresistive composites: A review," *Sensors*, vol. 14, no. 3, pp. 5296–5332, 2014.
[15] H. Zhao, K. O'Brien, S. Li, and R. F. Shepherd, "Optoelectronically innervated soft prosthetic hand via stretchable optical waveguides," *Science robotics*, vol. 1, no. 1, p. eaai7529, 2016.
[16] R. S. Dahiya, G. Metta, M. Valle, and G. Sandini, "Tactile sensing—from humans to humanoids," *IEEE transactions on robotics*, vol. 26, no. 1, pp. 1–20, 2009.
[17] R. Pfeifer, M. Lungarella, and F. Iida, "Self-organization, embodiment, and biologically inspired robotics," *science*, vol. 318, no. 5853, pp. 1088–1093, 2007.
[18] L. Seminara, P. Gastaldo, S. J. Watt, K. F. Valyear, F. Zuher, and F. Mastrogiovanni, "Active haptic perception in robots: a review," *Frontiers in neurorobotics*, vol. 13, p. 53, 2019.
[19] F. Iida and S. G. Nurzaman, "Adaptation of sensor morphology: an integrative view of perception from biologically inspired robotics perspective," *Interface focus*, vol. 6, no. 4, p. 20160016, 2016.
[20] M. Stevens, "Sensory ecology, evolution, and behavior," *Curr. Zool*, vol. 56, pp. 1–3, 2010.
[21] T. Yang, D. Xie, Z. Li, and H. Zhu, "Recent advances in wearable tactile sensors: Materials, sensing mechanisms, and device performance," *Materials Science and Engineering: R: Reports*, vol. 115, pp. 1–37, 2017.
[22] D. F. Gomes, P. Paoletti, and S. Luo, "Generation of gelsight tactile images for sim2real learning," *IEEE Robotics and Automation Letters*, vol. 6, no. 2, pp. 4177–4184, 2021.
[23] P. Pagliuca and S. Nolfi, "The dynamic of body and brain co-evolution," *Adaptive Behavior*, vol. 30, no. 3, pp. 245–255, 2022.
[24] K. Sugiura, H. Kawakami, and O. Katai, "Simultaneous design of the sensory morphology and controller of mobile robots," *Electrical Engineering in Japan*, vol. 172, no. 1, pp. 48–57, 2010.

[25] M. Shimojo, "Mechanical filtering effect of elastic cover for tactile sensor," *IEEE Transactions on Robotics and Automation*, vol. 13, no. 1, pp. 128–132, 1997.

[26] G. Vásárhelyi, M. Adám, E. Vázsonyi, I. Bársony, and C. Dücső, "Effects of the elastic cover on tactile sensor arrays," *Sensors and Actuators A: Physical*, vol. 132, no. 1, pp. 245–251, 2006.

[27] J.-J. Cabibihan, S. S. Chauhan, and S. Suresh, "Effects of the artificial skin's thickness on the subsurface pressure profiles of flat, curved, and braille surfaces," *IEEE Sensors Journal*, vol. 14, no. 7, pp. 2118–2128, 2013.

[28] L. Costi, P. Maiolino, and F. Iida, "How the environment shapes tactile sensing: Understanding the relationship between tactile filters and surrounding environment," *Frontiers in Robotics and AI*, p. 180, 2022.

[29] U. Culha, U. Wani, S. G. Nurzaman, F. Clemens, and F. Iida, "Motion pattern discrimination for soft robots with morphologically flexible sensors," in *2014 IEEE/RSJ International Conference on Intelligent Robots and Systems*, pp. 567–572, IEEE, 2014.

[30] H. Wegiriya, N. Sornkarn, H. Bedford, and T. Nanayakkara, "A biologically inspired multimodal whisker follicle," in *2016 IEEE International Conference on Systems, Man, and Cybernetics (SMC)*, pp. 003847–003852, IEEE, 2016.

[31] N. Sornkarn and T. Nanayakkara, "Can a soft robotic probe use stiffness control like a human finger to improve efficacy of haptic perception?," *IEEE transactions on haptics*, vol. 10, no. 2, pp. 183–195, 2016.

[32] L. He, N. Herzig, T. Nanayakkara, and P. Maiolino, "3d-printed soft sensors for adaptive sensing with online and offline tunable stiffness," *Soft Robotics*, 2022.

[33] Q. Qi *et al.*, "Wrinkled soft sensor with variable afferent morphology: Case of bending actuation," *IEEE Robotics and Automation Letters*, vol. 5, no. 3, pp. 4102–4109, 2020.

[34] H. Yamashita, Z. Wang, S. Hirai, K. Shibuya, *et al.*, "Wrin'tac: Tactile sensing system with wrinkle's morphological change," *IEEE Transactions on Industrial Informatics*, vol. 13, no. 5, pp. 2496–2506, 2017.

[35] L. Costi, A. Tagliabue, P. Maiolino, F. Clemens, and F. Iida, "Magneto-active elastomer filter for tactile sensing augmentation through online adaptive stiffening," *IEEE Robotics and Automation Letters*, vol. 7, no. 3, pp. 5928–5933, 2022.

[36] J. Hughes, L. Scimeca, P. Maiolino, and F. Iida, "Online morphological adaptation for tactile sensing augmentation," *Frontiers in Robotics and AI*, vol. 8, 2021.

[37] N. H. Nguyen *et al.*, "Tactile compensation for artificial whiskered sensor system under critical change in morphology," *IEEE Robotics and Automation Letters*, vol. 6, no. 2, pp. 3381–3388, 2021.

[38] J. Konstantinova, G. Cotugno, P. Dasgupta, K. Althoefer, and T. Nanayakkara, "Palpation force modulation strategies to identify hard regions in soft tissue organs," *PloS one*, vol. 12, no. 2, p. e0171706, 2017.

[39] E. Hamid, N. Herzig, S.-A. Abad, and T. Nanayakkara, "A state-dependent damping method to reduce collision force and its variability," *IEEE Robotics and Automation Letters*, vol. 6, no. 2, pp. 3025–3032, 2021.

[40] H. Arakawa and R. S. Erzurumlu, "Role of whiskers in sensorimotor development of c57bl/6 mice," *Behavioural brain research*, vol. 287, pp. 146–155, 2015.

[41] D. Silvera-Tawil, D. Rye, and M. Velonaki, "Artificial skin and tactile sensing for socially interactive robots: A review," *Robotics and Autonomous Systems*, vol. 63, pp. 230–243, 2015.

[42] S. Saeedvand, M. Jafari, H. S. Aghdasi, and J. Baltes, "A comprehensive survey on humanoid robot development," *The Knowledge Engineering Review*, vol. 34, 2019.

[43] R. S. Dahiya, P. Mittendorfer, M. Valle, G. Cheng, and V. J. Lumelsky, "Directions toward effective utilization of tactile skin: A review," *IEEE Sensors Journal*, vol. 13, no. 11, pp. 4121–4138, 2013.

[44] M. J. Pearson, B. Mitchinson, J. C. Sullivan, A. G. Pipe, and T. J. Prescott, "Biomimetic vibrissal sensing for robots," *Philosophical Transactions of the Royal Society B: Biological Sciences*, vol. 366, no. 1581, pp. 3085–3096, 2011.

[45] C. Zhao, Q. Jiang, and Y. Li, "A novel biomimetic whisker technology based on fiber bragg grating and its application," *Measurement Science and Technology*, vol. 28, no. 9, p. 095104, 2017.

[46] N. F. Lepora, N. Burnus, Y. Tao, and L. Cramphorn, "Active touch with a biomimetic 3d-printed whiskered robot," in *Conference on Biomimetic and Biohybrid Systems*, pp. 263–275, Springer, 2018.

[47] M. Kaboli, R. Walker, G. Cheng, *et al.*, "In-hand object recognition via texture properties with robotic hands, artificial skin, and novel tactile descriptors," in *2015 IEEE-RAS 15th International Conference on Humanoid Robots (Humanoids)*, pp. 1155–1160, IEEE, 2015.

[48] L. Scimeca, J. Hughes, P. Maiolino, L. He, T. Nanayakkara, and F. Iida, "Action augmentation of tactile perception for soft-body palpation," *Soft robotics*, vol. 9, no. 2, pp. 280–292, 2022.

[49] L. He, N. Herzig, S. de Lusignan, L. Scimeca, P. Maiolino, F. Iida, and T. Nanayakkara, "An abdominal phantom with tunable stiffness nodules and force sensing capability for palpation training," *IEEE Transactions on Robotics*, vol. 37, no. 4, pp. 1051–1064, 2020.

IOP Conf. Series: Materials Science and Engineering 1292 (2023) 012008 doi:10.1088/1757-899X/1292/1/012008

Heterogeneous Underwater Swarm of Robotic Fish

S. Iacoponi[1], M. Hanbaly[1], A. Infanti[1] ,
B. Andonovski[1] , N. Mankovskii[2], I. Zhilin[2],
F. Renda[1], C. Stefanini[1,3], G. De Masi[1,2]

Khalifa University[1], Tecnology Innovave Institute[2], Sant'Anna School of Advanced Studies[3]

E-mail: saverio.iacoponi@ku.ac.ae

September 2022

Abstract. Underwater robotics is prevalent today in the field of undersea surveys. The current single-agent approach struggles to cover surveys of large areas, especially when visual analyses are required. Applications of underwater multi-robots fleets need to manage the coordination of the agents, which in turn stresses the already limited capabilities of underwater localization and communication. We propose a concept for a multilayered control architecture, which leverages visual recognition and swarm behavior to negotiate the coordination requirements of the single agent while retaining a significant level of control over the fleet of vehicles as a whole. This is obtained by including multiple control layers, paired with a heterogeneous composition of vehicles.

1. Introduction and Motivation

Undersea exploration and survey are of great interest for numerous fields in research and commercial applications [1]. As one of many examples, monitoring of the underwater ecosystem, particularly the health of the coastal shallow water areas, impacts commercial needs such as the fishing industry[2], [3], oil&gas infrastructures' development, civil structures (particularly harbors), development of new shipping lines. Sites of biological, naturalistic, and archaeological value still present open challenges concerning exploration and preservation effort [4] [5].

Due to the complexity of human underwater exploration, even when limited to shallow waters, robots are often employed in those types of surveys [6]. The main limitation of studies and surveys in such locations are determined by the challenges of the underwater environment: In particular, communication [7] and localization [8] represent the main concern for Underwater Autonomous Vehicles (AUV). Water, and especially seawater, is a natural barrier to most electromagnetic frequencies. Those include most, if not all, wireless electromagnetic communication channels used above water. Thus, today's state of the art for underwater communication is focused on acoustics, which is limited to the order of a few kbits/s of bandwidth at most. This characteristic of water affects also the localization, because GNSS (Global Navigation Satellite Systems) is unreachable underwater. Coupling the absence of GNSS with underwater currents and free-floating body dynamics, the estimation of absolute position becomes extremely challenging.

One less discussed challenge of the underwater world, is represented by the limited range of visibility. This is a byproduct of the absorption of sunlight from the water and particle suspensions. The first leads to progressive loss of natural light to complete darkness with the increase of depth. Particle suspension and chemical composition of water determine the absorption and scattering properties of the medium [9]. Those parameters strongly affect the visibility range, even considering the use of artificial lights, and typical values of effective visibility range could vary from tens of meters to under one meter depending on the conditions. Traditional underwater robotics bypasses the issue, relying mostly on sonar and acoustic technology when performing wide areas surveys. However, the amount of information that can be extracted by a sonar survey is less detailed and of different content from what can be obtained by a close visual inspection.

The most promising approach to improve the performance of visual underwater surveys is the implementation of a multi-robot approach, specifically a swarm approach [10, 11]. The first obvious advantage lies in the possibility to parallelize the job among multiple agents: this not only allows for lower operating times and costs, but enables wide area studies when the time frame for the data acquisition impacts the measurements (for example: when studying fauna active only at a certain time of the day, or chemical proprieties with a 3d pattern or vertical stratification). Moreover, the swarm allows tasks' allocation among the different AUVs, having different tasks performed at the same time independently or collaboratively by the AUVs [12]. This also allows more flexibility during the mission, especially when the environmental conditions are dynamically changing or external factors impose a change of the mission. Furthermore, underwater robots are susceptible to failure due to the difficult environment: a distributed approach offers significantly more robustness to a mission as the loss of functionality of single vehicles does not impede the mission itself . In addition, using a multi-robot approach opens new opportunities to challenge the issue of localization and path reconstruction. In this work, we envision the use of swarm control algorithms, where the control rules of each agent are not reliant on accurate localization of themselves or of the other agents, but only on the application of simple rules based on neighbor detection and minimal data sharing. By fusing the information of the single agents, a collective localization can be then achieved. A more traditional top-down control is then added to allow the operator to interface with the swarm. The different control layers that constitute the proposed architecture of our system are described below.

In this paper, we present a conceptual architecture of a fleet of underwater and surface robots for wide-range underwater surveys, that rely on swarm behavior to coordinate all the robots.

2. General Platforms description and functionalities

A full description of the platforms is presented in [13]. The robotic swarm is composed of the large majority of fish-robots. Tab 1 summarizes the main characteristics of such platforms. Those fish-robots are AUVs (Autonomous Underwater Vehicles), capable of freely moving underwater. Each of those robots mounts a set of three cameras that offer up to a 270-degree FOV (Field of View). The Fish-robots are then subdivided into two classes: basic and advanced. The main variation of the advanced robot is that it features an acoustic modem for communications (Succorfish V3). Differently from most AUVs, these models are not mounting any kind of sonar or DVL (Doppler Velocity Logger) or USBL (Ultra Short Base Line); those instruments are usually part of most AUV architectures and are essential in the process of localization and odometry. However, their cost, size, and complexity, are not fit to be mounted on small and expendable platforms as required for the single swarm agent.

A USV (Unmanned Surface Vehicle), named Floater is added to the swarm. The Floater features an array of sensors, along with GNSS localization, acoustic modem, downward pointing underwater camera, and wireless communication with the surface operators (see Tab 2). From the Floater, a second platform can be deployed. This second platform, called Sinker, does not have self-locomotion and it is dependent on the Floater. The two platforms are connected by a tether. A Pulley system allows for autonomous deployment and retrieval of the sinker. The

Table 1. Main characteristic of the fish-robot Platform.

Sizes	45 cm length, 25 cm height, 15 cm width
Dry weight	4.1 Kg
Actuation	3 motors. 1 Vertical, 2 horizontal
Vision	Frontal Camera, 2 lateral cameras
Sensors	Pressure, IMU 9DoF, Temperature, Turbidity, Electric Sense, Echosounder for distance measure from the bottom
Communication	Succorfish V3 Acustic Modem, WiFi and Bluetooth (only for outside-water communication)

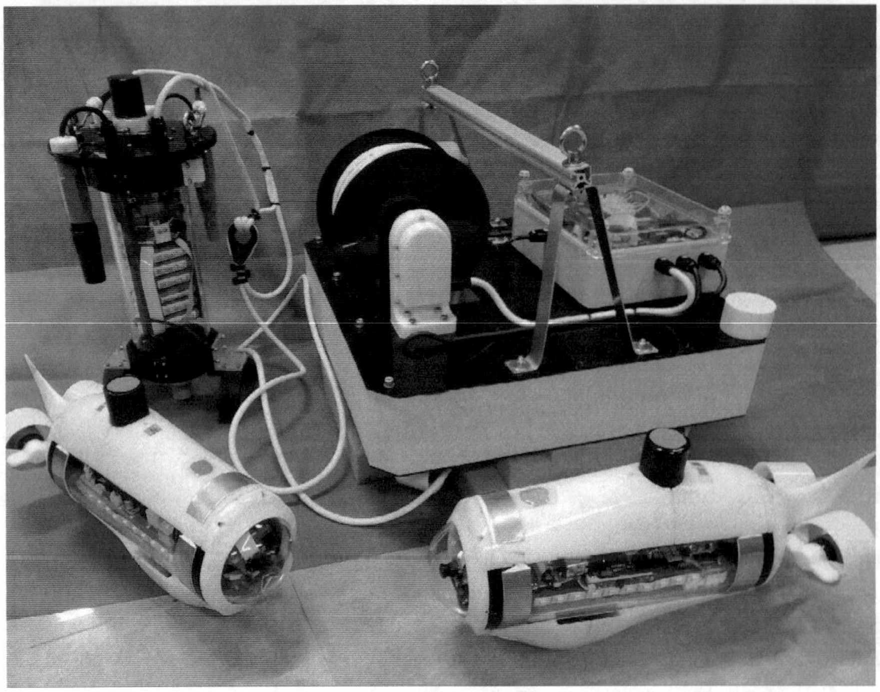

Figure 1. Picture of the swarm agents: on the left back the Floater platform, connected though a tether to the Sinker platform on the left. In the front 2 of the Fish-Robots are visible. Note that, since they are equipped with acoustic modem (visible on top), those agent can act as local group leaders.

Sinker features an array of sensors and modules as well, along which an acoustic modem and 360 FOV array of cameras (see Tab 3). Fig. 1 shows the swarm agents.

2.1. Communication

The communication among the different platforms play a crucial role. The final overall scenario considers an operator based on a control station able to be connected to the swarm underwater. This allows the operator to act as being present in the underwater environment, observe the environment and change the mission of the AUVs.

The communication between the floater and the base station is based on WiFi technology, while the communication between floater and sinker in guaranteed by an ethernet cable. More interesting and open to research are the communications channel between the floater and the fish and among the fish. While most underwater communication is handled via the acoustic

Table 2. Main characteristic of the Floater Platform

Sizes	50cm length, 45cm height, 50cm width
Dry weight	10.7 Kg
Actuation	4 motors O-configuration
Vision	1 down-pointing camera cameras
Sensors	IMU 9DoF, Temperature, Turbidity
Localization	GPS, GNS
Communication	Succorfish V3 Acustic Modem, Mobile Net WiFi and Bluetooth

Table 3. Main characteristic of the Sinker Platform

Sizes	22cm diameter, 45cm height
Dry weight	5.2Kg
Actuation	Pulley system
Vision	4 cameras, 360 degrees FoV total
Sensors	Pressure, IMU 9DoF, Temperature, Turbidity, Electric Sense, Echosounder for distance measure from the bottom, Chemical Probes (Ph, etc..)
Communication	Succorfish V3 Acustic Modem, Wifi (outside water)

channel, the proposed system requires some alternative modes of communication. The main drawbacks that the acoustic system present in our case are the limited bandwidth and the long communication delay: The acoustic channel is shared among all agents, so only one communication at a time can be transmitted at any moment for the whole swarm of robots. The acoustic modem selected for the project allows for approximately 64 bytes per second, which is close to the performance of most acoustic transceivers available. Possible solutions to overcome these limitations intrinsic to the acoustic channel rely on pre-processing of data before transmission, as for example using encoding-decoding schemes. To achieve a higher rate of communication different modes of communication have been implemented. Optical communication has been already implemented for underwater in [14] by using blue light wavelength; in this case bitrates of order of Gb/s can be achieved, with a maximum range, in optimal conditions, of 500 meters. For this purpose, blue light LEDs can be integrated; alternatively, other frequencies could be used for our case, such as high-infrared, as the communication distance required is less than 10 meters.

A hybrid (multimodal) method (optical and acoustic) is the best solution [15]. This way high bitrates can be achieved at short ranges, and the fish can still communicate when the members of the swarm move out of range of optical communication. Other means of communication also become available, still at a research stage, such as electric field emission and sensing, using the technology presented in [16], or low-frequency radio communication. Electric sense has been integrated on the new version of the fish platform, mainly for perception purposes. Another avenue that opens thanks to the proximity of the agents, is related to vision: the lateral line made by RGB LEDs can be used to broadcast static messages by position and color, beyond the more obvious purpose of guaranteeing visual recognition and allowing estimate of the relative position over time of neighbors.

3. Multi-robot control architecture

In this section, we propose a system control architecture for our heterogeneous Underwater Swarm. We approach the problem by subdividing it into layers.Each layer is described below, while Fig.2 present a synthetic summary of each layer structure.

3.1. Single Agent Layer

The first layer is represented by the fish-robot isolated. Each robot carries a series of sensors that allow for a minimal level of control. The depth control, heading control, and approximate surge control are handled at this level. Inertial and Pressure sensors provide feedback for a closed loop control on the first parameters, while surge velocity must be controlled open loop at this layer. However, considering the large number of platforms sharing the same hydrodynamics and propulsion system, it is reasonable to invest in an in-depth characterization of the command-to-velocity model to allow for fairly reliable control. An in-depth study of such a model is being conducted for our platforms in [17].

3.2. Local Group Layer

The local group layer interests a group of robot-fish that work in close proximity. Each group is composed of at least one advanced fish-robot. In this layer, acoustic communication is not used for coordination or control. The agents are capable of seeing each other in their FOV and are close enough that the water proprieties do not affect the capability of each agent to see or recognize each other. On this layer, visual broadcast communication (LED strips) can be implemented, as well as short-range communication. Onboard software, with the application of Neural Network algorithms, is able to identify close neighbor relative positions appearing in the FOV as well as roughly estimate their relative heading. Multiple swarm-like algorithms can be implemented using relative position and internal odometry deviated from the single agent layer. This allows for different behaviors and formations to be implemented and swapped according to mission and situation. The local group uses the alternative short-range communication channels presented above to share the behavior to be applied.

3.3. Super Group Layer

The supergroup layer focuses on the implementation of acoustic communication and ranging. TDoA (Time Differential of Arrival) can be used to localize the different local group leaders (Advanced fish-robot). Acoustic communication allows sending commands to local groups, effectively using each local group as a single agent in a traditional multi-robot approach. This represents also the lower level of control accessible to the human operator. The collected TDoA is locally used for odometry refinement, but more importantly for post-mission trajectory reconstruction.

3.4. Floater and Sinker

The Floater and Sinker platforms can be considered a separated layer that interacts transversely with all other layers. The Floater and sinker have their own shared single agent layer. On top of that, they can use fast internal communication, GNSS self-localization, and vision to accurately track the 3D absolute position of the Sinker. The Sinker can be recognized by close neighbors and support local groups by sharing information with the operator, thanks to the cabled connection to the surface. Furthermore, a close neighbor can be identified visually and properly localized via triangulation, when appearing in the FOV of the Sinker and Floater downward camera. Floater and Sinker also act as separate acoustic nodes to further refine the agents' odometry and trajectory reconstruction. Finally, Floater and Sinker act as a bridge with the human operator.

3.5. User Layer

Thanks to the Floater, a human operator can send directives and receive feedback from underwater agents. It can send specific instructions to a local group or send general instructions

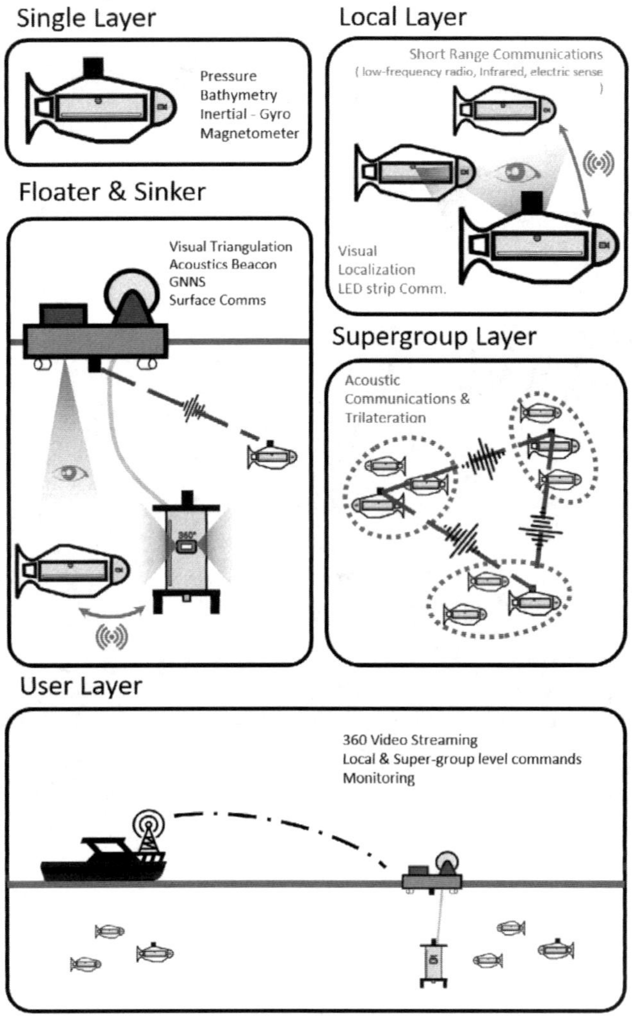

Figure 2. A representation of the different layers, and the types of localization and communication strategies implemented at each level. Starting from single layer, with only basic sensing, local group dominated by visual, super-group depending acoustic, user layer interfacing from the surface, and floater sinker acting as bridge among the different layers.

to be executed globally. The Sinker 360 view paired with a fast connection enables continuous visual feedback that allows for a third-person view of the swarm. We envision here a Virtual Reality set to be used to enable a full virtual immersion.

4. Conclusions

We envision for a multilayered control approach for an underwater swarm of robots to enable several new types of mission profiles. A localized use of swarm control coupled with higher level multi-agent allows to negotiate an uncertain localization of single agents while retaining top-level control and feedback of the system. The multi-agent approach allows for a later trajectory refinement by comparing relative distances and ranging measurements collected during the mission, which in turn allows more accurate spacial localization of all the readings and measurements collected. In Fig. 3 we have a conceptual representation of the swarm agents

Figure 3. Rendering of a potential mission scenario: Floater on the surface deploying the sinker to monitor two local grops of the agents each, busy monitoring the area.

acting in a mission.

A large number of agents enable wide-scale visual inspection as well as chemical survey mapping. Potential missions may include seabed algae and coral mapping for sea-health inspection, fish-counting surveys for fishing industries, patrolling of fish farms, archaeological sites virtual reconstruction, underwater mine clearance operations, preliminary and routine inspection for underwater constructions and facilities, including pipes and cable systems, and collaboration in Search and Rescue mission.

Acknowledgements

This work is part of the project "HSURF: Underwater Heterogeneous swarm of robots," funded by the Technology Innovation Institute (UAE) and developed with Khalifa University (Project No.8434000392).

References

[1] R. Bogue, "Underwater robots: a review of technologies and applications," *Industrial Robot: An International Journal*, 2015.

[2] M.-J. Rochet and V. M. Trenkel, "Which community indicators can measure the impact of fishing? a review and proposals," *Canadian Journal of Fisheries and Aquatic Sciences*, vol. 60, no. 1, pp. 86–99, 2003.

[3] J. A. Holmes, G. M. Cronkite, H. J. Enzenhofer, and T. J. Mulligan, "Accuracy and precision of fish-count data from a "dual-frequency identification sonar" (DIDSON) imaging system," *ICES Journal of Marine Science*, vol. 63, pp. 543–555, 01 2006.

[4] M. Reggiannini and O. Salvetti, "Seafloor analysis and understanding for underwater archeology," *Journal of Cultural Heritage*, vol. 24, pp. 147–156, 2017.

[5] A. Bowens *et al.*, *Underwater archaeology: the NAS guide to principles and practice.* John Wiley & Sons, 2011.

[6] J. Yuh and M. West, "Underwater robotics," *Advanced Robotics*, vol. 15, no. 5, pp. 609–639, 2001.

[7] C. M. Gussen, P. S. Diniz, M. L. Campos, W. A. Martins, F. M. Costa, and J. N. Gois, "A survey of underwater wireless communication technologies," *J. Commun. Inf. Sys*, vol. 31, no. 1, pp. 242–255, 2016.

[8] X. Su, I. Ullah, X. Liu, and D. Choi, "A review of underwater localization techniques, algorithms, and challenges," *Journal of Sensors*, vol. 2020, 2020.

[9] C. Gabriel, M.-A. Khalighi, S. Bourennane, P. Leon, and V. Rigaud, "Channel modeling for underwater optical communication," in *2011 IEEE GLOBECOM Workshops (GC Wkshps)*, pp. 833–837, IEEE, 2011.

[10] A. Luvisutto, A. Al Shehhi, N. Mankovskii, F. Renda, C. Stefanini, and G. De Masi, "Robotic swarm for marine and submarine missions: Challenges and perspectives," in *to appear in AUV 2022, Singapore*, IEEE, 2022.

[11] J. Connor, B. Champion, and M. A. Joordens, "Current algorithms, communication methods and designs for underwater swarm robotics: A review," *IEEE Sensors Journal*, vol. 21, pp. 153–169, Jan 2021.

[12] P. Zahadat and T. Schmickl, "Division of labor in a swarm of autonomous underwater robots by improved partitioning social inhibition," *Adaptive Behavior*, vol. 24, 03 2016.

[13] S. Iacoponi, G. J. Van Vuuren, G. Santaera, N. Mankovskii, I. Zhilin, F. Renda, C. Stefanini, and G. De Masi, "H-surf: Heterogeneous swarm of underwater robotic fish," in *to appear in OCEANS 2022, Hampton Roads*, IEEE, 2022.

[14] F. Hanson and S. Radic, "High bandwidth underwater optical communication," *Applied optics*, vol. 47, no. 2, pp. 277–283, 2008.

[15] I. V. Zhilin, O. M. Bushnaq, G. De Masi, E. Natalizio, and I. F. Akyildiz, "A universal multimode (acoustic, magnetic induction, optical, rf) software defined radio architecture for underwater communication," in *The 15th International Conference on Underwater Networks & Systems*, WUWNet'21, (New York, NY, USA), Association for Computing Machinery, 2021.

[16] J. Zheng, C. Huntrakul, X. Guo, C. Wang, and G. Xie, "Electric sense based pose estimation and localization for small underwater robots," *IEEE Robotics and Automation Letters*, vol. 7, no. 2, pp. 2835–2842, 2022.

[17] M. El Hanbaly, S. Iacoponi, A. Infanti, G. De Masi, C. Stefanini, and F. Renda, "Inverse dynamics modelling as a state estimation aid for underwater robots navigation," in *Submitted in ICRA 2023*, IEEE, 2023.

IOP Conf. Series: Materials Science and Engineering 1292 (2023) 012009 doi:10.1088/1757-899X/1292/1/012009

Task-Mediated Design: A latent space for dexterous soft robotic design

Josh Pinskier and David Howard

Robotics and Autonomous Systems Group, CSIRO Data61
Brisbane, Australia

E-mail: `josh.pinskier@csiro.au`

Abstract. Soft robotics has created a paradigm shift within robotic grasping and manipulation. Rather than using complex control policies and path planning methods to undertake grasp objects, soft robots use their embodied intelligence to deform around an object and securely hold it. Yet despite major advances in the field in recent years, there remains an absence of practical methods for designing new high performing soft manipulators. In this perspective, we propose a novel task-mediated approach to soft manipulator design and discuss its application to robust and dexterous manipulation. The task-mediated approach first abstracts the manipulation task into a characteristics latent space, then designs a soft robot to minimise the distance between the actual performance and characteristic latent variables. In conjunction with state-of-the-art generative design tools and simulators, this approach can accelerate the development of bespoke and general purpose manipulators

1. Introduction

Soft robots are prized for their ability to adapt their shape and gait to perform numerous tasks in diverse environments, especially deformable and fragile. The potential capabilities of mobile soft robots are frequently compared to natural invertebrates like octopuses and worms, which can reconfigure their bodies to move through cavities and adapt their locomotion to their terrain [Majidi(2014), Laschi et al.(2012), Rus and Tolley(2015)]. However, despite the rapid growth in soft robotics research, artificial analogues of these animals remain a distant vision. A similar trend has emerged in robotic grasping and manipulation, where inspiration has regularly been drawn from natural grasping methods such as geckos' adhesive feet and chameleons' tongues [Pang et al.(2022), Hao and Visell(2021)]. Yet soft end-effectors remain largely based on anthropomorphic fingers [Hao and Visell(2021)], and despite major technological improvements neither soft nor rigid robots currently come close to matching the gold standard dexterity of human hands. Consider common human manipulation tasks: picking up a soft fruit, unscrewing a bottle-cap, buttoning a shirt, eating with chopsticks, pulling a trigger. Each of these tasks require both a different hand configuration and different force magnitudes and directions, and most require a coordinated sequence of actions to successfully perform the task. Although there are numerous antropomorphic robotic hands, soft grippers, and grasp planners on the market, these routine tasks remain a challenge for robots.

In industries such as healthcare and agriculture there is an acute need for dexterous, soft grasping and manipulation. So what's the holdup and what needs to be done to bring us closer to natural capabilities?

In this paper we discuss the hierarchy of design goals for robotic end-effectors, with an emphasis on soft robotic grasping, and propose a task-mediated design approach to bridge the gap between robotic and natural grasping and manipulation capabilities. Here manipulation can be taken to mean a task involving any combination of grasping, reorienting and reconfiguring an object or environment.

1.1. Embodied Design or Manipulation Policies

A philosophical divide has emerged in recent years between the communities researching conventional (rigid) robotics and embodied/soft robotics. The rigid robotics community sees the problem of robotic grasping primarily as a control policy problem. That is, if a human hand the gold standard in manipulation, then the research problem is how to teach robots to perform human manipulation strategies. Embodied and soft robotics takes a different perspective, which emphasises the connection between the robot's design and control, and the task. Rather than assuming that a well designed controller can overcome design flaws, it sees the manipulation problem as finding the best combination of tool design and controller (i.e body and brain) for a task. By embodying intelligence in the tool design, simple control strategies can robustly perform complex tasks, reducing the burden on high-level control policies and task planning. For example: monkey tails and eagle talons are both high performing natural grippers, which are simpler to operate than human hands.

Consider the example of harvesting a ripe strawberry, which we will refer to throughout the remainder of the paper. The process for harvesting a strawberry is:

(i) Grasping the strawberry's stem
(ii) Gently pulling the berry away from the plant and neighbouring berries
(iii) Rotating the berry with the stem held stationary
(iv) Placing it in a storage container

This process must be completed without damaging any berries or the plant, and must be able to be performed in the cluttered plant environment. Hence at a minimum, a candidate gripper must be able to exert the required grasping forces/moments whilst minimising the forces exerted on non-grasped fruit. Ideally it should also minimise the amount of infrastructure required to perform the task (sensors, actuators, computation, etc)

Rigid robotics assumes that with sufficiently precise controller, an actuated end-effector can grasp the berry without damaging it or neighbouring berries. In contrast, the communities of embodied intelligence and soft robotics would instead consider the tool's morphology and materials first, to see whether its design can improve task performance and reduce the controller complexity. For example soft fingertips increase grasp robustness and reduce the risk of damage.

To be clear, the concepts are not mutually exclusive, but the two approaches naturally lead to very different research prioritics. One focuses on training control policies to create generic, multi-functional end-effectors, whilst the other optimises the tool's design and control for high-performing application specific designs.

1.2. Design Challenges

Embodied intelligence design is not a simple task. To design high performing robots, the entire robot and its task must be considered holistically and cooptimised. The resulting problem has a very-high dimensional feature space, making the process of encoding features and identifying high-performing regions difficult.

Hence, the conventional robotics design approach has been to dramatically reduce the feature space by assuming stiff linkages with discrete joints. Whilst convenient, this vastly reduces the quality of results. It is now well understood that distributed actuation and flexible materials is essential for interacting with deformable objects and unstable terrain [Zhu et al.(2022)]. However, despite the rapid growth of the field, creating new high-performing soft robots remains a slow and laborious exercise. Our ability to generate new designs is impeded by the complex and non-linear design space, lack of high-quality simulators, complex manufacturing process, and limited sensing capabilities.

Notable methods to overcome these issues include voxel-based soft robotic evolution [Cheney et al.(2013)], differentiable simulation [Ma et al.(2021)], in materio evolution [Howard et al.(2022b)], multi-level evolution [Howard et al.(2019)], topology optimisation [Boddeti et al.(2021), Zhang et al.(2019)], and bioengineered robots [Kriegman et al.(2020)].

A more in-depth discussion of these challenges is beyond the scope of this work, but detailed reviews of the state of the art and current challenges in soft robotic design can be found at [Pinskier and Howard(2022), Chen and Wang(2020), Xavier et al.(2022)].

Here, we propose an overarching framework to develop new soft grippers, which is compatible with any of these design methods.

We hope it can be used not just to design novel grippers, but also to answer major open research questions in the field, such as:

(i) Can we make multi-functional soft manipulators with human-like dexterity?
(ii) How can we efficiently design soft robots/end-effectors for a specific task?
(iii) What strategy is best to perform a complex series of tasks? Is it better to use a single multi-functional tool, or a set of bespoke ones? How many?
(iv) How can we efficiently simulate tasks involving interactions between deformable objects and robots whilst capturing their material properties and actuation? Can the simulation to reality gap be closed by incorporating experimental data?

2. Task-Mediated Design

In most design optimisation/evolution methods, the design problem is specified by first considering the selection of material(s), actuation (real or approximated) and a basic description of task (terrestrial or aquatic locomotion, grasping, etc). With the exception of a small number of targeted studies [Collins et al.(2018), Howard et al.(2022a)], the detailed task requirements and environment in which the task is being performed is only considered superficially (smooth or rough ground, hard or soft object).

This simplificiation is a necessary requirement for tractible optimisation in a high-dimensional space, but leads to design spaces far removed from the actual design problem.

We invert this design paradigm and instead propose an design approach which considers the task as the primary optimisation feature, transforms that into a latent task space, then designs around the latent variables.

That is rather than considering manufacturing/design capabilities first (or their approximations) then working forwards to a design, we first consider the task and environment and work backwards to a design.

Hence, we propose a two-stage design process for task-mediated design. In the first stage the task is quantitatively defined and characterised; then in the second, a design is formulated which best approximates the quantitative task definition.

Whilst superficially, this may appear synonymous with inverse design, task-mediated design is far more general. It is not simply designing to reach a set of desired properties, but instead captures the essence of a task or operating environment and find an optimal design for that goal. It determines which forces and displacements are desirable and which are tolerable, and use these as a basis for automated design. Whilst we focus on manipulators in this work, we believe the reduced dimension framework will enable robots to evolve for a broad array of tasks.

An example of the task definition process is illustrated in Figure 1. The act of picking a berry is abstracted to an optimal 4D force array (x,y,z,t) representing the set of forces with picks the fruit with minimal disturbance. The design task is then to find a manipulator which best generates these forces within the required geometric constraints.

This 4D task definintion is suitable for design optimisation, but additional features may also be added to capture higher-level design goals like robustness and dexterity. In the remainder of this work we discuss the tiers of design goals, and task-mediated design can aid in designing up the hierarchy.

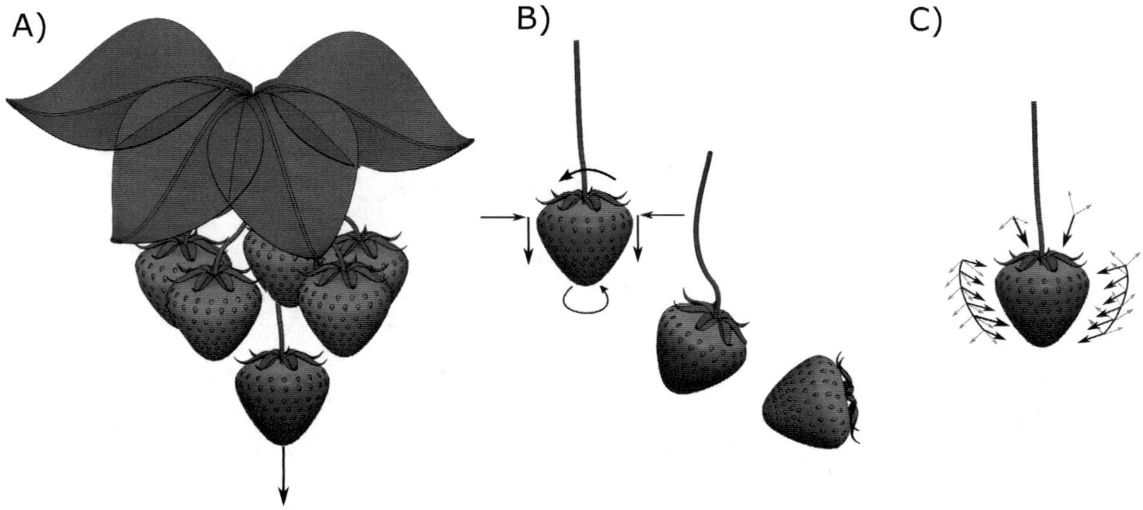

Figure 1. Quantifying the strawberry picking task. (A) The objective of the task is to pick the middle strawberry by pulling it away from the plant with minimal disturbance to the fruit and surrounding foliage. (B) The ideal way to do this is to rotate the berry as it is pulled away from the plant, gently breaking the stem. This requires a set of forces and torques to be applied to firmly hold the berry and generate the motion. (C) Using the berry's material properties (stiffness, coefficient of friction, etc), an optimal 3D force vector can be determined which generates the required motion whilst minimising the deformation of the berry

3. Design Goal Hierarchy

3.1. Task-Mediated Optimisation

The hierarchy of design is illustrated in Figure 2. The most basic design objective is to optimise a gripper for a single object. Whilst ideal for standardised and repetitive pick and place tasks, it is inherently fragile. Small perturbations in the object or the tool position lead to grasp failure. Consider an 'optimal' gripper for a strawberry, it may look something like a 2 part mould of the strawberry, which perfectly matches its geometry and is able to surround the berry to perfectly distribute forces. Whilst this 'imprint' gripping is a good strategy for a uniform object, it will fail to pick berries with even small differences in shape or orientation. For example, the simulated grasp success of imprint grippers fell from 78% to 70% under a 10 deg rotation [Ha et al.(2020)].

The task embodied design process to optimise a single design is a relatively straightforward 2-step process to first identify the task's characteristic load array, then design a gripper to approximate the array.

(i) Configure the task scene in a simulation environment (e.g. finite element or physics simulator), such that all relevant objects are present and their stiffness and friction, and

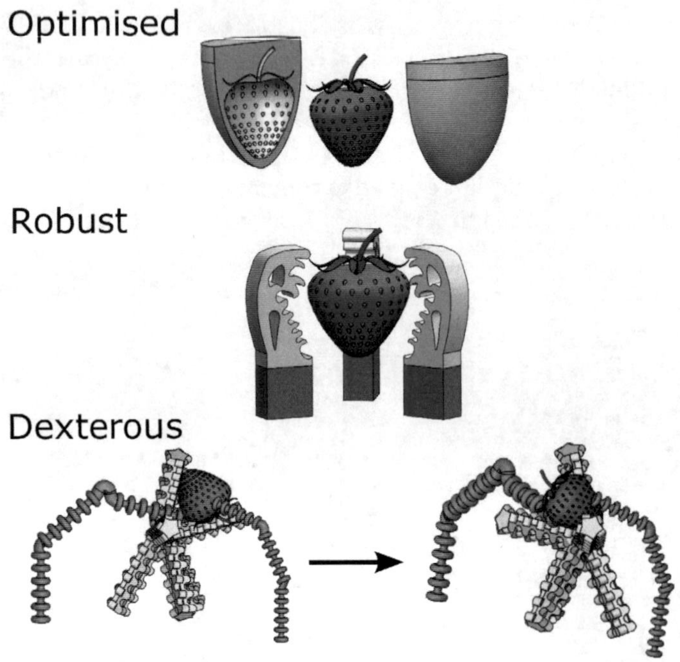

Figure 2. Hierarchy of design goals: optimisation is useful, but being robust to shape and positioning uncertainly is better, and being able to robustly perform multiple movements is ideal

stress and limits are captured. Then evolve the 4D load array to find a grasp is found which meets the requirements.

(ii) Generate and evaluate designs using a generative tool to find a valid approximation of the load array. That is, minimise the norm of distance between the applied and desired loads.

3.2. Task-Mediated Robust Design

Robustness design aims to make the design insensitive to these small changes, such that the manipulator can grasp a whole class of objects (e.g any strawberry) and tolerate positioning errors.

Designing for robustness is inherantly more challenging than optimality, as the entire operating envelope must be considered in the design. This can be achieved either by simulating across a large number of conditions, or by giving the design sufficient flexibility to be insensitive to small errors. Fit2form design is an example of the former approach, it used a generative design algorithm to optimise a bespoke (rigid) end-effector to robustly grasp a specified object using a parallel plate style gripper [Ha et al.(2020)]. The cost function contained terms which maximised grasp stability and robustness, allowing it to generate several robust grasp strategies even without a deformable end-effector. In soft robotics, robust embodied grasping is not even considered explicitly, but rather assumed as a byproduct of their inherent conformability [Ellis et al.(2022)].

To progress the field, robustness must be considered in the design of all soft manipulators. In practice, this means identifying the bounds of performance as well as the ideal grasp pose.

Incorporating this into our task-mediated design framework requires an additional dimension to be added to the cost function, representing the grasped object. The feature array is then a 4D force, plus the object index (x,y,z,t,n). Grasp sensitivity is captured by perturbing the optimised loads. To be valid a grasp must be able to be varied within specified bounds and still complete the task.

Here we see the benefit of the task-mediated approach: the actual design process does not change to accommodate robust design, it is only the task definition which changes. Hence, the method can be layered on top of any existing design strategy. It is fully compatible with sophisticated design strategies like multi-level evolution and adaptive optimisation [?, Pinskier and Howard(2022)]

3.3. Dexterous design

Atop the design hierarchy is dexterous design, in which the tool is able to perform a stable grasp with sufficient flexibility in its motion to accommodate significant shape and positioning uncertainty [Ciocarlie and Allen(2008)], and has the ability able to perform multiple or complex tasks. For example, a dexterous gripper may be able to perform the entire strawberry harvesting sequence, or pick and pack strawberries with a single tool.

This top tier approaches the goals of rigid robotics, to have single multi-functional devices. Given a fixed design it poses dexterous manipulation as a task to identify the required set of forces to reorient and object and translate that to manipulator links [Okamura et al.(2000)]. However, it is still build around the embodied design goals. It is the combination of material, morphology, and control which achieve the task rather than relying on control.

Task-mediated design provides a framework to generate new dexterous designs by simply adding one more dimension to the feature space, representing the task being performed. The task extraction procedure is then repeated across the set of tasks to find the characteristic array across the set of tasks, and objects.

4. Conclusion

In this perspective we introduce the notion of task-mediated design, a 2-step design process. It augments the traditional generate-design-test-regenerate methodology with the latent task-space. This will enable higher quality designs to be found more efficiently by capturing critical features and reducing the search dimension. We hope task-mediated design will help accelerate the design and development of novel high-performing soft manipulators and mobile soft robots.

References

[Boddeti et al.(2021)] Narasimha Boddeti, Tien Van Truong, Vincent S. Joseph, Thileepan Stalin, Theo Calais, Shien Yang Lee, Martin L. Dunn, and Pablo Valdivia y Alvarado. 2021. Optimal Soft Composites for Under-Actuated Soft Robots. *Advanced Materials Technologies* 2100361 (2021), 2100361. https://doi.org/10.1002/admt.202100361

[Chen and Wang(2020)] Feifei Chen and Michael Yu Wang. 2020. Design Optimization of Soft Robots: A Review of the State of the Art. *IEEE ROBOTICS AUTOMATION MAGAZINE* December (2020), 27–43. https://doi.org/10.1080/03772063.2020.1842159

[Cheney et al.(2013)] Nick Cheney, Robert Maccurdy, Jeff Clune, and Hod Lipson. 2013. Unshackling Evolution: Evolving Soft Robots with Multiple Materials and a Powerful Generative Encoding. In *GECCO 2013*. Amsterdam.

[Ciocarlie and Allen(2008)] Matei T. Ciocarlie and Peter K. Allen. 2008. On-line interactive dexterous grasping. *EuroHaptics '08: Proceedings of the 6th international conference on Haptics: Perception, Devices and ScenariosJune 2008* (2008), 104–113. https://doi.org/10.1007/978-3-540-69057-$3_1$1

[Collins et al.(2018)] Jack Collins, Wade Geles, David Howard, and Frederic Maire. 2018. Towards the targeted environment-specific evolution of robot components. In *GECCO '18: Genetic and Evolutionary Computation Conference*. Kyoto, 61–68.

[Ellis et al.(2022)] David Rostin Ellis, Martin Philip Venter, and Gerhard Venter. 2022. Generative Design Procedure for Embedding Specified Planar Behavior in Modular Soft Pneumatic Actuators. *Soft Robotics* 9, 3 (2022), 552–561. https://doi.org/10.1089/soro.2020.0013

[Ha et al.(2020)] Huy Ha, Shubham Agrawal, and Shuran Song. 2020. Fit2Form: 3D Generative Model for Robot Gripper Form Design. In *4th Conference on Robot Learning (CoRL 2020)*. Cambridge, MA, USA. arXiv:2011.06498v1 https://fit2form.cs.columbia.edu/

[Ellis et al.(2022)] David Rostin Ellis, Martin Philip Venter, and Gerhard Venter. 2022. Generative Design Procedure for Embedding Specified Planar Behavior in Modular Soft Pneumatic Actuators. *Soft Robotics* 9, 3 (2022), 552–561. https://doi.org/10.1089/soro.2020.0013

[Ha et al.(2020)] Huy Ha, Shubham Agrawal, and Shuran Song. 2020. Fit2Form: 3D Generative Model for Robot Gripper Form Design. In *4th Conference on Robot Learning (CoRL 2020)*. Cambridge, MA, USA. arXiv:2011.06498v1 https://fit2form.cs.columbia.edu/

[Hao and Visell(2021)] Yufei Hao and Yon Visell. 2021. Beyond Soft Hands: Efficient Grasping With Non-Anthropomorphic Soft Grippers. *Frontiers in Robotics and AI* 8, July (2021), 1–8. https://doi.org/10.3389/frobt.2021.632006

[Howard et al.(2019)] David Howard, Agoston E. Eiben, Danielle Frances Kennedy, Jean Baptiste Mouret, Philip Valencia, and Dave Winkler. 2019. Evolving embodied intelligence from materials to machines. *Nature Machine Intelligence* 1, 1 (2019), 12–19. https://doi.org/10.1038/s42256-018-0009-9 arXiv:1901.05704

[Howard et al.(2022a)] David Howard, Josh Kannemeyer, Davide Dolcetti, Humphrey Munn, and Nicole Robinson. 2022a. *Assessing Evolutionary Terrain Generation Methods for Curriculum Reinforcement Learning; Assessing Evolutionary Terrain Generation Methods for Curriculum Reinforcement Learning*. Vol. 1. Association for Computing Machinery. arXiv:2203.15172v1 https://doi.org/10.1145/nnnnnnn.nnnnnnn

[Howard et al.(2022b)] David Howard, Jack O'Connor, Jordan Letchford, James Brett, Therese Joseph, Sophia Lin, Daniel Furby, and Gary W. Delaney. 2022b. Getting a Grip: in Materio Evolution of Membrane Morphology for Soft Robotic Jamming Grippers. *2022 IEEE 5th International Conference on Soft Robotics, RoboSoft 2022* (2022), 531–538. https://doi.org/10.1109/RoboSoft54090.2022.9762197 arXiv:2111.01952

[Kriegman et al.(2020)] Sam Kriegman, Douglas Blackiston, Michael Levin, and Josh Bongard. 2020. A scalable pipeline for designing reconfigurable organisms. *Proceedings of the National Academy of Sciences of the United States of America* 117, 4 (2020), 1853–1859. https://doi.org/10.1073/pnas.1910837117

[Laschi et al.(2012)] Cecilia Laschi, Matteo Cianchetti, Barbara Mazzolai, Laura Margheri, Maurizio Follador, and Paolo Dario. 2012. Soft robot arm inspired by the octopus. *Advanced Robotics* 26, 7 (2012), 709–727. https://doi.org/10.1163/156855312X626343

[Ma et al.(2021)] Pingchuan Ma, Tao Du, John Z. Zhang, Kui Wu, Andrew Spielberg, Robert K. Katzschmann, and Wojciech Matusik. 2021. DiffAqua: A Differentiable Computational Design Pipeline for Soft Underwater Swimmers with Shape Interpolation. *ACM Transactions on Graphics* 40, 4 (2021). https://doi.org/10.1145/3450626.3459832 arXiv:2104.00837

[Majidi(2014)] Carmel Majidi. 2014. Soft Robotics: A Perspective - Current Trends and Prospects for the Future. *Soft Robotics* 1, 1 (2014), 5–11. https://doi.org/10.1089/soro.2013.0001

[Okamura et al.(2000)] Allison M. Okamura, Niels Smaby, and Mark R. Cutkosky. 2000. Overview of dexterous manipulation. *Proceedings - IEEE International Conference on Robotics and Automation* 1 (2000), 255–262. https://doi.org/10.1109/robot.2000.844067

[Pang et al.(2022)] Chohei Pang, Qicheng Wang, Kinwing Mak, Hongyu Yu, and Michael Yu Wang. 2022. Viko 2.0: A Hierarchical Gecko-Inspired Adhesive Gripper With Visuotactile Sensor. *IEEE Robotics and Automation Letters* 7, 3 (2022), 7842–7849. https://doi.org/10.1109/LRA.2022.3183249 arXiv:2204.10082

[Pinskier and Howard(2022)] Joshua Pinskier and David Howard. 2022. From Bioinspiration to Computer Generation: Developments in Autonomous Soft Robot Design. *Advanced Intelligent Systems* 4, 1 (2022), 2100086. https://doi.org/10.1002/aisy.202100086

[Rus and Tolley(2015)] Daniela Rus and Michael T. Tolley. 2015. Design, fabrication and control of soft robots. *Nature* 521, 7553 (2015), 467–475. https://doi.org/10.1038/nature14543

[Xavier et al.(2022)] Matheus S. Xavier, Charbel D. Tawk, Ali Zolfagharian, Joshua Pinskier, David Howard, Taylor Young, Jiewen Lai, Simon M. Harrison, Yuen K. Yong, Mahdi Bodaghi, and Andrew J. Fleming. 2022. Soft Pneumatic Actuators: A Review of Design, Fabrication, Modeling, Sensing, Control and Applications. *IEEE Access* 10 (2022), 59442–59485. https://doi.org/10.1109/access.2022.3179589

[Zhang et al.(2019)] Hongying Zhang, A. Senthil Kumar, Feifei Chen, Jerry Y.H. Fuh, and Michael Yu Wang. 2019. Topology optimized multimaterial soft fingers for applications on grippers, rehabilitation, and artificial hands. *IEEE/ASME Transactions on Mechatronics* 24, 1 (2019), 120–131. https://doi.org/10.1109/TMECH.2018.2874067

[Zhu et al.(2022)] Jihong Zhu, Andrea Cherubini, Claire Dune, David Navarro-Alarcon, Farshid Alambeigi, Dmitry Berenson, Fanny Ficuciello, Kensuke Harada, Jens Kober, Xiang LI, Jia Pan, Wenzhen Yuan, and Michael Gienger. 2022. Challenges and Outlook in Robotic Manipulation of Deformable Objects. *IEEE Robotics and Automation Magazine* September (2022), 67–77. https://doi.org/10.1109/MRA.2022.3147415 arXiv:2105.01767

Embodied Intelligence Conference 2022 (EI-2022) IOP Publishing
IOP Conf. Series: Materials Science and Engineering 1292 (2023) 012010 doi:10.1088/1757-899X/1292/1/012010

Imitation and reinforcement learning to control soft robots: a perspective

Muhammad Sunny Nazeer,[1, 2] **Cecilia Laschi,**[3] **and Egidio Falotico**[1,2]

[1] The BioRobotics Institute, Scuola Superiore Sant'Anna, Pisa, Italy
[2] Department of Excellence in Robotics and AI, Scuola Superiore Sant'Anna, Pisa, Italy
[3] Department of Mechanical Engineering, National University of Singapore, Singapore

E-mail: muhammadsunny.nazeer@santannapisa.it

Abstract. It would not be an over-stretch to appraise soft robotics as one of the most sizzling topic in this era, owing to its unparalleled performance in terms of exhibited deformability, soft-interaction and luscious dexterity. In spite of that, we see minimal applications of this class around us. We believe the reason roots in the gap between design and fabrication of the soft robots and the capability to control them. The remedy to this imbalance, we believe, is a necessity for a climacteric rise in the capability to control them. The control, however poses challenges due to virtually infinite degrees of freedom of soft robot design, in terms of modelling the behavior of the robot, modelling its interaction with the environment or the environment's effect on the robot. These challenges provoke the preponderant need for solutions that either can work with the inaccurate models of the robot obtained or avoid it altogether. To answer what can be used with these limitations we present here two potential classes of algorithms from machine learning capable of performing well under just the listed conditions: reinforcement learning (RL) can provide a solution in the form of a policy to control a platform with sophisticated dynamics even when the presented environment to train said policy has questionable accuracy, while Imitation Learning (IL) trains a policy while avoiding the need of a model or its constituent parts altogether and a policy is learned based on the teachings of an expert.

1. Introduction

The motivation for soft robotics stems from the biological systems due to their soft structure being able to exist and function efficiently in difficult and highly unstructured environments [1]. So, robots designed and fabricated under this inspiration tend to offer compliance, deformability and adaptive nature best suited, as initially intended, for unpredictable environments. Yet it is uncommon to find such a platform playing a major role in our society even though sophisticated platforms have already been designed and fabricated that can imitate the biological systems that inspired its design [2]. It is important to note here that this field has been around for almost two decades now. The hindrance, arguably exists in the current capability to model and control them.

For that matter, various researchers have managed to present modelling techniques able to capture their approximate behavior or behavior under constrained settings. The most notable among the modelling solutions are constant curvature [3] or piecewise constant curvature models [4], Cosserat rod theory [5] based models, FEM based models with slow computations [6] or SOFA Framework [7] with good accuracy [8] and low computational power using model order reduction [9] and finally machine learning based data-driven models [10], [11].

IOP Conf. Series: Materials Science and Engineering 1292 (2023) 012010 doi:10.1088/1757-899X/1292/1/012010

While most of the modelling techniques mentioned above tend to capture the behavior exhibited by the underlying soft robot, the accuracy of that behavior remains somewhat questionable. For example by imposing constant curvature constraint, while it is useful, the proposed solution still remains indecisive under non-negligible external loading. Similarly, Cosserat rod theory though improves the accuracy in the behavior, the underlying partial differential equations seem sensitive to noise in the underlying parameters. Moreover, the computational cost of these partial differential equations increase drastically with added modularity in the soft robot. Similar trend is seen with FEM based approaches as well. This class seems to work much better than the previously proposed approaches however, the computational cost hinders the real-time control of the robots or implementation of learning based approaches for control. On a similar note, data-driven approaches are more applicable with control methodologies based on learning, nevertheless they pose challenges of their own. For example, modelling the interaction with the environment requires dataset that is not easily available and the artificial neural networks trained to capture the soft robots performance tend to accumulate error over a finite horizon while working in closed loop.

So, though all the presented approaches tend to deal with the modelling problems differently, there is a single aspect missing in all, related to capturing global behavior of soft robots due to their virtually infinite degrees of freedom. It majorly impacts the number of solutions proposed to control them. All of the above models have been used for control purposes as in [4]. [4] uses piecewise constant curvature based models to achieve curvature and bending control, [12] uses Cosserat theory based simulation environment with RL for following a trajectory or tracking a goal in 3D space, [13] and [14] uses SOFA (FEM-based simulator) for position control with and without model order reduction, respectively, and finally data-driven models of the soft robots to elicit trajectory control using learning based adaptive control [15], open-loop dynamic controller [16] and dynamics model-based RL [17, 18].

These proposed solutions are typically effective in controlled environments. In this context, a controlled environment refers to a situation where the actual performance of the soft robot closely matches its modelled behavior. However, this assumption may not be entirely practical, as it requires the model to capture the global behavior of the soft robot, including any potential factors that may affect its performance over time. And there are several factors that may potentially alter the soft robot's performance, such as changes in material properties resulting from factors like elastomeric properties, variations in the external environment, aging of the material, and wear and tear. More often that not, these factors are either ignored or assumed consistent from the control derivation step to the application step.

To ensure the practical applicability of any control solution for soft robots, it is essential to account for these factors. To deal with these challenges and present solutions able to control soft robots with desired accuracy, we may look for the alternative solutions. These approaches should be either independent of the modeled behavior or robust enough to adapt if there is a significant difference between the modeled and actual behavior of the robot. We highlight two potential classes of algorithms able to address aforementioned issues: imitation learning (IL) and reinforcement learning (RL).

2. Imitation Learning for Soft Robots

Imitation learning is a class of algorithms where a desired task is demonstrated by an expert [19]. The expert here can be a human, a pre-trained or incrementally improving policy, a synthetic/learning agent, a demonstration by another robot or a biological system or even a video. Obtained demonstrations are then used to learn a policy (for underlying task in the demonstrations) to be elicited on the platform of choice.

In the literature, this class has been used extensively with a variety of applications, for example learning to imitate repetitive human arm movements on a rigid humanoid robot [20], autonomous and agile navigation of a ground vehicle [21] learned in online mode by imitating the expert states to continuous steering and throttle commands (here the expert is a model-predictive controller that uses models of expensive and advanced sensors), learning to manipulate objects [22] in a sophisticated environment with rigid manipulators using demonstrations acquired from virtual reality via teleoperation, and many more. This class is known to do much with a limited dataset.

While fewer demonstrations tend to work well for rigid robots to learn the underlying task and still be able to generalize over a minutely varying environment (as single shot learning schemes in [23, 24]), it may not necessarily be same for soft robots due to following factors: 1) task demonstrations recorded, where human experts are involved, may vary significantly from each other for the same task due to flexible shape and absence of fixed reference frame unlike rigid robots. 2) Soft robots tend to undergo varying kinematics upon interaction with the external objects. 3) Finally, stochastic behaviour exhibited by soft robots due to varying material properties. These varying properties result in ever-falling generalization capability. We consider all these factors when proposing final solutions in the progressing sections.

There are two practical ways to implement imitation learning on soft robots: scenario 01 - kinesthetic learning and scenario 02 - learning from observation. Kinesthetic learning is where the expert manipulates the soft robot. Expert demonstrations in terms of state-action pairs are recorded to train a policy. The policy, in return, predicts actions for the soft robot to replicate expert states. The expert task-demonstrations are elicited on the platform of choice with [25] or without expert intervention[26] depending on the level of difficulty of the task and platform at hand. The general algorithmic flow is as shown in the Fig. 1. The examples of this approach may include human teaching a soft robot to fish, perform a surgery, or produce a certain pattern/trajectory in 3D space or on an object, etc.

The second scenario, on the other hand, can be imagined where a robot learns to do a task by observing the expert. The expert do not physically interact with the soft robot here. The features are extracted from the expert demonstrations of a desired task. It is important to note here that the demonstrations may be in a space different than the robot's own configuration, action or operation space. The examples of this scenario can be a soft robot learning to flip burgers based on a video demonstration of a human, or soft robotic arm attempting to follow sequential series of motions as observed from a biological system like octopus arm movements being replicated on an octopus-inspired soft robot, or an elephant trunk movements on an elephant trunk-inspired soft arm, etc. The general algorithmic flow of this approach is as shown in the Fig. 2.

In the first scenario of using IL for soft robots, the gathered observations and associated actions can be used to derive a direct policy, utilizing behavior cloning techniques as described in Chapter 03 of the book by Osa et al. [27]. This policy can be improved in real-time using policy optimization, without the need to explicitly learn about the subject platform, the operating environment, or the task itself. However, in the second scenario, an additional mapping, also known as domain adaptation map [28], is required to transform the expert demonstrations into a space or frame replicable by the subject platform. This is because the soft robot and the expert may not only lie in different frames but may also have different morphologies and action spaces. Therefore, the expert demonstrations need to be transformed to match the specific requirements of the subject platform, allowing the IL algorithm to learn and improve the soft robot's performance.

In the context of soft robotics, the second scenario is considered more favorable. We (as a soft robotics community) strive to mimic biological systems, which require knowledge transfer from one system to another, such as transferring the motion of a real octopus to an octopus-inspired robot, as discussed by Kim et al. [1]. While the general appearance of the soft robot and the biological system may be similar, the actuation capability can differ significantly. Returning to

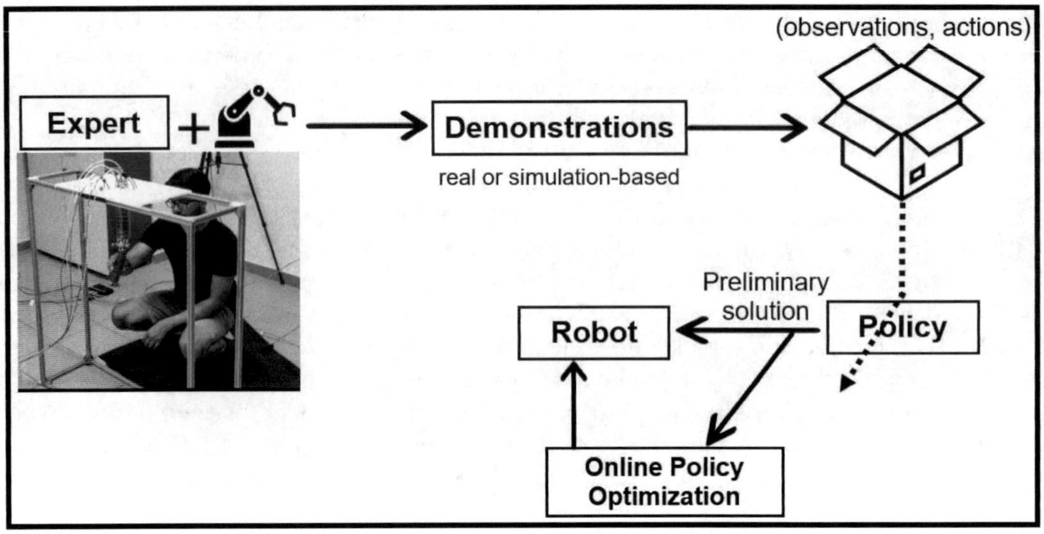

Figure 1: Scenario 01 - Kinesthetic expert demonstrations are recorded using a motion capture system. Behavior cloning can be applied to the expert's state-action pairs to train a preliminary policy, which can be further optimized using policy optimization approaches.

the first scenario, it appears more feasible to run this scheme on rigid robots, as they have a fixed shape, a reference frame to start from, and joint movement that is repeatable, ignoring the influence of human factors. The demonstrations thus accumulated tend to relate closely without significant variability. However, if such a scheme is applied to soft robots, the absence of a fixed shape or reference frame and virtually infinite degree of freedom, exacerbated by the influence of human experts, results in a significantly varying distribution of desired task demonstrations. Vanilla behavior cloning may not be sufficient to learn the policy for the underlying task. Furthermore, the kinematics or geometry of a rigid robot does not change if it comes into contact with another object. Such an assumption does not necessarily hold for soft robots, as their properties can be impacted by external factors, leading to deviations from the expected behavior.

The challenges posed by both scenario 01 and 02 combined present issues for policy derivation for soft robots, even after domain adaptation in scenario 02. Vanilla behavior cloning is not sophisticated enough to handle these challenges, requiring a more incremental and data-driven approach, such as the Data Aggregation approach or DAgger. Both scenarios can implement this approach with appropriate modifications, as highlighted in Figs. 1 and 2. Though such a learning scheme may be able to learn the underlying task from the distribution of demonstrations, it tends to under-perform when faced with a state not seen before. The policy only has knowledge from the expert demonstrations, and it is impossible to cover all possible situations that the policy may face during application, especially when a high degree of accuracy is desired. Given that IL is not known for exhibiting robust behavior in changing environmental settings, we move on to the next class of algorithms that are best known for robust behavior under varying conditions: RL.

3. Reinforcement Learning for Soft Robots

This class of machine learning algorithms provides a policy that is learned from a training environment based on trial and error. The training environment can be a simulation engine, a mathematical formulation, or a data-driven machine learning-based model that describes the

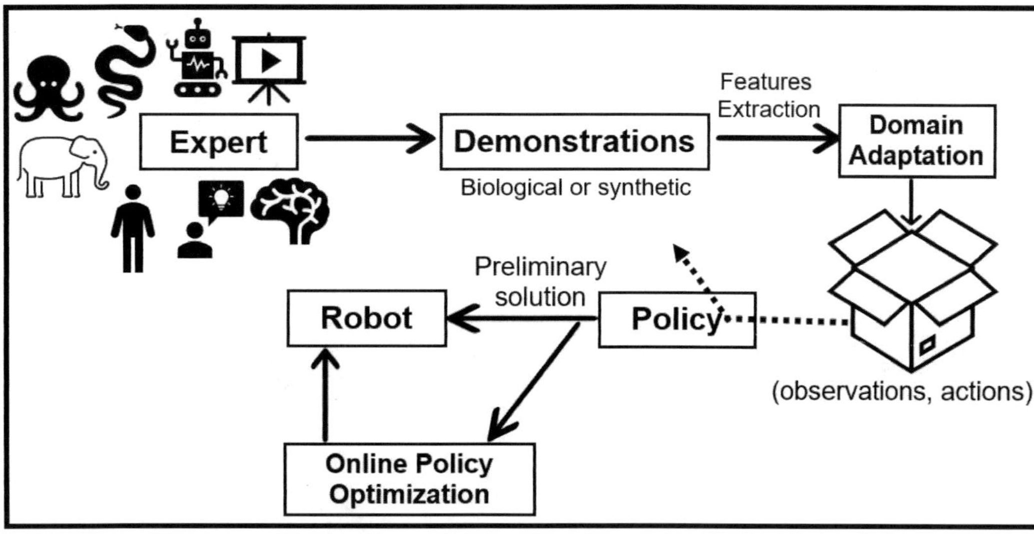

Figure 2: Scenario 02 - The desired behavior to replicate is observed from an expert, which could be a video, a different robot, a human agent, any other biological system performing the desired behavior, or even a hand-crafted idea. Next, features are extracted from these observations, and they are mapped into a domain that can be replicated by the robot in question, using a domain adaptive map. Finally, a policy can be trained and optimized as in Fig. 1.

approximate behavior of the platform under consideration, including encompassing modules that affect/complement the robot's behavior. These encompassing modules may include, but are not limited to, the reward/penalty/objective function that describes the task at hand, the interaction of the subject platform with objects/humans/robots in the environment, or physical constraints that affect/hinder the desired task.

On the other hand, a learning agent is an algorithm (under the RL umbrella) that attempts to learn a policy capable of performing the desired task by taking random actions and observing how the training environment responds to it in terms of a reward/penalty. Such signals, in turn, motivate future actions [29] to maximize the reward or minimize the penalty. Actions taken in this way finally converge to the desired actions that can perform the desired task. However, such a policy is acquired after a significant number of iterations have been performed. This signifies a rather obvious point that such a scheme cannot be executed directly on the robot, as performing tens of thousands of iterations to reach the desired results is not a practical approach from a safety and/or resource perspective. Therefore, such an environment needs to be created in a non-physical domain as highlighted in Fig. 3.

When it comes to the various components that make up the training environment, such as the robot behavior model, interaction model, and the effect of the external environment on the robot model, researchers have attempted to present solutions describing them over the past decade for soft robots, as found in literature [3, 4, 5, 6, 7, 8, 9, 10, 11, 15, 18]. However, the level of accuracy achieved remains a challenge and an ongoing topic of interest. This introduces an additional hurdle when implementing the trained policy in the real environment, acquired from the training environment, as the training environment only holds an approximation of the real platform's behavior. The discrepancy between the simulation environment (proposed models) and the real world is famously known as the simulation-to-reality (sim2real) gap, which will persist until the behavioral gap is closed. Depending on the training environment, this gap may be known differently, such as the training2real gap or machine2real gap, but serving the same

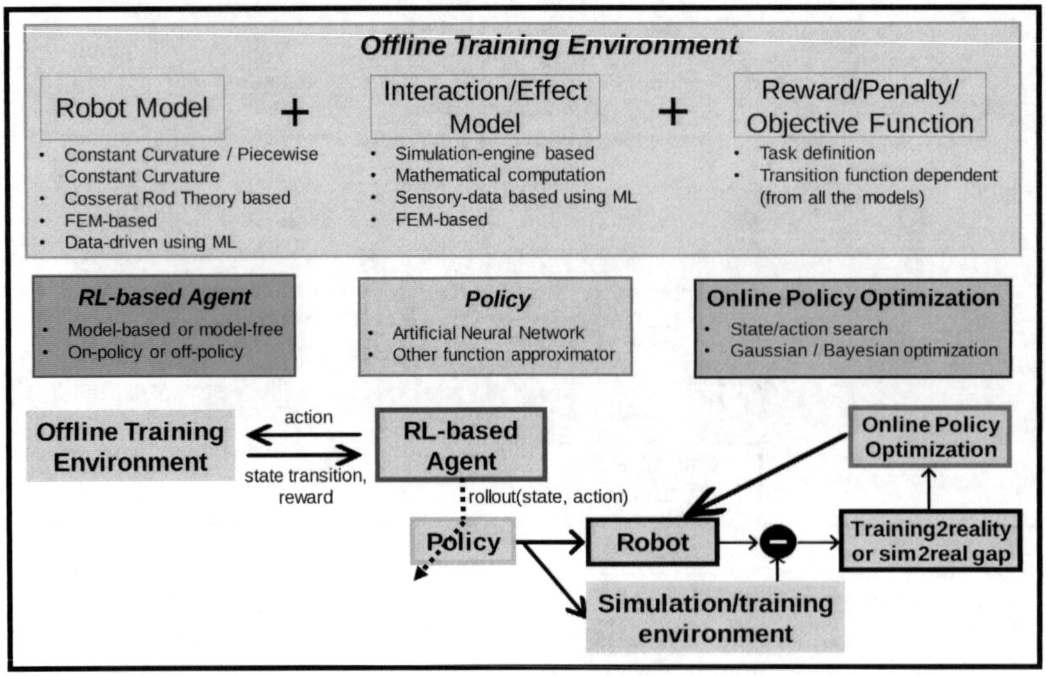

Figure 3: This figure highlights the definition of various components involved in the training environment for a policy trained using RL-based algorithms. Policies obtained from offline training environments often underperform when tested on the actual robot, particularly for robots with complex dynamics, such as high DoF soft robots. This discrepancy between the simulation/training environment and the actual environment is commonly referred to as the simulation-to-reality gap and typically requires an additional optimization step with the physical robot.

hurdle nonetheless. Additional optimization is thus required to implement the offline-trained policy in the real environment. The complete map for RL is shown in Fig. 3.

At the same time, the amount of time and computational resources required for optimization depend on several factors such as the complexity of the robot or the task at hand, the technique being used, and the level of discrepancy between the simulation environment and the real environment. This requirement becomes more intensified with the performance exacerbation, particularly when the soft robot exhibits stochastic behavior or undergoes changes due to material properties. As the learning agent explores its environment during the random search to learn the optimal sequential series of actions for performing the desired task, the resulting solution tends to be quite robust under varying conditions.

On a different note, it should be mentioned that it may not always be feasible to obtain all the necessary models for the offline training environment, such as models for interaction with an external object with unknown properties or for modeling the impact of external constraints on robot behavior. Additionally, creating a reward function for every type of desired task may introduce additional challenges. For instance, consider the challenge of deriving a reward/penalty/objective function to tie a knot, fish, or flip a burger. Such tasks may be learned more effectively from demonstrations.

The conundrum of choosing between a time-consuming vs fast solution, a computationally exhaustive vs relatively low computational burden approach, a robust vs failure-under-eluded-states approach, and a complicated hand-crafted reward function vs demonstrations-only approach, etc., can be mitigated by combining the two approaches. This leads us to the proposed scheme of combining IL and RL for a robust and fast solution.

4. Combination of IL and RL for Soft Robots

Looking back, we realize that RL introduces robusticity while IL produces a fast solution for quite sophisticated tasks. So, a combination of both schemes will instigate producing a preliminary policy using the expert demonstrations and optimizing said policy in online mode using an exploration-based optimization (using RL) to bolster unstructured and varying environment conditions. Similar solution is proposed in [25] where the author introduced a synthetic agent based on DQN (off-policy RL algorithm) to support the policy where it may fail under eluded states as shown in fig 4a.

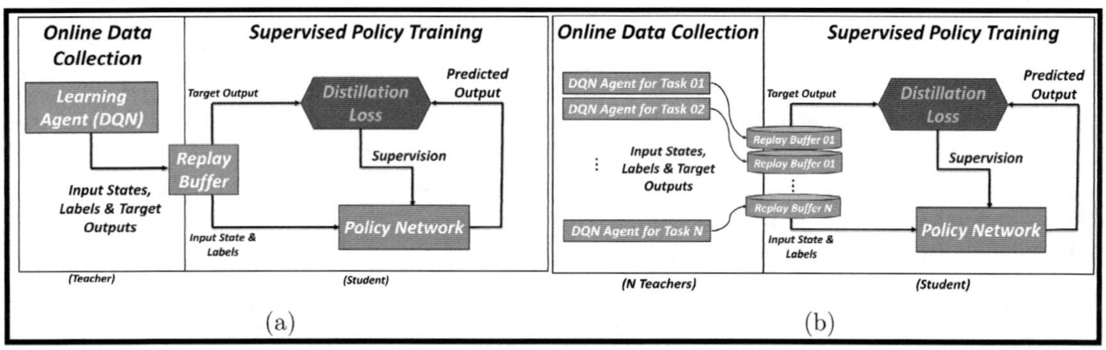

Figure 4: The block diagrams in (a) and (b) illustrate the systemic flow of the algorithm proposed by Rusu et al. in [25]. The goal is to train a student policy using policy distillation, where knowledge is distilled from a DQN agent in online mode. The DQN agent acts as an expert for the under-training student policy and provides demonstrations for each encountered state. In (b), N DQN agents act as experts for N different tasks and provide demonstrations for each task in an associated buffer. The knowledge is then distilled from all the DQN agents in online mode to train a student policy capable of performing N tasks.

A question arises regarding the exploration based on a search-based approach: where do the objective function comes out of for tasks where hand-crafted reward/penalty functions are challenging to design? In such situations, there is another class of algorithms known as Inverse Reinforcement Learning (IRL), which is surveyed in [30]. IRL algorithms learn a reward/objective function from expert demonstrations, observed behavior, a learned policy, or directly from an expert. The obtained objective function can then be used for policy optimization. However, there is an underlying assumption that the expert demonstrations are the optimal way to perform the task, which may not always be the case. Nonetheless, the obtained objective/reward function from the demonstrations can produce a solution for the desired task that is as good as the demonstrations themselves, and can also be robustified (as discussed in Chapter 04 of [27]) for unstructured environments. The RL agent, as made abundantly clear in Sec. 3, introduces robustness to the proposed policy for a desired task by using a reward function and search-based algorithms. So, why limit such a solution to merely exploring and intervening during eluded states only when used with IL? At the same time, as also indicated in Sec. 2, the policy learned simply from demonstrations may lack generalization capability due to the high-dimensionality of the observation space and continuous action space of the platform under consideration (soft robots in this case) or show distribution shift due to error accumulation [31] in the policy learned solely through supervised learning on the demonstrations. Under such conditions, an RL agent may be encouraged to train a policy that can match the expert demonstrations over a long finite horizon based on comparative incentives. It is of paramount importance to note that no reward function was hand-crafted for the task or extracted from the demonstrations; instead, the incentives are based on how closely the agent can match the expert demonstrations over a finite period (as presented in [31]).

In addition to learning a single task quickly and making it resilient to a varying and unstructured environment, it is also possible to train a single policy to perform multiple tasks. For example, consider Fig. 4a, where a trained DQN agent serves as the expert for a student policy training on a single task. Instead of using a single DQN agent for a specific task, we can have multiple experts, each associated with an independent task as shown in Fig. 4b. The demonstrations from each expert can be gathered in a different buffer and the student policy can be trained episode-by-episode with episodes coming from different buffers sequentially. This approach can result in a policy that can perform multiple tasks and is robust enough to work under varying conditions, thanks to the intervention of independent RL agents, as also demonstrated in [25].

5. Challenges and Future Work

Soft robots tend to exhibit stochasticity in their performance, so the viability of the IL-RL combination remains a topic of interest and a subject of our ongoing research. We believe that the policy obtained from the IL-RL solution can work under varying conditions. However, if the stochasticity is incorporated into this variation, the adaptability of this solution may degrade. This is because we are relying solely on the search capabilities of an RL algorithm to adapt, which was trained based on a single training signal from a continuous-state environment. Although a knowledge base such as the one suggested in Fig. 4b, where multiple signals are being used to train a single policy, could introduce adaptability with significantly varying environments, the forever exploring RL-agents would again require more interactions to learn if the environment has changed since the training is running in parallel.

If we try to train sequentially with one type of environment and then another and so on, we will end up with a problem known as catastrophic forgetting, where the policy learns a new task but either underperforms or forgets how to do the previous task as it overwrites the policy weights. Previously learned task in a very different environment may also be construed as a new task. For such scenarios, intelligently exploring or effectively using the gathered knowledge base for learning a new task or adapting previously learned tasks in a very new environment is required. Continual learning is a class of algorithms that deals with such issues and may prove to be an exciting area to explore for future research in the current problem.

6. Conclusion

In this paper, we have discussed the potential of two main classes of algorithms in machine learning, IL and RL, to find optimal control solutions for soft robots. We have shown that each approach has its own strengths and limitations, and that combining them can provide a more comprehensive solution that is both efficient and robust under varying and unstructured environments. We have further highlighted the importance of addressing the issue of stochasticity in soft robots control, which can pose a challenge to the adaptability of IL-RL solutions. We further propose employing Continual learning for the adaptability problem as it can help to overcome this challenge and enable soft robots to learn and adapt to new tasks and environments effectively. Overall, our perspective on combination of IL and RL for soft robots control highlights their potential to reform the applicability of soft robots for sophisticated tasks. We hope that this paper will inspire further research in this exciting area and lead to the development of new and innovative approaches for controlling soft robots.

Acknowledgements
We would like to acknowledge the collaborative effort between the BRain-Inspired Robotics Lab (BRAIR Lab) at The BioRobotics Institute and the Soft Robotics lab at the National University of Singapore that made this work possible. It is funded by the European Union's Horizon 2020 Research and Innovation Programme in the framework of the project SMART (Soft, Self-responsive Smart Materials for Robots) under Marie Skłodowska-Curie Actions (MSCA), Innovative Training Network (ITN) with grant agreement no. 860108.

References

[1] S. Kim, C. Laschi, and B. Trimmer, "Soft robotics: A bioinspired evolution in robotics," *Trends in biotechnology*, vol. 31, 04 2013.

[2] C. Laschi, M. Cianchetti, B. Mazzolai, L. Margheri, M. Follador, and P. Dario, "Soft robot arm inspired by the octopus," *Advanced robotics*, vol. 26, no. 7, pp. 709–727, 2012.

[3] B. Jones and I. Walker, "Kinematics for multisection continuum robots," *IEEE Transactions on Robotics*, vol. 22, no. 1, pp. 43–55, 2006.

[4] C. Della Santina, A. Bicchi, and D. Rus, "On an improved state parametrization for soft robots with piecewise constant curvature and its use in model based control," *IEEE Robotics and Automation Letters*, vol. PP, pp. 1–1, 01 2020.

[5] D. C. Rucker and R. J. Webster III, "Statics and dynamics of continuum robots with general tendon routing and external loading," *IEEE Transactions on Robotics*, vol. 27, no. 6, pp. 1033–1044, 2011.

[6] M. Pozzi, E. Miguel, R. Deimel, M. Malvezzi, B. Bickel, O. Brock, and D. Prattichizzo, "Efficient fem-based simulation of soft robots modeled as kinematic chains," 05 2018, pp. 1–8.

[7] F. Faure, C. Duriez, H. Delingette, J. Allard, B. Gilles, S. Marchesseau, H. Talbot, H. Courtecuisse, G. Bousquet, I. Peterlik, and S. Cotin, "SOFA: A Multi-Model Framework for Interactive Physical Simulation," in *Soft Tissue Biomechanical Modeling for Computer Assisted Surgery*, ser. Studies in Mechanobiology, Tissue Engineering and Biomaterials, Y. Payan, Ed. Springer, Jun. 2012, vol. 11, pp. 283–321. [Online]. Available: https://hal.inria.fr/hal-00681539

[8] E. Coevoet, T. Morales-Bieze, F. Largilliere, Z. Zhang, M. Thieffry, M. Sanz Lopez, B. Carrez, D. Marchal, O. Goury, J. Dequidt, and C. Duriez, "Software toolkit for modeling, simulation and control of soft robots," *Advanced Robotics*, vol. 31, pp. 1208–1224, Nov. 2017. [Online]. Available: https://hal.inria.fr/hal-01649355

[9] O. Goury and C. Duriez, "Fast, generic and reliable control and simulation of soft robots using model order reduction," *IEEE Transactions on Robotics*, vol. 34, no. 6, pp. 1565 – 1576, 2018. [Online]. Available: https://hal.archives-ouvertes.fr/hal-01834483

[10] T. George Thuruthel, F. Renda, and F. Iida, "First-order dynamic modeling and control of soft robots," *Frontiers in Robotics and AI*, vol. 7, 2020. [Online]. Available: https://www.frontiersin.org/articles/10.3389/frobt.2020.00095

[11] T. G. Thuruthel, E. Falotico, M. Cianchetti, and C. Laschi, "Learning global inverse kinematics solutions for a continuum robot," in *ROMANSY 21 - Robot Design, Dynamics and Control*, V. Parenti-Castelli and W. Schiehlen, Eds. Cham: Springer International Publishing, 2016, pp. 47–54.

[12] M. Gazzola, L. Dudte, A. McCormick, and L. Mahadevan, "Forward and inverse problems in the mechanics of soft filaments," *Royal Society open science*, vol. 5, no. 6, p. 171628, 2018. [Online]. Available: https://doi.org/10.1098/rsos.171628

[13] F. Largilliere, V. Verona, E. Coevoet, M. Sanz-Lopez, J. Dequidt, and C. Duriez, "Real-time control of soft-robots using asynchronous finite element modeling," in *2015 IEEE International Conference on Robotics and Automation (ICRA)*, 2015, pp. 2550–2555.

[14] O. Goury, B. Carrez, and C. Duriez, "Real-time simulation for control of soft robots with self-collisions using model order reduction for contact forces," *IEEE Robotics and Automation Letters*, vol. 6, no. 2, pp. 3752–3759, 2021.

[15] F. Piqué, H. T. Kalidindi, A. Menciassi, C. Laschi, and E. Falotico, "A learning-based approach for adaptive closed-loop control of a soft robotic arm," in *Proceedings of the I-RIM 3D Conference, online*, 2020, pp. 10–12.

[16] T. G. Thuruthel, E. Falotico, M. Manti, and C. Laschi, "Stable open loop control of soft robotic manipulators," *IEEE Robotics and Automation Letters*, vol. 3, no. 2, pp. 1292–1298, 2018.

[17] T. G. Thuruthel, E. Falotico, F. Renda, and C. Laschi, "Model-based reinforcement learning for closed-loop dynamic control of soft robotic manipulators," *IEEE Transactions on Robotics*, vol. 35, no. 1, pp. 124–134, 2019.

[18] A. Centurelli, L. Arleo, A. Rizzo, S. Tolu, C. Laschi, and E. Falotico, "Closed-loop dynamic control of a soft

manipulator using deep reinforcement learning," *IEEE Robotics and Automation Letters*, vol. 7, no. 2, pp. 4741–4748, 2022.

[19] A. Hussein, M. Gaber, E. Elyan, and C. Jayne, "Imitation learning: A survey of learning methods," *ACM Computing Surveys*, vol. 50, 04 2017.

[20] T. Asfour, F. Gyarfas, P. Azad, and R. Dillmann, "Imitation learning of dual-arm manipulation tasks in humanoid robots," in *2006 6th IEEE-RAS International Conference on Humanoid Robots*, 2006, pp. 40–47.

[21] Y. Pan, C.-A. Cheng, K. Saigol, K. Lee, X. Yan, E. Theodorou, and B. Boots, "Imitation learning for agile autonomous driving," *The International Journal of Robotics Research*, vol. 39, p. 027836491988027, 10 2019.

[22] T. Zhang, Z. McCarthy, O. Jow, D. Lee, X. Chen, K. Goldberg, and P. Abbeel, "Deep imitation learning for complex manipulation tasks from virtual reality teleoperation," 2017. [Online]. Available: https://arxiv.org/abs/1710.04615

[23] C. Finn, T. Yu, T. Zhang, P. Abbeel, and S. Levine, "One-shot visual imitation learning via meta-learning," 2017. [Online]. Available: https://arxiv.org/abs/1709.04905

[24] A. Bonardi, S. James, and A. J. Davison, "Learning one-shot imitation from humans without humans," *IEEE Robotics and Automation Letters*, vol. 5, no. 2, pp. 3533–3539, 2020.

[25] A. Rusu, S. Gómez, C. Gulcehre, G. Desjardins, J. Kirkpatrick, R. Pascanu, V. Mnih, K. Kavukcuoglu, and R. Hadsell, "Policy distillation," 11 2015.

[26] P. Oikonomou, A. Dometios, M. Khamassi, and C. S. Tzafestas, "Task driven skill learning in a soft-robotic arm," in *2021 IEEE/RSJ International Conference on Intelligent Robots and Systems (IROS)*. IEEE, 2021, pp. 1716–1723.

[27] T. Osa, J. Pajarinen, G. Neumann, J. A. Bagnell, P. Abbeel, and J. Peters, 2018.

[28] A. Farahani, S. Voghoei, K. Rasheed, and H. R. Arabnia, "A brief review of domain adaptation," 2020. [Online]. Available: https://arxiv.org/abs/2010.03978

[29] R. S. Sutton and A. G. Barto, *Reinforcement Learning: An Introduction.* Cambridge, MA, USA: A Bradford Book, 2018.

[30] S. Arora and P. Doshi, "A survey of inverse reinforcement learning: Challenges, methods and progress," 2018. [Online]. Available: https://arxiv.org/abs/1806.06877

[31] S. Reddy, A. D. Dragan, and S. Levine, "Sqil: Imitation learning via reinforcement learning with sparse rewards," 2019. [Online]. Available: https://arxiv.org/abs/1905.11108

IOP Conf. Series: Materials Science and Engineering 1292 (2023) 012011 doi:10.1088/1757-899X/1292/1/012011

How to Fool Your Robot: Designing Exploitable Sensory Systems

David Hardman and Fumiya Iida
Bio-Inspired Robotics Lab, University of Cambridge, UK.
E-mail: dsh46@cam.ac.uk, fi224@cam.ac.uk

Abstract. Based on real-world interactions in our lives and in the lives of our ancestors, humans have developed a multitude of psychological, social, and reflexive actions for efficient living. We consider the integration of similar behaviours into embodied robots through the design of their sensory systems, evaluating their impact through a novel lens: how magicians exploit these human behaviours in order to fool their spectators into experiencing impossible events. We explore the consequences of designing agents which can experience magic effects, and argue that such design facilities lifelike actions.

1. Introduction

Over the last 20 years, researchers have become increasingly interested in the psychology of magic and the psychological effects which have been empirically discovered and exploited by magicians [1, 2]. In particular, misdirection - the intentional control of a spectator's focus - has proven a fruitful area for exploration and categorisation of psychological phenomena [3]. Magicians exploit learned social cues, psychological behaviours, and optical illusions [4, 5, 6, 7], alongside a conscious suspension of disbelief similar to that employed when watching films or reading books [8]. The effect emerges from a mismatch between predicted and observed events [9], much like the psychology of humour [10, 11].

During a similar time frame, the fields of robotics and artificial intelligence (AI) have continued to grow exponentially. Researchers have explored the uses of AI in performance-based creativity tasks including comedy, acting, and painting [12, 13, 14]. Few examples exist for magic performances: Zaghi et al. train deep neural networks to track and analyse simple coin tricks using object permanence [15], whilst Williams and McOwan use optimisation techniques to design the most deceptive puzzles [16, 17]. However, magic relies on a knowledge and experience of physical laws, which is difficult for AIs to understand without a form of embodiment. Magicians exploit the assumptions and affordances which we learn and develop from real-world interactions [18], based on the information received by our sensory systems. Though this is most commonly visual and audio information, other senses are involved in similar processes, with pickpockets exploiting a form of tactile change blindness [19].

By considering the effect of magicians' techniques on physical robotic sensory systems, we set the stage for future developments of embodied AI. Examining the behaviours which lead to being fooled sheds light on biomimetic actions [20, 21, 22], thought processes [23], expectations [24], and perception [25], enabling the development of more lifelike robots.

In this work we assume that our robotic agent has finite processing and storage capabilities, and is not omniscient i.e. it has access only to the information collected by its sensors. Under these constraints, we focus on the practicalities of implementation, and how the placement of imperfections in our system could lead to our robot being fooled by conjuring techniques. To evaluate this, we must first define what we mean by 'fooling': Section 2 explores a number of suggestions. From here, we categorise and explore a selection of phenomena which lead to this experience in humans, discussing the practicalities and implications of reproducing these behaviours in our robotic designs.

2. Fooling Metrics

2.1. Probabilistic Approaches

Ideally, we aim for a quantitative measure of 'being fooled' when watching a magic performance. Lamont and Wiseman [4] qualitatively consider a magician's spectator to be fooled when they perceive the effect and not the method. This is a good starting point for our considerations: the effect is a physically impossible set of events which the magician encourages by controlling and manipulating the available sensory data [26, 27]. If our agent chooses this interpretation over the genuine actions then we consider it fooled; however, it is difficult to assign quantitative thresholds as to when this occurs, or whether one agent is more fooled than another.

With this in mind, a number of probabilistic models have been developed, building upon the popular 'Bayesian brain' model in which predictive coding deduces the likely source of incoming information [28, 29, 30]. This is achieved using prior knowledge and real-world experiences which, in order for a robot to have equivalent knowledge, must be similarly experienced or explicitly programmed. Grassi and Bartels [9] consider predictive coding in the context of magic & magicians' techniques, in which magicians aim to maximise the prediction error between an anticipated prior and the observed sensory input, manipulating these inputs alongside the spectator's memory-based expectations. In this way, fooling can be explicitly quantified as the Shannon surprise between the predicted and observed distributions [31]. However, this contrasts with the 'impossibility' models in that fooling corresponds to a near-zero, but still modelled probability. Additionally, Grassi and Bartels note that *'while not all surprises are experienced as magic, all magic is experienced as surprise'*. This suggests that there are additional contributors to the experience of magic and its enjoyment; Caffarrati et al. observe strong neural responses to magic tricks using scalp EEG recordings [32].

Nonetheless, throughout this work we consider a Bayesian framework, choosing to focus on the numerous factors which lead to this erroneous prediction, rather than those which evoke an emotional response from its occurrence. This framework helps to model a multitude of psychological effects which magicians have learned to exploit: for example, an observer believing that a bag already contains a coin will believe this even more strongly upon hearing a 'clink' when the bag is tapped with a solid pen. To our Bayesian agent, this aural information transforms the prior - a distribution skewed in favour of a coin's presence - into a more skewed posterior. When the bag is then seen to be empty, the effect is the same: the coin has vanished from the bag. However, by further skewing the distribution (using a technique referred to by magicians as a 'convincer'), the surprise between the anticipated event (*'the coin is definitely in the bag'*) and the visual information of an empty bag is larger, leading to our agent being more fooled. Similarly, the model explains the difference in performing a trick to an audience of magicians or to an audience primed to be inherently suspicious: these experiences lead to different priors which require more observable evidence to shift in favour of effect over method [24, 5].

In these examples, the requirement of embodiment is not necessarily clear. A camera and microphone which feed into a network trained for object permanence could also be fooled by this series of actions, despite having little real-world experience. However, convincers occur to various extents throughout the entire effect: the dynamics of the coin as it leaves the magician's hand follow the expected parabolic arc; the bag deforms at the moment the coin hits its interior, with exactly the shape and magnitude expected of the perceived fabric and coin weight/velocity; the expected sound is that which we expect from the ballpoint pen which the spectator is so used to using in their everyday life. Beyond extreme cases, a spectator does not consciously consider all of these variables, but each sensory input which matches their real-world expectations serves to shift the posterior distribution further from the final effect. As the tricks become more complex, these expectations become increasingly important in augmenting the effect: the magician relies on exploiting both the lower-level priors (e.g. *'the bag now contains a £1 coin'*) and the hyperprior understandings of the world learned from experience of physical laws (e.g. *'a £1 coin will make a clink when tapped through a felt bag with a ballpoint pen'*). Embodiment is therefore necessary in developing this broad set of hyperpriors, but also in the experience of many tricks: we are especially surprised when a magician makes an object vanish from inside our own fist since we are able to constantly feel and confirm its presence up until the final moment. Furthermore, we wish to consider the implementation of imperfect embodied agents: these may still have gaps in their hyperprior knowledge, as discussed in Section 3.3. Whilst the idealised Bayesian model considers that an agent can consider *every* possible outcome in parallel, these gaps can lead to the observed outcomes being unaccounted for by the model; the agent does not know whether the outcome was likely or unlikely. Most robot control algorithms would consider an unknown state of this type to be highly undesirable, aiming to use the available sensory information to return to known conditions. Whether this state can be equated with the human experience of 'being fooled' or just 'being confused' is beyond the scope of this discussion. In Sections 3 & 4 we consider the sensory underloads & overloads which magicians exploit to induce fooling or confusion, and how their techniques are likely to affect an imperfect agent. Before this, we briefly examine the existing literature which categorises these techniques.

2.2. Existing Taxonomies

Broadly speaking, magicians control the sensory information perceived by their spectators using misdirection: a topic which has been widely discussed and debated in the magic literature. A popular manual [33] describes the principle: *'wherever you direct their attention, the audience will look there'*. Nelms [8] subdivides misdirection into that of the mind, attention, and eye, whilst Corinda [34] distinguishes between misdirection achieved by speech and achieved by action. A number of more thorough taxonomies and categorisations of misdirection techniques have been developed [35, 36, 4, 37]: the reader is directed to Kuhn et al. [38] for a thorough analysis and comparison. At the highest level, this work separates perceptual, memory, and reasoning misdirection. Whilst all three levels contribute to the Bayesian framework described above, we focus here on the perceptual, which relates most strongly to an embodied agent's sensory collection and filtering. This is subdivided into attentional and non-attentional forms: using psychological phenomena to control what the spectator should and shouldn't perceive, respectively. Sections 3 & 4 take a selection of techniques from each category, examining the arguments for and against designing an agent which is vulnerable to each. Finally, Section 5 briefly considers the 'ruse/feigning' sublevel of reasoning misdirection, in which an agent is presented with false information.

3. Dealing with Incomplete Information

The first case to consider is that in which our agent is not presented with sufficient sensory information to strongly reach a particular conclusion i.e. the Bayesian posterior has a large standard deviation, rather than a sharp peak. Examples include: concealment, in which an object is obscured from view and the observer assumes its location; and speed, in which visual information is presented faster than the framerate of the agent (equivalent to saying that *'the hand is quicker the eye'*).

An idealised Bayesian agent could simply consider all available scenarios at once, whilst keeping track of and updating their associated probabilities. However, the true parallel tracking of every possibility requires infinite bandwidth, and a practical implementation must instead ‘fill in the gaps’ to reduce the search space to a discrete size. The same is true with humans: we can’t focus on every method, and instead take cues from the magician’s behaviours and actions to fill in the gaps in our sensory inputs and draw our own conclusions. For example, we regularly rely on gestures and gaze direction in our social interactions to ‘fill in’ unknown information. If we have been told that there is a cat stuck in a tree which is obscured from our viewpoint, we deduce its location by following the direction in which a crowd of people looks and points. Though it is still possible that the cat is elsewhere, or that the cat does not even exist (see Section 5), we instinctively use these social cues as an input to reduce the standard deviation of our Bayesian posterior.

Similarly, if we missed the moment at which a magician picked up a coin, such that it could be in either hand, we instinctively focus on the clenched fist towards which the magician looks and points. Provided this is not done consciously, we soon forget to consider that the coin might not be in this hand, experiencing surprise when this is shown to be the case.

To be similarly fooled, our agent must have learned to instinctively recognise and follow these social cues; else they experience minimal surprise when the hand is shown empty. As we strive to develop robots which can socially interact with humans, these are the type of developments which shift the robot out of the ‘uncanny valley’ [39] and into the realm of lifelike interactions. Conversely, we are not fooled by these same actions if we realise their deliberateness: if the magician intently stares at one fist and rigidly points with the other, we grow so suspicious as to shift the posterior in favour of the pointing hand [20]. It is not as clear that a socially-adept robot would implement this suspicion threshold unless it had been explicitly programmed against deception. It is therefore straightforward to program a robot which is never fooled by these actions - neglecting all social cues - or which is always fooled - believing all perceived social cues - but much more difficult to produce a lifelike balance between the two. The choice strongly depends on application: companionship robots could be designed to follow all social cues, whilst a safety-critical application might be less trusting.

3.1. Dynamic Expectations

Another exploitable shortcut made by our brains is the constant prediction of dynamics. We anticipate a ball’s parabolic arc as soon as it is thrown, based on our prior observations of rigid body mechanics. We can approximate the position at which the ball should be caught just by observing its release, though in practice we continue to observe the trajectory in order to make closed-loop prediction updates.

Kuhn and Rensink [40] demonstrate how these short-term predictions can be exploited in a magic effect, causing a ball to seemingly vanish despite never being thrown, in the same way a dog begins to chase a tennis ball which has not been released. Before the vanish, a number of ordinary throws serve to prime the spectator’s dynamic expectations, leading to reports of the ball vanishing mid-air. Interestingly, eye tracking shows that this deception does not occur in the oculomotor system, with the authors hypothesising that ‘*the illusory effect is caused by covert redirection of the attentional spotlight to the predicted position of the ball*’. In a robotic implementation, this corresponds to the predicted dynamics of an object being weighted more heavily than the visual input of its motion. This might be beneficial if the visual systems operate with low frame rates, such that the current position could significantly deviate from the last observed position. Trajectory predictions are highly useful in a number of dynamic robots, and high surprise levels could be temporarily caused by a mismatch in prediction and observation, before the posterior is shifted to the ball remaining in the hand. Small mismatches could even be used to prevent fooling: if a thrown object does not follow the expected trajectory, an agent might infer the presence of threads or magnets which are supposed to remain hidden.

The interesting question is then perhaps whether there are benefits to designing an agent which foucses on predictions and disregards its sensory input to experience the ball vanishing mid-air. Since Kuhn and Rensink find priming to increase the likelihood of this effect, we can consider a bio-inspired energy-efficient robot which lowers the processing power designated to repeated observations: if it has seen and predicted multiple throws of the ball, it could lower its frame rate during future throws. Magicians have known and exploited similar human behaviours for decades: In a chapter on controlling attention, Nelms [8] notes how '*monotony kills interest*'.

3.2. Amodal Completion

Camí et al. [37] note the influence of unconscious amodal completion in deception: that is, perceiving an entire object despite its partial concealment. As well as forming the basis of many optical illusions, this effect can be exploited by magicians: if we see a pen being held horizontally by its centre, we visualise the concealed middle of the pen, based on our prior experience and knowledge of biros. If our embodied agent has learned the concept of a pen, its onboard vision system can use the available information to assign a high probability to the object being a pen, rather than two separate halves or a painted piece of wood. These deductions are vital in a robot which navigates and maps its surroundings based on the available information; it should not need to see every side of a chair to deduce the presence of a chair, assuming the floor space which it occupies. For any level of task-efficiency, the robot must rely on both its previous experience of physical interactions with common objects and of visual training data rather than reconstructing these assumptions from scratch. Magicians are well-aware of a similar necessity in humans, often relying on everyday objects being overlooked in order to facilitate deception [33].

Conversely, amodal absence can be exploited: if most of an empty hand is visible, we find it easy to imagine that the entire hand is empty. This effect has already been shown to be a contributor to road accidents [41], suggesting that similar psychological assumptions should not be programmed into self-driving cars or safety-critical applications.

3.3. Hyperprior Incompleteness

The final example of incomplete information which we consider is that of an incomplete or incorrect background knowledge. In the Bayesian framework, Grassi et al. [9] refer to 'deeply held hyperpriors' such as physical laws: we experience a high surprise if our sensor inputs suggest that these hyperpriors are violated. Robots do not necessarily need to follow the same hyperpriors as us, depending on their environmental niche [42]: an ice-hockey playing robot might experience high surprise if it encounters an object which does not glide smoothly along the floor, but its restriction to a 2D plane might mean that it requires no hyperpriors of gravitational laws, and experiences no impossibility when witnessing a levitation.

Lamont and Wiseman [4] identify a 'lack of scientific knowledge' as a potential area of exploitation for magicians: early fraudulent séances used phosphorescent paints to create ghostly apparitions which would be unlikely to phase modern audiences [43]. More recently, advances in self-healing materials might prove pivotal in the development of new methods for 'cut-and-restored' magic effects. Any agent without knowledge of these materials could be fooled by their demonstration - robots would need to be kept up to date with scientific breakthroughs to avoid this kind of deception. Similarly, Kuhn et al.'s taxonomy [38] considers 'false assumptions about magic' as a sublevel of reasoning misdirection.

On a psychological level, it is well observed that we often behave less randomly than we expect, such as the presence of biases when asked to select a random number [44]. Magicians use this in their favour to increase the likelihood of an impossible experience [34], though it seems unlikely that robotic spectators would be exploitable in this way: a robot can easily generate a random playing card, without a bias towards the Ace of Spades or Queen of Hearts.

4. Dealing with an Information Excess

We next consider the opposite case: rather than making assumptions to fill in the gaps from missing information, we now imagine that there is too much information for our robot to process in real-time, and that it must decide which sensory inputs to ignore, simplify, or not examine in depth. This threshold is of course dependent on the physical hardware - we assume that any practical implementation must have a finite limit, and cannot be omniscient. We briefly consider the effect of memory, or stored information, before looking at bio-inspired methods of sensory filtering.

4.1. Memory

We assume that our agent's on-board memory is sufficiently limited to prevent the storage of all sensory data for later processing, it being more efficient to store key moments, images, or conclusions.

According to Kuhn et al.'s taxonomy [38], memory misdirection can be subdivided into forgetting and mis-remembering, which can occur on different timescales. A thorough analysis of the role of memory in the reconstruction and recollection of magic performances is well beyond the scope of this work, though it is empirically observed that spectators of a magic effect will tend to augment the impossibility of their experience during recollections [4]. Unless an agent holds an incomplete or incorrect hyperprior (Section 3.3), a perfect storage of all available information should eventually enable deduction of the method used through replaying, especially with the knowledge of the final effect. This relates to the common advice for magicians to never repeat a trick to the same audience. We focus here on instantaneous fooling i.e. the initial discrepancy between prediction and observation at the moment of the effect.

4.2. Sensory Filtering

We consider sensory filtering - the separation of useful information from available background noise - to occur in two parts. The first is physical and largely refers to sight, in which the vision system (a mounted camera or our head and eyes) can be redirected to select the perceived area. The second considers information which is physically received, but psychologically neglected: humans perform much of this filtering unconsciously, which could provide clues in the design of lifelike robots.

Much of the physical choices made for gaze direction are contextual, such as looking directly at a task being performed or a person to whom we are speaking. Within this area of interest, magicians search to direct a spectator's gaze towards or away from a specific stimulus, much as with the social cues discussed in Section 3. Being asked a question or called by name can induce a brief moment of eye contact, which is exploited at the moment of a secret move [45]. Social robots without these inbuilt reflexive actions might seem distant or unengaged with human interactions. However, tasks which require extreme focus should not involve these reflexes, requiring some kind of hierarchy-based architecture for prioritisation.

Amongst others, movements play a large part in directing visual attention and in inducing change blindness [46, 47]. Pickpockets differentiate between fast linear and slower curvilinear gesture trajectories in order to control attention [19], whilst magicians rely on larger gestures to cover smaller secret movements [8]. Again, the duplication of these behaviours in our robots would help them seem more natural. This alone is not a reason for implementation - these behaviours would also emerge from a bio-inspired algorithm which tracked objects which it believed to be of the highest interest, based on context, motions, history, and social cues. Choosing which objects to touch can also be grouped under our physical filtering category:

physically interacting with an object provides details about its size, shape, and material properties. If an agent's fitness function is based solely on reducing uncertainty then these interactions would be constant. This is not what we see in humans, indicating a number of other factors in this motor control: social conformity, lack of interest, or belief that we already know the object's properties (Section 3.2).

The post-perception filtering of information can be both conscious - such as straining to hear a friend's conversation in a crowded room - or unconscious, such as failing to feel the glasses resting on your head during a prolonged search. A notable example is that of change blindness - the unawareness of of a stimulus's visual change, despite its obviousness when attention is drawn to it. Its effect is widely studied beyond the scope of this work, but we briefly note how it has been observed to result from transient interruptions such as blinking, eye movements, or flickering [48, 49]. Kuhn et al. investigate a spectator's experience of a magic trick in which the method is clearly visible, using social cues to draw attention away from this area [50]. If we continue to develop and incorporate bio-inspired robotic vision systems which filter the incoming information into regions and objects of interest, we would expect our systems to be similarly susceptible to change blindness in real-time if the system's rules are exploited to conceal stimuli.

Post-perception filtering need not necessarily be visual. Tactile information is similarly filtered - Macknik et al. analyse the pickpocket's exploitation of certain aspects of our somatosensory systems during the theft of a watch [19]:

> *"To steal a watch directly from the wrist of a mark, the pickpocket might first squeeze the wrist while the watch is still on (invoking contrast gain adaptation). This has two effects. First, it makes a high contrast somatosensory impression that adapts the touch receptors in the skin, making them less sensitive to the subsequent light touches that are required to unbuckle and remove the watch. second, the high contrast impression leaves behind a somatosensory after image, giving rise to the illusion that the watch is still on after it has been removed".*

From this, we see that available tactile information - an object in contact with the skin - is filtered and ignored by the brain, given an initial stimuli by the pickpocket. This is in part achievable becuase of the wrist's insensitivity compared to the rest of the hand and fingertips: it is easier to steal a watch than a wedding ring. This is due to the tactile information which we receive being pre-filtered by the morphological distribution of sensors over our bodies [51]. In robotic systems, sensor morphology similarly impacts the available data. Robots with highly sensitive hands might learn similar grasping techniques to us, whilst the relocation of all tactile sensors to one finger would certainly lead to different developments and pickpocketing techniques.

5. Dealing with False Information

One sublevel of Kuhn et al.'s reasoning misdirection is 'ruse/feigning' in which a magician deliberately presents their spectator with false information. As previously discussed, the level to which an agent trusts the information presented to it should be highly application-dependent. An agent which knows it is watching a magician must assign some level of uncertainty to statements to avoid being fooled, else the statement *'this is an ordinary pack of cards'* immediately assigns a zero probability to any kind of gimmicked deck. We can imagine such intrinsic dubiousness being present in any robot which regularly interacts with the public, even if only through acceptable action thresholds [52]: *'You must kill all humans.'*

False information is not necessarily verbal - see Section 2.1's discussion on convincers in the Bayesian framework. The example here focuses on false information which directly impacts our assumptions (*'there is a coin in the bag'*), though this can also be indirect: if we observe a magician struggle to cleanly shuffle a pack of cards we quickly discount the possibility of intricate sleight-of-hand moves. Avoiding these effects in robotic software design is a matter of assigning uncertainty to all information received and inferred. This is commonplace in many systems which employ Kalman filtering to counter the effects of measurement noise or sensor damage. Our considerations raise a difficult question in these systems: how do we determine that our sensors are damaged, and that a series of improbable events - deception - have not instead been correctly observed?

6. Conclusions

Using the actions and behaviours which allow us to perceive magic effects, we consider the design and filtering of a robot's visual and somatosensory information in the context of morphological design, probablisitic frameworks, and psychological reflexes. We see how physical information selection emerges from social cues, the incorporation of which would help to bring our designs out of the uncanny valley and towards natural interactions.

Post-perception, psychological studies provide a very useful basis for developing 'unconscious' filtering rules in the face of hardware limitations, though we must incorporate an awareness of the potential for these to be exploited. In this way, Bayesian frameworks enable explicit considerations of the uncertainties associated with conclusions, observations, and assumptions; their magnitudes must be application specific, often compromising between social conformity and information certainty. As we continue to develop more general-purpose and complex designs, a robot's perception of magic effects can provide a useful set of benchmarks as to which end of the spectrum we operate.

Acknowledgements

This work was supported by the SHERO project, a Future and Emerging Technologies (FET) programme of the European Commission (grant agreement ID 828818), and by EPSRC DTP EP/R513180/1.

References

[1] G. Kuhn, *Experiencing the impossible : the science of magic.* 2019.

[2] M. L. Tompkins, "A science of magic bibliography," Mar 2021.

[3] G. Kuhn, P. Kingori, and K. P. Grietens, "Misdirection-magic, psychology and its application,"

[4] P. Lamont and R. Wiseman, *Magic in theory : an introduction to the theoretical and psychological elements of conjuring.* University of Hertfordshire Press, [paperback edition]. ed., 2005.

[5] R. Wiseman and E. Greening, "'it's still bending': Verbal suggestion and alleged psychokinetic ability," *British Journal of Psychology*, vol. 96, pp. 115–127, 2 2005.

[6] K. Eykholt, I. Evtimov, E. Fernandes, B. Li, A. Rahmati, C. Xiao, A. Prakash, T. Kohno, and D. Song, "Robust physical-world attacks on deep learning models," 7 2017.

[7] R. M. Williams and R. V. Yampolskiy, "Optical illusions images dataset," 9 2018.

[8] H. Nelms, *Magic and Showmanship: A Handbook for Conjurers.* Dover Magic Books, Dover Publications, 1969.

[9] P. R. Grassi and A. Bartels, "Magic, bayes and wows: A bayesian account of magic tricks," *Neuroscience and Biobehavioral Reviews*, vol. 126, pp. 515–527, 7 2021.

[10] J. Goldstein, *The Psychology of Humor: Theoretical Perspectives and Empirical Issues.* Elsevier Science, 2013.

[11] R. Martin and T. Ford, *The Psychology of Humor: An Integrative Approach.* Elsevier Science, 2018.

[12] M. A. Boden, "Creativity and artificial intelligence," *Artificial Intelligence*, vol. 103, no. 1, pp. 347–356, 1998. Artificial Intelligence 40 years later.

[13] N. T. Fitter, H. Knight, N. Martelaro, and D. Sirkin, "What actors can teach robots," in *Proceedings of the 2017 CHI Conference Extended Abstracts on Human Factors in Computing Systems*, CHI EA '17, (New York, NY, USA), p. 574–580, Association for Computing Machinery, 2017.

[14] T. Veale, *Your Wit Is My Command: Building AIs with a Sense of Humor.* MIT Press, 2021.

[15] R. Zaghi-Lara, M. Ángel Gea, J. Camí, L. M. Martínez, and A. Gomez-Marin, "Playing magic tricks to deep neural networks untangles human deception,"

[16] H. Williams and P. W. McOwan, "Magic in the machine: A computational magician's assistant," *Frontiers in Psychology*, vol. 5, 2014.

[17] H. Williams and P. W. McOwan, "Magic in pieces: An analysis of magic trick construction using artificial intelligence as a design aid," *Applied Artificial Intelligence*, vol. 30, pp. 16–28, 1 2016.
[18] J. Gibson and L. Carmichael, *The Senses Considered as Perceptual Systems.* Houghton Mifflin, 1966.
[19] S. L. Macknik, M. King, J. Randi, A. Robbins, Teller, J. Thompson, and S. Martinez-Conde, "Attention and awareness in stage magic: turning tricks into research," *Nature Reviews Neuroscience*, vol. 9, 2008.
[20] F. Phillips, M. B. Natter, and E. J. Egan, "Magically deceptive biological motion - the french drop sleight," *Frontiers in Psychology*, vol. 6, 2015.
[21] E. Garcia-Pelegrin, C. Wilkins, and N. S. Clayton, "Investigating expert performance when observing magic effects," *Scientific Reports*, vol. 12, 12 2022.
[22] A. Hergovich, K. Gröbl, and C. C. Carbon, "The paddle move commonly used in magic tricks as a means for analysing the perceptual limits of combined motion trajectories," *Perception*, vol. 40, pp. 358–366, 2011.
[23] C. Thomas and A. Didierjean, "Magicians fix your mind: How unlikely solutions block obvious ones," *Cognition*, vol. 154, pp. 169–173, 9 2016.
[24] E. Garcia-Pelegrin, A. K. Schnell, C. Wilkins, and N. S. Clayton, "Exploring the perceptual inabilities of eurasian jays (garrulus glandarius) using magic effects," vol. 118, p. 2026106118, 2021.
[25] R. J. Wiseman and T. Nakano, "Blink and you'll miss it: The role of blinking in the perception of magic tricks," *PeerJ*, vol. 2016, 2016.
[26] W. Smith, F. Dignum, and L. Sonenberg, "The construction of impossibility: A logic-based analysis of conjuring tricks.," *Frontiers in Psychology*, vol. 7, 2016.
[27] R. A. Rensink and G. Kuhn, "A framework for using magic to study the mind.," *Frontiers in Psychology*, vol. 5, 2015.
[28] K. Friston and S. Kiebel, "Predictive coding under the free-energy principle," *Philosophical Transactions of the Royal Society B: Biological Sciences*, vol. 364, no. 1521, pp. 1211–1221, 2009.
[29] R. P. N. Rao and D. H. Ballard, "Predictive coding in the visual cortex: a functional interpretation of some extra-classical receptive-field effects," *Nature Neuroscience*, vol. 2, pp. 79–87, Jan 1999.
[30] T. Parr, G. Rees, and K. J. Friston, "Computational neuropsychology and bayesian inference," *Frontiers in Human Neuroscience*, vol. 12, 2018.
[31] P. Baldi and L. Itti, "Of bits and wows: A bayesian theory of surprise with applications to attention," *Neural Networks*, vol. 23, no. 5, pp. 649–666, 2010.
[32] H. Caffaratti, J. Navajas, H. G. Rey, and R. Q. Quiroga, "Where is the ball? behavioral and neural responses elicited by a magic trick," *Psychophysiology*, vol. 53, pp. 1441–1448, 9 2016.
[33] H. Tarbell, *The Tarbell Course in Magic: Volume 1 (lessons 1 to 19).* Louis Tannen, 1953.
[34] T. Corinda, *Thirteen Steps to Mentalism.* D. Robbins & Company Incorporated, 1926.
[35] A. Ascanio, "Consideraciones sobre la misdirection," 1964. Misdirection 1, 4-6.
[36] J. Bruno, *Anatomy of Misdirection.* Createspace Independent Pub, 1978.
[37] J. Camí, A. Gomez-Marin, and L. M. Martínez, "On the cognitive bases of illusionism," *PeerJ*, vol. 8, 2020.
[38] G. Kuhn, H. A. Caffaratti, R. Teszka, and R. A. Rensink, "A psychologically-based taxonomy of misdirection," *Frontiers in Psychology*, vol. 5, 2014.
[39] M. Mori, K. F. MacDorman, and N. Kageki, "The uncanny valley [from the field]," *IEEE Robotics Automation Magazine*, vol. 19, no. 2, pp. 98–100, 2012.
[40] G. Kuhn and R. A. Rensink, "The vanishing ball illusion: A new perspective on the perception of dynamic events," *Cognition*, vol. 148, pp. 64–70, 3 2016.
[41] V. Ekroll, M. Svalebjørg, A. Pirrone, G. Böhm, S. Jentschke, R. van Lier, J. Wagemans, and A. Høye, "The illusion of absence: how a common feature of magic shows can explain a class of road accidents," *Cognitive Research: Principles and Implications*, vol. 6, 12 2021.
[42] R. Pfeifer and C. Scheier, *Understanding Intelligence.* MIT Press, 2001.
[43] H. Tarbell, *The Tarbell Course in Magic: Volume 2 (lessons 20 to 33).* Louis Tannen, 1975.
[44] R. N. Shepard, D. W. Kilpatric, and J. P. Cunningham, "The internal representation of numbers," *Cognitive Psychology*, vol. 7, no. 1, pp. 82–138, 1975.
[45] C. L. Kleinke, "Gaze and eye contact: A research review.," *Psychological Bulletin*, vol. 100, pp. 78–100, 1986.
[46] J. W. Suchow and G. A. Alvarez, "Motion silences awareness of visual change," *Current Biology*, vol. 21, pp. 140–143, 1 2011.
[47] R. Yao, K. Wood, and D. J. Simons, "As if by magic: An abrupt change in motion direction induces change blindness," *Psychological Science*, vol. 30, pp. 436–443, 3 2019.
[48] J. K. O'Regan, R. A. Rensink, and J. J. Clark, "Change-blindness as a result of 'mudsplashes'," *Nature*, vol. 398, p. 34, 1999.
[49] J. K. O'Regan, H. Deubel, J. J. Clark, and R. A. Rensink, "Picture changes during blinks: Looking without seeing and seeing without looking," *Visual Cognition*, vol. 7, pp. 191–211, 2000.
[50] G. Kuhn, B. W. Tatler, and G. G. Cole, "You look where i look! effect of gaze cues on overt and covert

attention in misdirection," *Visual Cognition*, vol. 17, pp. 925–944, 2009.

[51] H. Mori and Y. Kuniyoshi, "A human fetus development simulation: Self-organization of behaviors through tactile sensation," in *2010 IEEE 9th International Conference on Development and Learning*, pp. 82–87, 2010.

[52] R. Clarke, "Asimov's laws of robotics: Implications for information technology-part i," *Computer*, vol. 26, p. 53–61, dec 1993.

IOP Conf. Series: Materials Science and Engineering 1292 (2023) 012012 doi:10.1088/1757-899X/1292/1/012012

Metaphorising and bayesian inference according to biology of cognition and enaction

Amaranta Valdés-Zorrilla[1,2], Juan-Carlos Letelier[3] and Jorge Soto-Andrade [1,2]

[1] Department of Mathematics, Faculty of Science, University of Chile, Santiago, Chile
[2] Centre for Advanced Research in Education, University of Chile, Santiago, Chile
[3] Department of Biology, Faculty of Science, University of Chile, Santiago, Chile

E-mail: amvaldes@uchile.cl, letelier@uchile.cl, sotoandrade@uchile.cl

Abstract. The most accepted paradigm of brain function sustaining mathematical thinking is the "Brain as Computer" metaphor. Since the early 1970s, however, a new viewpoint based on the Biology of Cognition, created by Humberto Maturana and Francisco Varela, has slowly positioned itself as a challenger to the computer metaphor. Here, we interpret the notion of metaphorising in mathematics in the context of Biology of Cognition. Specifically, we introduce the fundamental concept of Structural Coupling, which is the mechanism by which living systems create the "objects" populating their niche by exploiting correlations in the never-ending Perception ↻ Action loop. Furthermore, we show that how a living system decides which action to perform is the outcome of a Bayesian Inference mechanism; therefore, randomness is fundamental to living systems and not just a consequence of the mere lack of sufficient information to compute the next action. Additionally, in Bayesian inference, the underlying conditional probability distribution $P(action|perception)$ changes by the very execution of every Perception ↻ Action loop, performing a biased random walk according to a Hebbian-like rule. With this rich set of concepts derived from cutting-edge biology, we show that Biology of Cognition is a good fit to understand mathematical metaphorisation.

1. Introduction

The most used metaphor to understand brain function is "The Brain is a Computer". This metaphor can be traced to the early years after the Second World War, where influential books (Cybernetics by Wiener [1]) and influential papers [2] introduced the idea of computing in neuroscience. The power of this metaphor, with its seductive ideas of Codes, Programs, Optimization, Decision Theory, Memory and Information made it the mainstream framework by the early 1970s. According to this widespread and often implicit metaphor, the brain is a machine surrounded by a sensory apparatus that almost perfectly transduces –like a one-to-one correspondence– the properties of external objects into internal data. Data, once produced, is interpreted as an (almost) perfect representation of the external object [3].

Following the Brain-Computer metaphor, once transduction is completed, the internal representation is manipulated using self-evident formal rules that reflect the dynamics of the objective world surrounding the organism. The sequence of rules is implemented by a Turing Automaton that a) recognises the external object, b) analyses the context of the object and

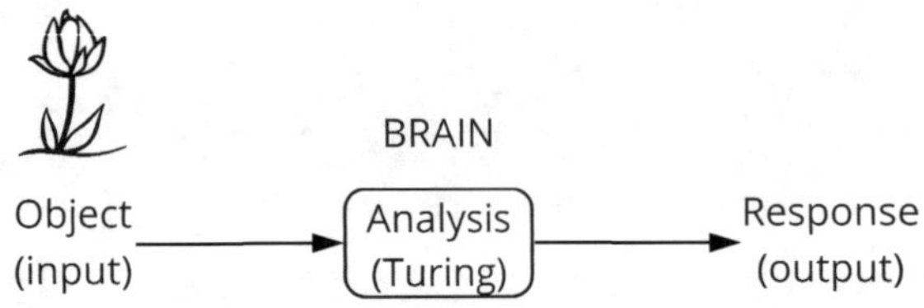

Figure 1. Sequence of interpretation of an object by a brain acting as an information processing device.

the circumstances where it is found, and c) designs an appropriate response (Figure 1). This is why the period 1970-2000 is rich in neuroscience papers that try to interpret the visual system as a simple telephoto system [4, 5] or a cooperative system where high-level image descriptors (i.e., a cat) are built from simpler visual elements (contrast, borders, edges, occlusion) [6]. These ideas reached wide acceptance as experimental research *seemed* to indicate, mainly in the visual and auditory systems, that the inner workings were similar to electronic devices built by humans (radars, sonars, correlators, linear predictors, etc.). Even in the motor system, a kind of population coding was found that, while different from normal circuits, was interpreted as an evolutionary adaptation of ideas of multivariate linear predictions [7]. Although the use of the computer metaphor, plus the idea of Information, plus the many experimental results of the period 1960-1990 were formidable, some obvious cracks appeared in that conceptual edifice. Namely, the neuroanatomy of the vertebrate's brain is very different from von Neumann's computers. And perhaps the main difference is the non-hierarchical connectivity of the brain. Indeed, in modern electronics, the cabling follows a simple logic: massive connections go from *sensors* to the processing area (usually a *CPU/RAM-memory* assembly) and from there, a few cables go to the *effectors* (motors) and only a few cables project back (from the CPU/RAM-memory assembly to *sensors* or from the *effectors* to CPU/RAM-memory). Brain connectivity is radically different as *feedback connections* are massive and ubiquitous (for example, in the mammalian visual system, for every visual axon connecting the lateral geniculate nucleus to the visual cortex *two* axons from the cortex return to the geniculate [8]).

But the fundamental reason why the computer metaphor has lost its previous pre-eminence, uncontested during the 70-80s, is that it has not helped to advance, in the sense of producing *deep* ideas, in neuroscience. Furthermore, the computer metaphor did clash, from its very beginnings, with cherished ideas in theoretical biology, primarily the notion that living systems construct their *Umwelt* (the world as it is experienced) by the sequence of actions they perform. The *Umweltian* viewpoint has been stated since the seminal works of Jakob von Uexküll in 1920s [9] and since its inception, it has steadily grown in acceptance.

The point of view derived from Biology of Cognition [10] is firmly anchored in the Umwelt lineage and it can be considered its most technically precise formulation as it uses ideas obtained from current neuroscience. From the outset, Biology of Cognition posits that the nervous system is not a computer or a machine *processing information* but rather it is an infinite spiral of *Perception* ↻ *Action* elementary acts where the main problem is *deciding the next action.* Thus, at its core, every living system is an unending stream of recursive Perception↻Action loops and not a device that interprets, using the immutable laws of logic, an objective external world [11, 12].

In our formulation, the Perception↻Action stream is recursive because at any given moment t, the perception at time t, denoted $Perception(t)$, discerned by a living system, is the direct result of the action previously performed, that is, the action at time $(t-1)$, denoted $Action(t-1)$, and in turn triggers $Action(t+1)$.(Figure 2). [1]

For example, if you perceive a lion ($Perception(t)$) and decide to approach it slowly

[1] For the sake of simplicity, we model for now our system dynamics in discrete time

($Action(t+1)$), your subsequent perception ($Perception(t+2)$) is different than if you decide to run away (alternative $Action(t+1)$). Here $Action(t-1)$ might have been "scanning the landscape with your binoculars".

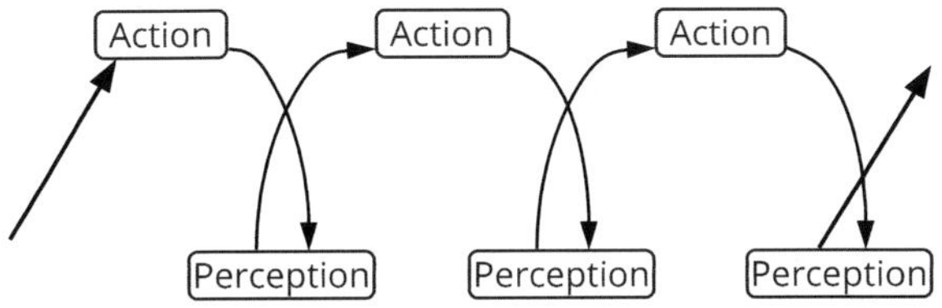

Figure 2. The infinite spiral of Action↻Perception Loops

We may write then

$$\rightarrow Perception(t-1) \rightarrow Action(t) \rightarrow Perception(t+1) \rightarrow Action(t+2) \rightarrow$$

where the arrows indicate "stochastic mappings", that is, for a given specific $Perception(t-1)$, say, $Action(t)$ is not uniquely determined: there are several possible options, each with a given probabilistic weight (whose total sum equals 1, of course). Notice that $Perception(t+1)$ differs from $Perception(t-1)$ because $Action(t)$ always results in a new situation (i.e., confrontation) of the organism with respect to its medium. Thus, the actual objective of living systems is not to understand "reality" but rather to modulate their perception-action spiral to maintain their internal (living) organisation invariant. According to this view, the main formal problem that living systems must solve is not to implement a "*capture* → *recognize* → *analyse* → *decide*" stream, but rather to create a Perception-Action spiral where the notion of *formal truth* is replaced by a notion of *coherence* between the living system and its circumstances.

2. The basic mechanism of Structural Coupling/Enaction

Structural Coupling, as first proposed by Maturana [11] and later refined by Varela [12] as Enaction, is the phenomenon by which living systems, as they undergo the never-ending perception↻action spiral, create the "objects"[2] they encounter, thus creating their ecological niche (Figure 3). "Objects" are an unavoidable side effect of the Perception↻Action loop when embodied by an autopoietic system.

The basic idea (see Figure 4) is that when an organism finds itself in recurrent interaction with its medium, the relationship that emerges from the perception-action loop is initially undefined (at $t = t_0$). But, as living systems are self-fabricating mechanisms with myriads of regulatory mechanisms, the re-occurrence of the interactions begins to change the structure of the living system as well as the interface with its medium (at $t = t_1$). If the history of mutual interactions is dense enough, in the sense of being recurrent, then the structure of the living system and of the interface become perfectly complementary (at $t = t_2$) and, for an observer not privy to the history of previous interactions, it appears that the living system has "adapted" to a *pre-existent* object. This misinterpretation is understandable because the observer is only witnessing the final stage of the process (t_2), not seeing the most important aspect of the phenomenon of Structural Coupling: that all objects are co-constructed relationships between the organism and its medium.

3. All Objects are always relational.

Our world is populated by a wide variety of *objects* that appear to us as given objective entities. An object is, however, defined by the list of behaviours that we can have towards it. For example,

[2] By "objects" we mean material-tangible in-animated objects, as well as living beings or even abstract notions.

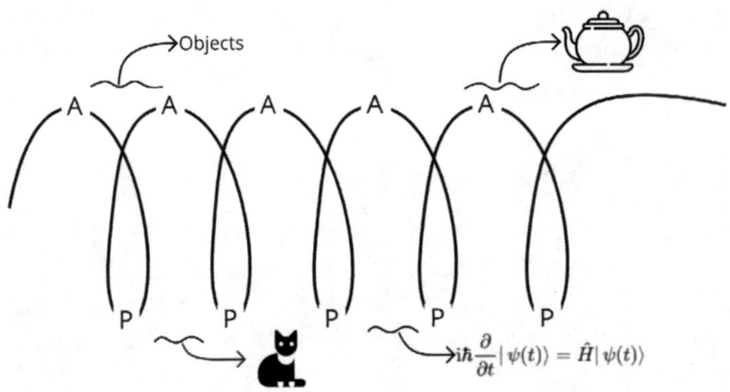

Figure 3. Every *Object* (from *teapots* and *cats* to *Schrödinger's equation*) are constructed through the perception↻action spiral as the result of a random walk in the Cartesian product space *Perception* × *Action*.

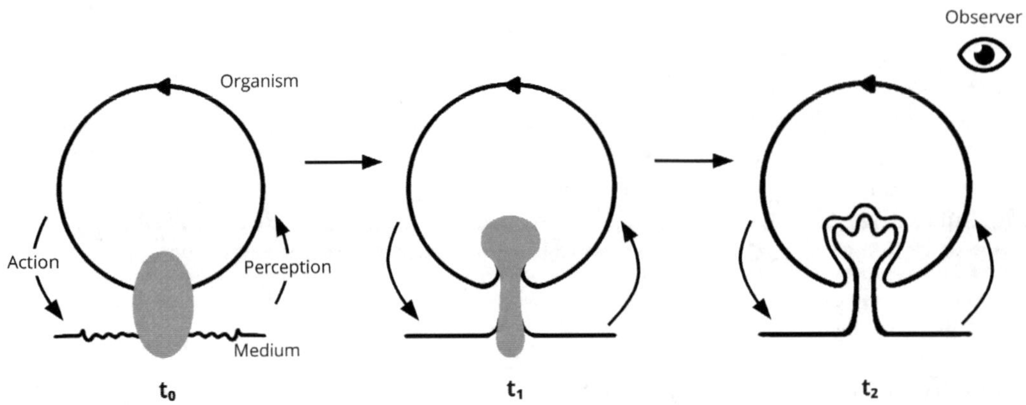

Figure 4. Objects are co-constructed relationships between the organism and its medium. Initially ($t = t_0$), the relation is undefined (grey blob), but as the recurrent interaction is maintained ($t = t_1$), the inner structure of the living system changes, as well as the interface with its medium. In well-developed interactions ($t = t_2$), the organism's structure has undergone a significant change along with the interface, and these changes are perfectly complementary. External Observers (upper right corner) could misinterpret the overall situation if they only focus on stage $t = t_2$ and not on the overall process; they might describe the situation as an *adaptation* of the organism to a pre-existent external object. Thus if the observer only considers the situation at $t = t_2$ he or she will miss the crucial fact that the *object* (the tree-like contour) started as an un-structured blob. *Structural Coupling* refers to the complete process of internal changes in the organism and changes in the interface organism/medium.

the object "cat" *is* the list of actions: "to feed the cat", "to play with the cat", "to comb the cat", "to brag about the cat",... This is also true for more abstract objects such as *vector spaces and linear transformations* in mathematics.

Thus, we posit that every object is a list of possible behaviours we can engage *vis à vis* this entity. Every object, either *real* or *formal* is defined relationally by our behaviours. This is true for us (*Homo sapiens*) as well as for bacteria [13].

4. The problem of Living System: What to do next?

As explained, a living system is a machine that relates perception to action. This relation is mediated by a metabolic organisation whose internal logic is implemented by a self-fabricating system – a concept that still eludes modern science. Indeed, as it has been explained in the last 30 years [10, 14], the "secret" of living systems is not that they behave like computers (they do not) or that they have genetic material (like DNA / RNA macromolecules, although they do). Their secret is that they fabricate the vast majority of their own components (Autopoiesis), and regulate the speed and nature of their internal process of transformations (i.e. the living system controls the kinetics of all biochemical reactions by producing and modulating all the catalysts –enzymes– in charge of facilitating the myriad of internal processes) [10, 15, 16].

As a consequence of the complex network of interlocking processes that constitute metabolism, when the sensory lamina (sensorium) of an organism encounters an "object" (y) in its milieu, a series of internal signals (Sy) are activated within metabolism. Crucially, the relation between "objects" and internal signals is not a simple one-to-one correspondence. Each encounter triggers a complex spectrum of internal signals, not all of which have the same temporal course. Some signals occur as soon as the object is encountered, while others are only slowly recruited. Thus, in the enactive view of the organism, the capture (or transduction) process is far more complex than a simple functional relation between an object and some limited set of internal variables [16, 17, 18]. From the point of view of the organism, a fundamental complexity arises as it must decide which course of action it must implement using as input variable the set of internal signals (S) that are not in direct one-to-one correspondence with the "external world". The organism, from this point of view, should not be considered as a simple computer but as **a self-fabricating Bayesian inference machine entangled with the world through Structural Coupling**. At all times, it must answer the following question:

> What is the probability (denoted $P(A_x|S_y)$) that I need to implement action A_x given that my internal signals are S_y?

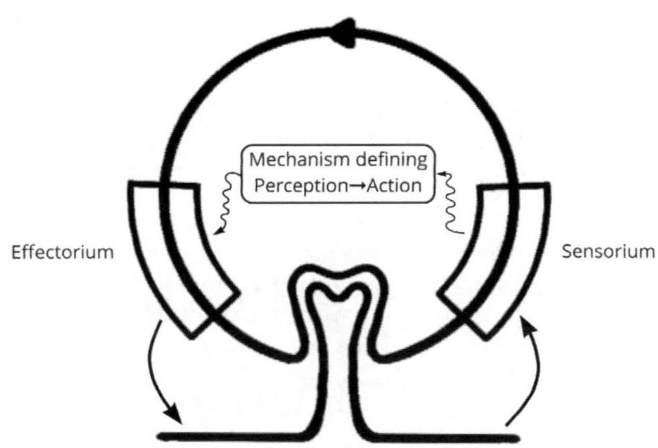

Figure 5. When a living system encounters a given *object* in its niche, many *internal signals* arise inside the metabolic network (arrow connecting the *Sensorium* to the internal mechanism defining the next action). These signals modulate the mechanism defining the next action that the living system must perform (arrow arriving to the *Effectorium*). The organism does not detect the "external reality out there" –a non existing entity– it only sees the internal changes as manifested by the spectrum of internal signals. Some of these internals signals however could be generated by the metabolic network bypassing the *Sensorium*.

In a compact form, we then have a *Conditional Probability Matrix* (denoted by CPM), associated with the organism, relating the set of signals (S_y) to the set of possible subsequent actions $(Ax)^3$, i.e.:

$$CPM = (P(A_x|S_y))_{x,y}$$

In this view of the organism (as a self-fabricating Bayesian inference machine entangled with the world through structural coupling), randomness is an intrinsic property of the world of the organism and not just simple ignorance of the factors controlling a perfectly deterministic situation (Fig 4 and Fig 5).

We can interpret the conditional probability matrix CPM as a *stochastic mapping*[4], call it Φ, which sends a signal vector S_y (triggered by the perception process) to the *random* action vector:

$$\Phi(S_y) = \sum_x P(A_x|S_y)A_x$$

Notice that our stochastic mapping Φ admits a sort of "dual", or "mirroring analogue" stochastic mapping Φ^*, going from actions to signals, defined by the "dual" conditional probability matrix:

$$CPM^* = (P(S_y|A_x))_{y,x}$$

whose (y, x)-entry can be interpreted as the probability that the action vector A_x triggers the signal vector S_y. In some sense, the matrix CPM^* can be seen as the "CPM of the medium of the organism" if we become aware that actions of the organism trigger inner signals of the medium and the other way around. It follows that by combining the stochastic mapping Φ, together with its dual Φ^*, we have a *random walk* in perception-action space. Indeed, Φ sends signals (and more generally random signals) to random actions, then Φ^* sends random actions back to random signals, which are sent to random actions by Φ^* and so on... We then have a random walk in the space of signals and actions, given by the iterated application of the stochastic mappings $\Phi, \Phi^*, \Phi, \Phi^*, \cdots$

To be more precise about this random walk, which intertwines a random walk on signals with another on actions, we need to take into account the temporal unfolding of the "object construction process" of our living being, via the infinite spiral of *perception* ↺ *action* loops of Figure 2. We address this in the following subsections.

4.1. A further problem for Living Systems: Updating of the Conditional Probability Matrix $P(Actions|Signals)$

Any living system must solve the problem:

Internal Signals (S) → Performed Action (A).

where the action performed must be consistent with the state of the surrounding medium. This viewpoint is valid for all living systems from bacteria to kittens and Homo Sapiens. The big difference is the size of the Conditional Probability Matrix (CPM). In bacteria, the medium can be catalogued by just a few circumstances (place with low food, place with high temperature, place with many bacteria, etc.), whereas a cat's medium has thousands of circumstances (master's home, it is August!, morning nap, mean dog running down the street, another nap, etc). Any Homo sapiens' medium, thanks to the amazing plastic abilities of neurones, and specially language, could reach a billion of circumstances (Marilyn Monroe, Alexander Grothendieck, Prime Number, Darwin, Password, Electron, Enzyme, Goldbach Conjecture, Gödel's theorem, red colour, Motherland, credit card, debt, inflation, UNIX, grades, tristesse,

[3] We often refer to the "signal *vector*" S_y and the "signal *vector*" A_x

[4] A *stochastic mapping* sends points from its domain to random points in its range, i. e., probability measures (also called probability distributions) on its range; the usual (deterministic) mappings correspond to the case where all these probability measures are Dirac measures.

The Beatles, CyberBullying, etc.). Furthermore, as internal signals are also products of the normal action of self-fabrication, it happens that internal states (i.e. those not triggered by specific encounters between organism and niche) also produce configurations of signals that are treated by the organism as being triggered by specific encounters. In other words, because of the many regulatory mechanisms as well as the complexity of the self-fabricating network, internal states are treated as external events because the only relevant fact is the variation in the spectrum of internal signals. In this sense, we agree with Seth's claim that we "hallucinate reality" [19].

4.2. Plasticity of the Bayesian Inference mechanism behind Enaction/Structural Coupling

As mentioned above, the main problem of an organism is to decide its next action after internally detecting a given signal S. Thus, we can think that the decision mechanism uses the Conditional Probability Matrix $P(Action|Signal)$ as the starting point to decide on the final action to implement. This is the core of Structural Coupling [20] and similar theories like the Free Energy Principle [21, 22].

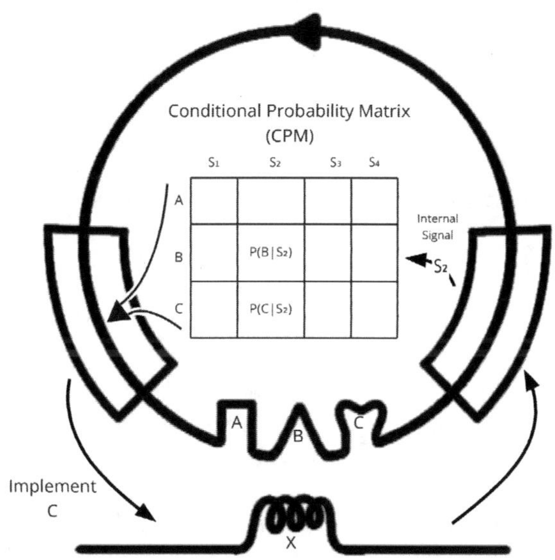

Figure 6. The Conditional Probability Matrix (CPM) of a Living System. Inside the metabolic network of any organism, from bacteria to *Homo sapiens*, there exists a mechanism whose operation is akin to a Bayesian probability model driven by a *Conditional Probability Matrix* (CPM), whose entries answer questions of the type: Which is the probability of executing action B, given that the internal signals are S_2. Notably, the matrix CPM is not static, it changes moment by moment according to the particular *Perception* ↺ *Action* spiral

To explain the following we start from a very simple case (see Figure 6). Suppose that an organism can implement 3 different actions (A, B, C) and detect 4 different signal configurations (S_1, S_2, S_3, S_4). Once an action has been implemented (imagine C) because a given signal configuration (S_2) was detected, the corresponding conditional probability $P(C|S_2)$ must be updated to reflect the increased strength (or effectiveness) of the association $S_2 \to C$.

Thus, we can write a simple evolution equation:

$$P(C|S_2)_{t+1} = P(C|S_2)_t + \Delta P_t$$

where ΔP_t represents the change in the probability $P(C|S_2)_t$ and can take many forms, such as

$$\Delta P_t \propto P(C) \cdot P(S_2),$$

somewhat akin to the Hebbian mechanism of synaptic plasticity [23].

If action C was effective, from the point of view of the organism, ΔP_t would be positive and, in that case, $P(C|S_2)_{t+1} > P(C|S_2)_t$ at the expense of one or more other *action/signal* couples, for which we would have had $P(action|signal)_{t+1} < P(action|signal)_t$.

Other rules for the variation of $P(C|S_2)_t$ can be proposed, but overall the important point is that the Conditional Probability Matrix $(CPM)_t$ at time t dynamically changes moment by moment according to the specific trajectory of $(\cdots action|internal\ signal, \ldots)$ undergone by the organism.

In other words, the transition probabilities of the random walk arising from the iterated application of the stochastic mappings $\Phi, \Phi^*, \Phi, \Phi^*, \cdots$ (or the corresponding conditional probability matrices $CPM, CPM^*, CPM, CPM^*, \cdots$ if you prefer) are *not stationary.*

Indeed, if a signal vector S_y arises in the metabolism of our living being at time t, the ensuing action A_x at time t+1 is probabilistically afforded by applying the stochastic mapping Φ at time t (or equivalently, the conditional probability matrix CPM at time t). We should then write,

$$A_x(t+1) = \Phi_t(S_y(t)) = (CPM)_t.S_y(t) = \sum_x P(A_x(t+1)|S_y(t))A_x(t+1)$$

Therefore, the temporal unfolding of our *perception* ↻ *action* random walk, starting, say, at an initial signal vector $S_y(t_0)$, would be given by the iterated application of $\Phi_{t_0}, \Phi^*_{t_0+1}, \Phi_{t_0+2}, \Phi^*_{t_0+3}, \cdots$ to $S_y(t_0)$. Of course, we could see this random walk as the intertwining, like two threads of the same cord, of the "pure perception" random walk given by $\Phi^*_{t_0+1} \circ \Phi_{t_0}, \Phi^*_{t_0+3} \circ \Phi_{t_0+2}, \cdots$[5] and the "pure action" random walk given by $\Phi_{t_0+2} \circ \Phi^*_{t_0+1}, \Phi_{t_0+4} \circ \Phi^*_{t_0+3}, \cdots$

Notice now that updating Φ_t and Φ^*_t (or equivalently, $(CPM)_t$ and $(CPM)^*_t$) in turn enacts a random walk in the space of conditional probability matrices. This random walk "hovers" above the above random walks in perception space and in action space, given by the suitable iterated application of our changing stochastic mappings. So, in fact, the latter are random walks with *randomly varying transition probabilities*, "second-order" random walks of sorts!

4.3. Metaphors as homologous paths in CPM space

In our view, Aristotle's classical description of metaphor [24] is just the tip of the iceberg of metaphorising as a biological cognitive process.

Indeed, the crucial role of metaphorising in mathematics and mathematics education has been increasingly acknowledged only in the last decades [25, 26, 27, 28, 29, 30, 31, 32, 33].
We are especially interested in the *poietic* role of "hallucinatory metaphors ": while Aristotle described metaphor as giving a thing a name that belongs to something else, we say that metaphorising is looking at something and seeing something else, i.e., "hallucinating", in the sense of Seth [19], for whom "reality" is a "controlled hallucination" of ours. Then, metaphorising enables us to fathom, grasp, or construct new concepts by taking advantage of our previous experiences. For example, when we look at the symmetric random walk of a frog on a row of stones in a pond and we wonder where it will be after a given number of jumps, we may "hallucinate", seeing the frog repeatedly splitting evenly into pieces instead of jumping. Or, engaging in an animal-friendlier hallucination, we may see one litre of fruit juice being shared

[5] Here "$\circ$" denotes as usual composition or concatenation of mappings, where $(g \circ f)(x) = g(f(x))$. Notice that $\Phi^*_{t_0+1} \circ \Phi_{t_0}$ its a mapping that goes from *Perceptions* to *Perceptions.*

equally among next neighbours in a row of drinkers, starting with one holding the whole litre. This allows us to *construct* the abstract notion of probability. Indeed, the probability of finding the frog on a given stone after n jumps can be understood as the amount of fruit juice a specific drinker has after n sharings. Initially, the problems of finding the frog and the distribution of juice after a given number of steps appear unrelated, but metaphorically they can be regarded as the "same". Metaphor can then be seen as the use of previous experience to explain current experience, as suggested by Tall's felicitous pun: "a metaphor is a met-afore" [30].

On the other hand, we can fathom the arising of metaphorising as a biological cognitive process, from our enactivist viewpoint, as follows. The interaction profile between an organism and its Umwelt is not a random collection of perceptions followed by certain actions, because the encounters between an organism and its medium follow regularities created by its ontogeny and phylogeny. In other words, the *local history of interactions* organism/medium contains strong correlations that are reflected in that the time series of internal signals $(..., S_{t-2}, S_{t-1}, S_t, S_{t+1}, ..)$ contains internal correlations reflecting the regularities in the interactions between the organism and its Umwelt. Thus, the random walk is not fully random as any effective $(Perception, Action)$ trajectory contains local regularities or correlations.

By way of example, let us consider two sequences of interactions, i.e. sequences of perception/action) I_1 and I_2 that contain some correlations in common; for example, they consist of petting a cat (I_1) and petting a dog (I_2), see Figure 7. In this case, random trajectories in CPM space induced by interactions I_1 and I_2 are themselves correlated. Thus, in some sense, each trajectory could be thought of as a metaphor of the other. This metaphorising could be embodied in several ways, from the simple equality of changes in CPM (i.e., for a given $P(A_i|S_k)$ the overall change in one interaction is proportional to the one in the other) to more complex forms where internal correlations in CPM are preserved in I_1 and I_2. .

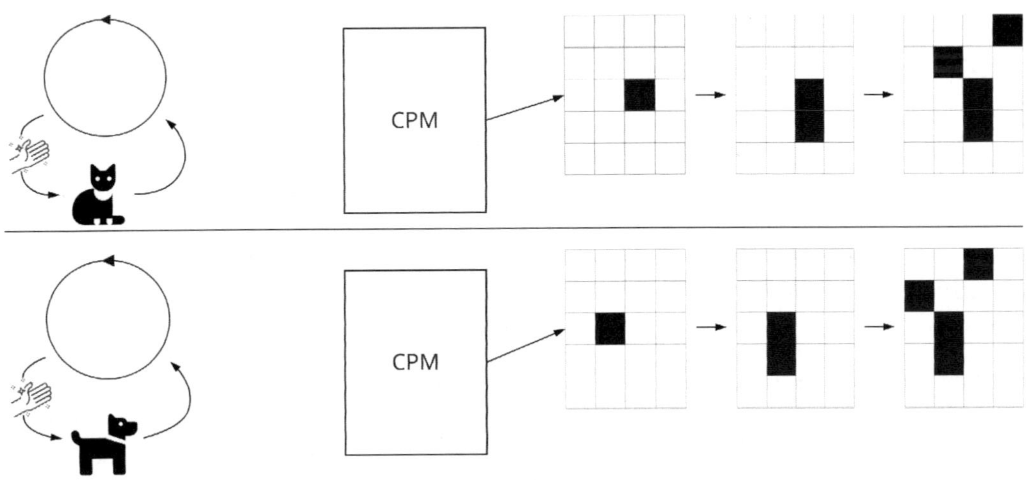

Figure 7. Metaphors and $CPMs$. Two interactions, sequences of *perception* ↻ *action* (upper-row: petting a cat, lower-row: petting a dog), transform the initial state of their corresponding CPM according to different, but correlated paths. In the end, the states of both CPM are different (compare the corresponding matrices), but subtle correlations exist between both matrices (the two black square patterns are not identical but "homologous"). In this case, we would say that each random walk is a metaphor of the other.

We remark that the same applies to more "abstract" circumstances, e. g., in the context of the learning of mathematics, when instead of two pets to fondle, we have two problems to solve. For example, the classical *ruin problem* for two players, in probability, and the symmetric random walk of our frog now along a row of six stones, say, in a tropical river, numbered 0 to 5, with the caveat that the end stones 0 and 5 are, in fact, alligator heads. Then, initial fortunes of 2 coins and 3 coins of the players, say, correspond to the frog starting from stone number 2. The learners wonder about the unfolding of the game or the fate of the frog. Each problem appears as a metaphor of the other, and both can in turn be metaphorised as the draining down of water through a suitable vertical network of ducts. See [34] for a detailed account of this metaphorisation.

5. Discussion: Self-reference and Metaphorising using Enactive Bayesian Cognition.

Our first-level introduction of the basic concepts of Enaction/Biology of Cognition uses two main ideas: a) Bayesian inference triggered by internal signals and b) plasticity of the Bayesian mechanism according to a Hebbian-like rule. Notice that our understanding of mathematical thinking based on these premises will be very different from the standard viewpoint based on applying perfect logic to a disembodied universe.

We summarise our take-homes below.

First, the Enactive viewpoint considers *generation of action* (rather than analysis) as the central property of a nervous system.

Second, the mechanism generating actions is intrinsically probabilistic, not a mere trick to hide our ignorance about the value of some systemic variables.

Third, the input to our mechanism is not a simple transduction of the external world. The input is an interplay of the confrontation of the nervous system (or organism) against its medium mediated by its own metabolism. This confrontation is always embedded in the task of self-fabrication. In some sense the input could be represented by a filtering of the medium by the metabolic network. Thus, the input to the inference mechanism contains many elements of the internal rules produced by the self-fabricating metabolism and is not in a one to one correspondence with the "external world", thus it is more complex than a *blurred* image of the external world.

Fourth, most actions can be triggered by purely internal events without any input from an external, pre-existing, reality.

Fifth, through the never-ending interplay between the mechanisms of Bayesian inference and Hebbian plasticity, a dynamics akin to a random walk arises. Thus, if we consider simultaneously the random walk generated by the aforementioned mechanism in the space of *Perceptions* and the associated random walk in the space of *Actions*, in some sense each random walk is a metaphor of the other. Although both walks exhibit a significant random component, they are strongly correlated to each other. Indeed, the corresponding sequences of perceptions and actions both converge to (the construction of) the same object in the Umwelt of the organism.

Sixth, internal metabolic states are handled by the organism as bona-fide "external object". The identification between internal states and objects (which are defined by recurrent perceptually guided actions) is crucial, as it is *the* mechanism we use to create, recursively, all sorts of abstract mathematical objects (number, function, Fermat last theorem, etc.). We then present the following diagrammatic description of the processes involved:

$$(\textit{Internal-State}_t \rightarrow \textit{Action}_{t+1} \rightarrow \textit{Internal-State}_{t+2}),$$

but since internal states are encapsulated as *Objects*, the above sequence could be written as:

$$(\textit{Object}_t \rightarrow \textit{Action}_{t+1} \rightarrow \textit{Object}_{t+2})$$

making clear that *Objects* are continuously created by the very fact that the organism handles internal states as external objects. Finally, we can synthetically rewrite this sequence as follows.

$$Action(Objects) = Objects$$

analogously to the standard functional notation $f(x) = y$, meaning that actions applied to objects give objects.

But since $Objects \longleftrightarrow Actions$, i.e. Actions and Objects are identified with each other, we can write the overall process as a system of *self-referential* equations [35] to foster the understanding of the intertwining between cognition and action [36, 37]"

$$Action(Action) = Action,$$
$$Perception(Perception) = Perception$$

The latter being entailed by the former because of the intertwining of perception and action, embodied in the recurring *perception* ↻ *action* loops.

Seventh, instead of considering the correlated random walks in perception space and in action space sequentially, as intertwined alternating random walks, we can fathom them simultaneously as a single random walk whose states are ordered pairs $(perception, action)$, so that $(perception(t), action(t+1))$ goes to $(perception(t+2), action(t+3))$, etc. This generates an overarching random walk in the Cartesian product of *Perception* space and *Action* space, whose transition matrix at time t is a $D \times D$ square block diagonal matrix, whose blocks are $CPM^*_{t+1} \cdot CPM_t$ and $CPM_{t+2} \cdot CPM^*_{t+1}$, with $D = dim(Perception) + dim(Action)$. The "2-dimensional" trajectories of this random walk provide a visualisation of our complex sequences of *perception* ↻ *action*. Metaphorisation appears then as similarity of these trajectories, as is the case of *Pet a Cat* and *Pet a Dog* in the toy example of Figure 7 (where $D = 4 + 5 = 9$).

In summary, we have presented the backbone of a theory explaining how living systems construct the objects that populate their niche. The main ideas are that organisms are self-fabricating systems and that they live inside a never-ending spiral relating perception to action. Furthermore, we claim that, in order to select the action to be executed among a *behavioral repertoire* each organism performs Bayesian inference on dynamic probabilities. This viewpoint has the advantage of explaining how a circumstance (an object or a mathematical result) can be viewed as the metaphor of another object. We see mathematical thinking then as centred on the world the organisms create and not on immutable rules of logic. Our perspective opens the door to build an organismic Artificial Intelligence based on actions rather than simple predictions based on oceanic amount of data. In a nutshell we present a novel manner to interpret mathematical work based on the premise that mathematics is the product of concrete self-fabricating living systems entangled with their Umwelt through Structural Coupling.

References

[1] Wiener N 1948 Cybernetics. Scientific American 179 14–9

[2] Shannon C E 1948 A Mathematical Theory of Communication. Bell System Technical Journal 27 379–423

[3] Baddeley R, Hancock P and Földiák P 2000 Information Theory and the Brain. Cambridge University Press

[4] Hubel D H and Wiesel T N 1959 Receptive fields of single neurones in the cat's striate cortex. J Physiol 148 574–91

[5] Wurtz R H 2009 Recounting the impact of Hubel and Wiesel. J Physiol 587 2817–23

[6] Marr D and Ullman S 2010 Vision: A Computational Investigation into the Human Representation and Processing of Visual Information. Cambridge, MA: The MIT Press.

[7] Georgopoulos A P, Kalaska J F, Caminiti R and Massey J T 1982 On the relations between the direction of two-dimensional arm movements and cell discharge in primate motor cortex. J. Neurosci. 2 1527–37

[8] Varela F J and Singer W 1987 Neuronal dynamics in the visual corticothalamic pathway revealed through binocular rivalry. Exp Brain Res 66 10–20

[9] von Uexküll J 1992 A stroll through the worlds of animals and men: A picture book of invisible worlds. Semiotica, 89(4):319–391.

[10] Maturana H and Varela F 1980 Autopoiesis and Cognition: The Realization of the Living. Boston Studies in the Philosophy and History of Science (vol 42), Reidel.

[11] Maturana H 1970 Biology of cognition. biological computer laboratory research report bcl 9.0. Urbana, IL: University of Illinois.

[12] Varela F, Thompson E and Rosch E 1991 The embodied mind: Cognitive science and human experience. MIT Press.

[13] Lyon P 2015 The cognitive cell: bacterial behavior reconsidered. Frontiers in Microbiology 6

[14] Rosen R 2005 Life Itself: A Comprehensive Inquiry Into the Nature, Origin, and Fabrication of Life. Columbia University Press

[15] Rosen R 1958 The Representation of Biological Systems from the Standpoint of the Category Theory. Bulletin of Mathematical Biophysics, 20:317-345.

[16] Cornish-Bowden A and Cárdenas M L 2020 Contrasting theories of life: historical context, current theories. In search of an ideal theory. Biosystems, 188:104063.

[17] Gánti T 2003 The Principles of Life. Oxford University Press.

[18] Hordijk W, Steel M and Dittrich P 2018 Autocatalytic sets and chemical organizations: Modeling self-sustaining reaction networks at the origin of life. New Journal of Physics 20

[19] Seth, A 2021 Being you: A new science of consciousness. Faber and Faber.

[20] Maturana H R, Varela F J 1992 The tree of knowledge: The biological roots of human understanding (Rev. ed). Shambhala.

[21] Friston K 2010 The free-energy principle: a unified brain theory? Nature reviews neuroscience, 11(2):127–138.

[22] Sajid N, Ball P J, Parr T and Friston K J 2021 Active inference: demystified and compared. Neural Computation, 33(3):674–712.

[23] Hebb D O 2005 The organization of behavior: A neuropsycho- logical theory. Psychology Press.

[24] Aristotle 1984 Poetics (trans: Bywater I). In J. Barnes (Ed.), The Complete Works of Aristotle: The Revised Oxford Translation. Princeton University Press.

[25] Abrahamson D, Gutiérrez J F and Baddorf A K 2012 Try to see it my way: The discursive function of idiosyncratic mathematical metaphor. Mathematical Thinking and Learning, 14, 55–80.

[26] Díaz-Rojas D, Soto-Andrade J and Videla-Reyes R 2021 Enactive Metaphorizing in the Mathematical Experience. Constructivist Foundations, 16, 265–274.

[27] English L D 1997 Mathematical reasoning: Analogies, metaphors, and images. L. Erlbaum Associates.

[28] Lakoff G and Nuñez R 2000 Where mathematics comes from. Basic Books.

[29] Manin Y I 2007 Mathematics as metaphor. Amer. Math. Society.

[30] McGowen M and Tall D 2010 Metaphor or Met-Before? The effects of previous experience on practice and theory of learning mathematics. J. Math. Behaviour, 29, 169–179

[31] Sfard A 2009 Metaphors in education. In: Daniels H, Lauder H, Porter J (eds) Educational theories, cultures and learning: a critical perspective. Routledge, New York, pp 39–50

[32] Soto-Andrade J 2014 Metaphors in Mathematics Education. In S Lerman (Ed.), Encyclopedia of Mathematics Education (pp. 447–453). Springer.

[33] Soto-Andrade J 2018 Enactive metaphorising in the learning of mathematics. In G. Kaiser, H. Forgasz, M. Graven, A. Kuzniak, E. Simmt, B. Xu (Eds.), Invited Lectures from the 13th International Congress on Mathematical Education (pp. 619–638). Springer.

[34] Soto-Andrade J, Diaz-Rojas D and Reyes-Santander P 2018 Random walks in the didactics of probability: Enactive metaphoric learning sprouts. In C. Batanero E. J. Chernoff (Eds.), Teaching and learning stochastics: Advances in probability education research (pp. 125–143). Springer.

[35] Soto-Andrade J, Jaramillo-Riveri S, Gutiérrez C and Letelier J 2011 Ouroboros avatars: A mathematical exploration of self-reference and metabolic closure. In T. Lenaerts, M. Giacobini, H. Bersini, P. Bourgine, M. Dorigo, R. Doursat (Eds.), Advances in Artificial Life ECAL 2011: Proceedings of the Eleventh European Conference on the Synthesis and Simulation of Living Systems (pp. 763–770). The MIT Press.

[36] Harvey I 2019 Circular causation, circular cognition: A tour around some common confusions. Artificial life, 25(4):334– 351.

[37] Bechtel W and Bich L 2021 Grounding cognition: heterarchical control mechanisms in biology. Philosophical Transactions of the Royal Society B, 376(1820):20190751.

Exploring Embodied Resources in Gaze in Human-Robot Collaborative Environments

Chapa Sirithunge[1], Kalhan S. Boralessa[2], W. H. Eranga[3], Buddhika P. Jayasekara[3], D. P. Chandima[3] and Manjula U. Hemapala[3]

[1] Department of Engineering, University of Cambridge, Cambridge CB2 1PZ, UK
[2] Department of Computer Science, Nottingham Trent University, Nottingham NG1 4FQ, UK
[3] Department of Electrical Engineering, University of Moratuwa, Katubedda 10400, Sri Lanka

E-mail: csh66@cam.ac.uk

Abstract. Among various types of embodied resources in humans, gaze, started with mutual gaze, plays a major role in embodied cognition. In addition to establishing relationships during interactions, gaze further portrays information about the level of engagement in a dyadic interaction. Hence the gaze and gaze-related behaviors such as averted gaze can be used as cues to make decisions regarding an interaction. This holds true for a human and a robot during human-robot interaction (HRI) as well. Hence proactive robots could evaluate human gaze as a parameter to achieve situation-awareness. In this work, we present the outcomes of several experiments aimed at evaluating such gaze behavior of human-human nonverbal interactions and other behaviors initiated as a result during dyadic interactions. The possibility of evaluating situations by such behavioral responses of individuals as cues, is also examined. We further compared the relationships between gaze behavior of humans during HRI and human-human interaction (HHI). We considered the properties of existence and aversion of gaze, as gaze behaviors during this study. Results of these experiments indicate interesting tendencies in verbal and nonverbal human behavior in initiating of an interaction in both HHI and HRI. The behavioral patterns related to gaze, observed during the study were analyzed using statistical methods and critical observations are highlighted. The potential of analyzing gaze behavior in displaying messages to the outside world during HRI is discussed.

1. Introduction

A robot's embodiment affects the its emotional relationship with human companions.Hence robotic bodies with adequate affective qualities must be designed to foster synergies between various aspects of interactivity [1] . 'Gaze' plays an important role in continuing communication in a co-operative fashion. Despite the complexity of human behavior, gaze can be stated as a simple, but prominent cue observed in a social human behavior [2,3]. Gaze is further considered a successful way of displaying the innermost intentions of engagement in an interaction to outside or the expected conversant [4]. With the growth of collaborative robots, their cognitive skills have to be improved adequately [5]. In such applications, robots received much higher acceptance from their human users. This is because most of the modern robots which we encounter today are specialized to perform a given task within a predefined model with limited constraints [6], hence lack social intelligence. Cognitively and emotionally intelligent robots are expected to have a general sense of social contexts than task-specific robots. Then only the robots will be able to assist their users without causing any disturbance. However, theoretical models in embodiment and cognition are less specific to explain a complete behavioral scenario at present [7]. Furthermore, users will not have to constrain their behaviors to adapt to the limitations in perceptive skills of a robot.

According to Arkin in [21], an intelligent agent has to be able to generate an abstract model of its current context and people in it before it acts. To achieve an adequate sense of social context, robots require a proper interpretation of the surrounding, especially its user [8]. Psychological theories such as the theory of mind (ToM) [9] and some previous studies in HRI identified various observable human cues related to movements and behavior that display human interest in a particular subject. Gaze, gestures, posture changes, involuntary movements, verbal responses and facial expressions to name a few. A method to estimate the attention of a user is proposed in [10]. The authors have used the pose and speeds of specific angular joints of a human to predict the nonverbal interaction demanding of a human user. These two cues were evaluated based on fuzzy logic to derive a user's interest level towards a robot. Even so, there can be other parameters that portray attention more clearly, especially when the users are not engaged in a task.

The human study performed in [11] was planned to investigate human responses towards an interaction with a robot in various circumstances. As indicated in this investigation, human reactions upon a robot-initiated interaction vary depending on their current activity. However several studies suggested that a gaze from a robots is interpreted differently than a gaze from a human [12]. However this work does not explore the characteristics of gaze behavior between proactive robots and humans. More recently, attempts have been made to understand the temporal characteristics of social gaze but little research has been conducted in a natural setting with two interacting participants [13].

Interactivity of an encounter was measured using four types of connection events incorporating gesture and speech: directed gaze, mutual facial gaze, conversational adjacency pairs and back channels according to [14]. The work explained here used initiator and responder gaze times to determine gaze and gaze is used with the other three parameters to determine user engagement during interaction. The trade-off between user's and robot's gaze were used as a cue to evaluate a certain situation in [15]. In [16], 3D positioning of human frustum and eye movement were continuously monitored to evaluate situation-awareness during a joint task. However tracking eye movement needs to move the robot to close vicinity. Therefore this evaluation cannot be performed at a distance prior to an interaction. Hence such robots may use 'gaze' instead of 'eye movement'. In addition, authors in [20] have identified gaze as a parameter which portrays the user engagement with social robots.

The duration or time in consistent gaze proved to be an important parameter in determining conversational attention in multi-user environments [17]. Findings of this work confirmed that the gaze directional cues of users could form a reliable source of input for the conversational systems that need to establish whether the user is speaking or listening to them. This further discusses that the predictive power of signals related to gaze may depend on the individual and the visual design of the overall conversational system. Mendez et al. in [18] proposes evaluating the diveristy of gaze patterns in older adults and young during HRI. The authors expect to compare differences in reaction times in response to gaze in both the age groups although the authors do not expect to see much difference in preferences for the robots between age groups. This can be stated as one research that will go in-line with the findings of our set of experiments in this regard.

An evaluation of personality traits in terms of gaze aimed at eye and body during HHI and HRI is explained in [19]. Results of the experiment confirmed that the participants gaze more at body in HRI than in HHI, and they gaze more at the partner's face in HHI than in HRI. The reasons authors guessed for these trends were that robot's face is only blinking by changing color of the eye, which does not provide any information and to understand the meaning of the gesture of the robot respectively. From the above examples, we could observe that many robotic systems developed so far have considered 'gaze' as a parameter to evaluate a situation between a human and a robot. Therefore we tried to find the role of 'gaze' during an interaction to find how important it is to evaluate this feature and which other facts can be explored by analyzing gaze. In this research we tried to find answers for the following questions.

1) How do humans use gaze-related features as actions during human-human interactions?

2) How similar the features are associated with human-robot interactions with these?

3) How similar are HHI and HRI in terms of gaze behavior?

For that, we conducted a series of experiments with human-human and human-robot dyads to find out the patterns in their gaze behavior and hence find answers to the above questions.

2. Understanding human behavior

Robots in human-robot collaborative environments will have to co-operate with humans successfully, to ensure their survival in the real world. Humans adopt various physical behaviors to display their intentions to outside. Changes in movements, pose, posture, gestures, nature of speech, to name a few. Understanding human intents through visual cues is one critical ability such robots will have to master. In this paper, we took an ethological approach to analyze how human gaze behavior changes in the presence of another human and how should we adopt gaze observations in human-robot interactions.

2.1. Perception of behavior

Out of the numerous cues which humans display, gaze which receives much higher attention during an interaction [22] is analyzed in this paper. Furthermore, gaze can be used as a cue which displays our intentions to the outside world [3]. Hence we intend to identify features related to gaze that are worth analyzing. We expect that such an analysis will facilitate developing cognitive models related to HRI in the future.

2.2. Timescales of human actions

According to the time scale of human actions given in [23], actions that take place within 100 milliseconds (ms) to 10 seconds (s) fall under the cognitive band of actions. In our work, we selected 'gaze' behavior which is also a cognitive act. Gaze can sometimes be an automatic behavior and sometimes, a controlled behavior [24] where the reaction time may take up to about 1.3 seconds. Hence we adopted a timescale to observe gaze behavior between human-human and human-robot dyadic interactions.

2.3. Understanding gaze behavior

Figure 1 illustrates a typical human-human encounter and the change in responder's gaze with time. Alongside this, three parameters involved in gaze: response time, duration, averted/not were analyzed during this study. These occur at different time scales when initiating an interaction. Here, the 'delay' represents the time taken by the responder to react, once a change in situation was detected. When considering gaze, the change in situation was the presence of an outsider who seeks attention. Here, the 'duration of gaze' relates to the entire time the responder looks at the approacher. The 'averted gaze' is when the responder returns his/her gaze back at his previous task. These features associated with gaze can say a lot to an outsider about the inner intents and priorities of the responder. Hence these three features of gaze during HHI and HRI have been investigated separately in our experiments.

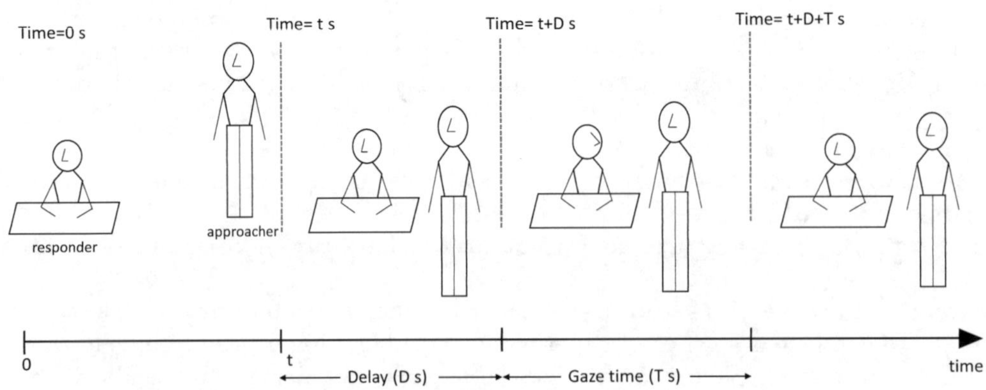

Figure 1. Phases in a typical encounter prior to an interaction are shown. At time= 0, the approacher intends to approach the target person: the responder. At t=t s, he/she approaches the responder. The responder notices his/her presence, but takes another D seconds to give attention. The responder looks at him/her for a duration of T seconds and averts his/her gaze to continue the work.

3. Experiments and results

3.1. Setting

The experiment was conducted in study areas and laboratories inside the university with the participation of 47 subjects in an age range of 25-51 years (mean of 38.6 and SD of 11.31). Nearly half of the participants (23) had no strong technical background. The experiment was conducted in 2 stages: HHI and HRI.

Initially participants were asked to be seated and read the guidelines of the experiments in the form of a document opened on a laptop on a table as shown in Figure 2. While the participant was reading these guidelines, a person previously known to them, but not family/ a relative, was allowed to approach him/her. The participant's reaction to the approacher were recorded. The role of the approacher was to walk towards the target person, without any verbal or nonverbal interaction. Here interactions refer to dialogs, gestures, facial expressions, etc. These observations were analyzed under stage 1: human-human interactive scenario, shortened into HHI. The same procedure was repeated in stage 2. But this time, the approacher was a robot. Stage 2 was conducted once a week over a period of 3 weeks to give the participants some time to adapt to social robots. We terminated the set of experiments once we observed several trends in HRI which were similar to HHI.

Stage 2: human-robot interactive scenario, shortened into HRI, was conducted over 3 weeks in 3 occasions. These occasions are HRI-week 1, HRI-week 2 and HRI-week 3 in which the same experiment was conducted in the first, second and third week respectively. The reason to conduct only stage 2 along 3 weeks is that human-robot interaction process was not normalized as human-human interaction even at present. Therefore we gave participants time to adapt to the human-robot collaborative environments. Therefore we repeated the same experiment thrice within 3 consecutive weeks. The only difference in stage 2 was that the approacher was the robot, wherein the approacher was another human in stage 1.

Therefore in the end, there were four occasions considered as follows.

- HHI- Both the approacher and the participant were humans.
- HRI (HRI-week 1, HRI-week 2, HRI-week 2)- The interaction took place between human participant and robot.

In all 4 occasions, the interaction was dyadic. We did not allow other people in the surrounding as human responses might have an influence from outsiders' presence [11]. The interaction stopped before any conversation between the two parties begins. In all the occasions, the features associated with the gaze: presence of gaze/averted gaze, delay in gaze and the duration of existed gaze, were recorded using a video camera set up in front of the setting. At a later stage, this video was analyzed to extract required parameters during the interaction. Two example scenarios from HHI and HRI from the experiment are shown in Figure 2.

3.2. Results of the experiments and discussion

Firstly, we analyzed whether gaze existed or not in a scenario and if existed, we recorded parameters or properties of existed gaze behavior. At the end, we looked for the existence of other responses towards an interaction between two participants in addition to gaze. An in-depth analysis of our observations is given in the following paragraphs.

Several labels have been used for the ease of referring the parameters related to gaze. These are mentioned below.

- Continued gaze- When the user looked at the approacher and continued looking until the interaction ends.
- Averted gaze- When the user looked at the approacher and averted gaze after a while.
- Delay- The time taken by the participant to look at the approacher after the person approached him/her.
- Duration- The duration of existed gaze before the participant averted gaze.

0 and 1 in the graphs denote non-existence and existence of a considered parameter respectively.

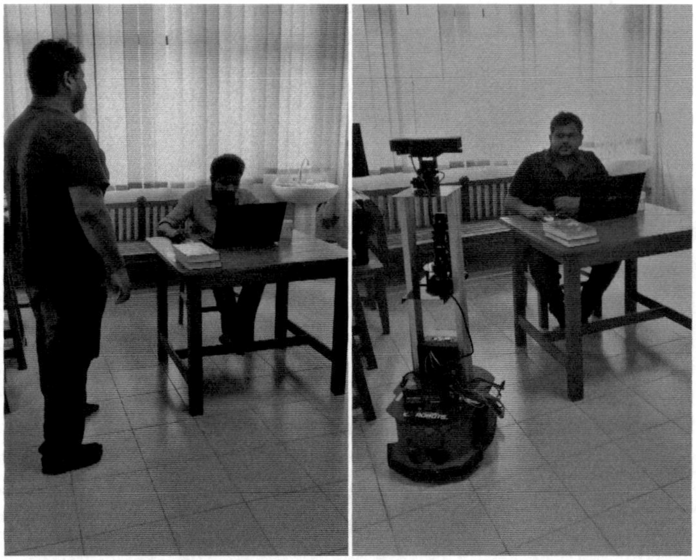

(a) (b)

Figure 2. Two scenarios during the experiment are shown. (a) HHI scenario is shown. Both the responder and approacher were humans. (b) An HRI scenario is shown. The approacher was a robot while the responder was human. In both the occasions, the responses generated by the responder after noticing the presence of the approacher were recorded.

IOP Conf. Series: Materials Science and Engineering 1292 (2023) 012013 doi:10.1088/1757-899X/1292/1/012013

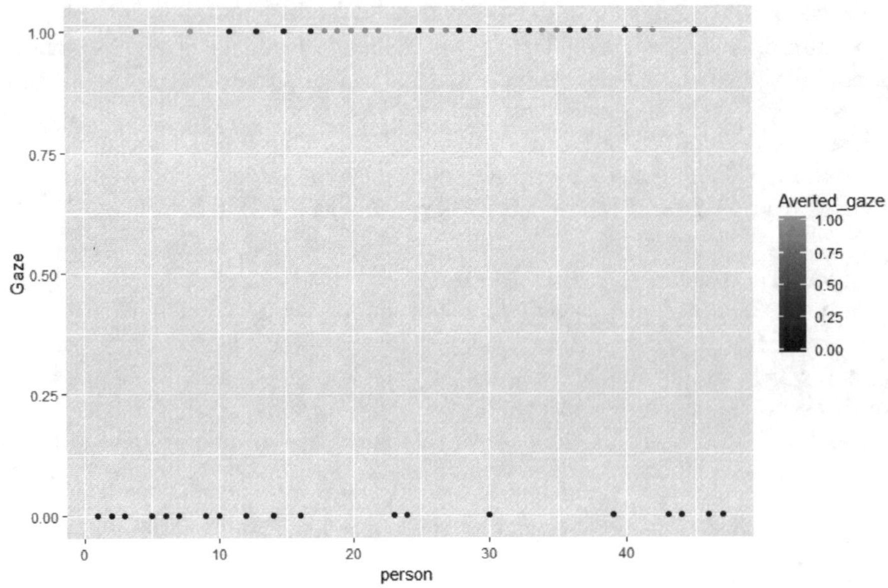

Figure 3. Existence or nonexistence of gaze during HHI is plotted against each user. If the user averted his/her gaze, such instances are marked in light blue while continued gaze is marked in dark blue.

Figure 3 shows the presence of gaze parameters during HHI. As seen from the graph, 26 persons out of the 47 considered (55%), maintained gaze with an approacher. Out of them, 14 averted their gaze and turned back to their work, which accounted for 29.8% in total. That is, 54% of individuals who maintained gaze, averted gaze after a while and continued their work. Even when at work, more than half of the participants (55%) maintained gaze with the approacher, although it is averted later. It can be considered a friendly gesture of paying attention towards the approacher Even so, more than half of them (54%) averted gaze and gave their attention back to work. However it is interesting that most individuals tried to be interactive and not to be disturbed at the same time.

The existence of continued/averted gaze as a percentage in each of the 4 occasions is shown in Figure 4. In HHI, this percentage was 55 and in HRI, these percentages were 81, 68, and 49 in the three consecutive weeks. In the first week of HRI, a higher percentage of individuals maintained gaze in week 1, while this reduced considerably (by 32%) by week 3. HRI was only slightly different from HHI in gaze behavior by week 3.

Considering scenarios with averted gaze, the duration of existed gaze is plotted against the average delay of the gaze for the four occasions in Figure 5. In the four occasions, the durations of gaze were recorded as 4.45, 3.36, 2.97 and 1.8 seconds respectively. Corresponding average delays in gaze were 1.06, 1.88, 3.23 and 3.35 seconds. The difference between the duration of gaze and the delay in maintaining gaze was maximum in HHI and HRI-week 3. The duration of gaze was higher in HHI, and the delay in maintaining gaze was higher in HRI-week 3. From this, it can be seen that, when humans are engaged, robots received much higher negligence from humans.

Figure 6 illustrates box plots showing the gaze delays in encounters where a continued gaze was present for the four occasions. First and third quartiles and the mean for the four occasions are marked on the box and whisker plots here. Outliers, minimum and maximum values recorded are marked on the plots as well. When the average delay of the four occasions were considered, HRI-week 1 recorded the least average delay in continued gaze. This value gradually increased

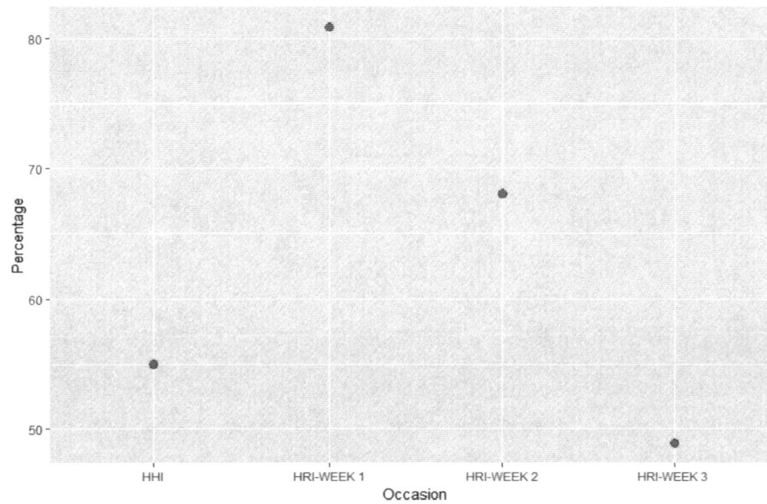

Figure 4. The percentage (%) of encounters with any kind of gaze (continued or averted), out of the total encounters considered, is plotted against the occasion: HHI, HRI-week 1, HRI-week 2 and HRI-week 3.

by week 3 and, HHI and HRI-week 3 resulted in close average delays in continued gaze. The trend in HHI and HRI in terms of gaze delay, at the end of the 3 weeks. A possible reason for the scatter of data in week 2 is that people are getting familiar with social robots in their environments starting from week 2. The scatter of data was again normalized in week 3. This can be the reason to observe negative delays in gaze as well. The unfamiliarity of humans with robots resulted in giving much more attention to robot's behavior when they approach.

Figure 7 plots several box plots to identify the behavior of gaze delay of encounters where a continuous gaze was present. First and third quartiles and the mean for each occasion are marked on the blox and whisker plots. Similar to the previous graphs, duration of averted gaze also recorded similar trends in HHI and HRI-week 3 in terms of the average duration and the

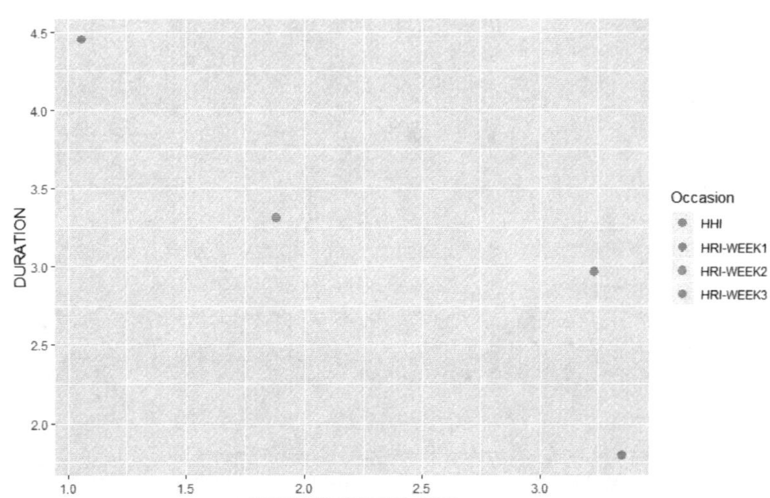

Figure 5. Of the total encounters where averted gaze could be observed, the average duration of existed gaze was plotted against the delay. All the units are in seconds (s).

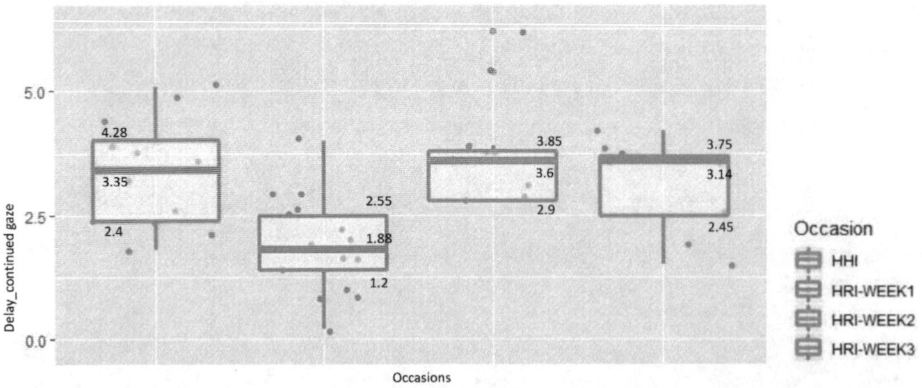

Figure 6. The delays of encounters where a continued gaze was present. The box plots are drawn to identify the behaviors of parameters and their evolution with time and experience.

pattern of data scatter. HRI-week 2 recorded data considerably scattered as in previous graphs. As seen from the box-plots, the trend in delays of averted gaze received numerically slightly different values in HHI and HRI-week 3.

When continued gaze delay and averted gaze delay were compared, the difference between HHI and HRI-week 3 is very small. A probable reason for this is if users are interested they paid attention for both robots and humans. But if they are not interested in an interaction, they will tend to neglect the robot for long. Even though the mean values of averted gaze duration in HHI and HRI- week 3 are close, the actual durations in HRI were scattered over a wide range of values. Familiarity and past experience with robots and personal opinions on robots can be stated as a probable reason for this observation. Humans still consider a robot as a 'machine' rather than a 'companion'. Such perceptions can also support this observed behavior.

In addition to gaze-related behaviors, several other responses towards the interaction were observed during the study. These were voice responses such as greetings (e.g: "Hey", "Hello, how are you?"), smile, changes in body pose and changes in hand pose. Figure 8 illustrates these

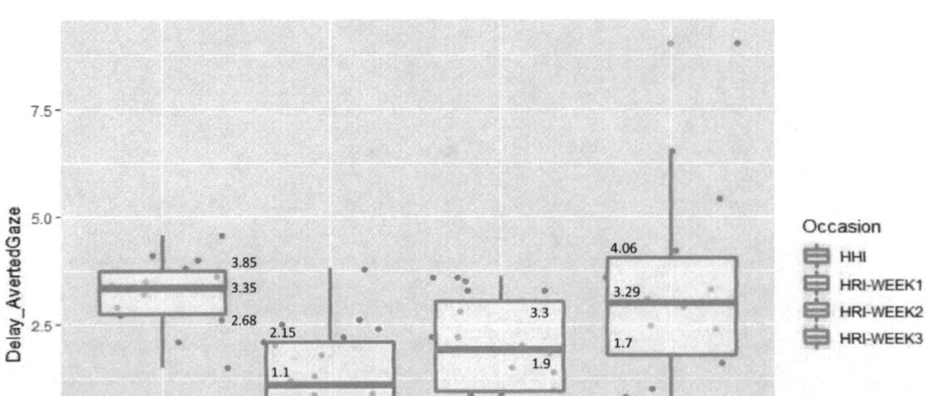

Figure 7. The delays of encounters where an averted gaze was present. The box plots are drawn to identify the behaviors of parameters and their evolution with time and experience. The markers are drawn to the scale of the duration of gaze.

behaviors adopted by the participants to react to the approacher, in addition to gaze. Out of the total participants, 44.68% used some kind of voice response to welcome the approacher. 36.17% smiled at the approacher while only 23.40% changed their pose voluntarily or involuntarily. 17.02% changed their hand pose during the encounter according to the graph. Therefore humans use several other responses to support an interaction even when engaged. Proactive robotic systems have a challenge in perceiving such behaviors as well as analyzing gaze behavior.

3.3. Finding answers

As mentioned in section 1, the objectives of this set of experiments were to find answers to following questions and the answers that were derived from the study are as follows.

1) How humans use gaze-related features as actions during human-human interactions?

During the study we observed the existence/nonexistence of gaze and if gaze existed, the delay and the duration of that gaze. These parameters associated with gaze were numerically measured as a part of the study. In addition, four other responses were observed; voice, smile, pose changes and hand movements. From the results, it could be observed that gaze was the most frequently used behavior during HHI, upon all these observed behaviors. Hence exploring gaze-related behaviors received much higher attention while analyzing the results.

2) How similar features were associated with human-robot interactions?

Once humans were comfortable with social robots, similar responses they use towards other humans could be observed towards robots as well. Comparison of results in Figure 8 and [9] testify to this. According to these two studies, when users were busy, robots received lesser attention than humans, as a result lesser responses as well. Somehow the behaviors observed in front of robots were only slightly different from those in front of humans, except in number.

3) How similar are HHI and HRI in terms of gaze behavior?

Out of the three parameters considered: continued gaze delay, averted gaze delay and averted gaze duration, we could observe similarities between HHI and HRI in continued gaze delay and averted gaze delay. Averted gaze duration during HHI and HRI showed a significant difference. Therefore we can say that there are dissimilarities between HHI and HRI in terms of gaze. It is important to identify these changes as an insight for the accurate cognitive models of future robot assistants. It is also important in evaluating the human acceptance of social robots as well.

The answers to these questions were critically discussed under results and discussion in section 3.2. According to the findings, not only gaze plays an important role during interactions, but also in HHI and HRI with some unique features in both. It will be helpful to identify where

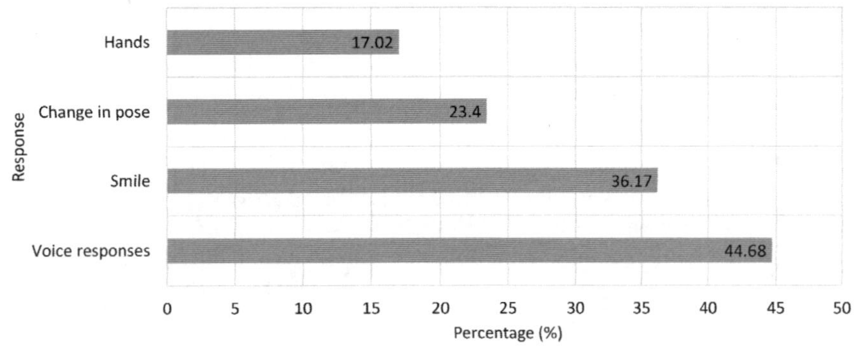

Figure 8. Responses observed from users during HHI, in addition to gaze are plotted against each individual.

gaze parameters differ in HHI and HRI in future applications of social robots.

3.4. Implications of the study

Gaze is an important cue which determines the interactivity of a situation. Hence considering gaze in decision-making can be proposed as a design guideline in future Robotics.

The second design guideline derived from our study is analyzing the properties associated with gaze, without considering the existence/nonexistence of gaze solely.

Human-robot interaction was not very different from human-human interaction in terms of gaze behavior. Therefore proactive robotic systems which are intended to initiate interactions with humans, can adopt human rules and norms related to gaze. This is the third design guideline proposed by this study.

Robots are deployed in social environments in different levels in different communities. In this study we considered younger and adult Asians originated in Sri Lanka where robots are rare in social environments. Therefore considering previous experience with the robot will be our forth design guideline.

In addition to the relationship between the two conversant, there can be other factors which contribute for gaze parameters to differ. For instance factors such as age group and gender might have an influence, at some level. Therefore considering some other factors which might be of interest, will be beneficial in decision making during HRI. This will be another implication derived from our experiments.

Even though gaze is a strong display of emotions during interaction, there are other factors which determine emotions associated with an encounter as well. Therefore multimodal perceptive skills will be required for a robot to perceive a situation entirely during social interaction. But considering gaze as an observable cue is crucial according to the results of this work. Development of such a perceptive ability will be the sixth design guideline derived from the study.

4. CONCLUSIONS

We conducted a set of experiments based on gaze behaviors between human- human and human-robot dyads extends through a period of few weeks. We analyzed similarities and differences in three parameters related to gaze: delay in maintaining any kind of gaze, duration of gaze and whether the gaze is averted or not. Results confirmed the fact that 'gaze' plays an important role in HHI as well as HRI with some observable variations. We could further discover that there are differences between HRI and HHI, with respect to the parameters associated with gaze. Therefore before imitating humanlike behaviors, it is important to identify how and why gaze parameters alter differently during these two scenarios. Hence our experiments combined both HHI and HRI in terms of gaze behavior to lay a justifiable basis for the development of multimodal approaches for a robot's cognition. In line with the previous studies, gaze proved to impact largely on the interactivity of a scenario than other behavioral responses. We evaluated and discussed how existence/non-existence of gaze, gaze time, and averted gaze differ in HRI compared to HHI. Our findings further suggest that averted gaze showed a significant difference during HHI and HRI upon the other parameters. These findings will help proactive robots in the future to make situation-cautious decisions in human-robot collaborative environments. Trends in human-human interaction that could be adopted in human-robot interaction are suggested as the outcomes of this work, helping future robots satisfactorily evaluate a situation in proactive HRI. In the future, we will explore the roles of other observable human cues which will have an impact on interactivity of a situation, except for gaze.

Acknowledgment

The authors are thankful for the participants who volunteered the study.

References

[1] Caroline Yan Zheng, Cherie Lacey, and Mark Paterson. "Affect and Embodiment in HRI." In Companion of the 2020 ACM/IEEE International Conference on Human-Robot Interaction, pp. 667-668. 2020.

[2] Sooha Park Lee, Jeremy B. Badler, and Norman I. Badler. "Eyes alive." In ACM transactions on graphics (TOG), vol. 21, no. 3, pp. 637-644, ACM, 2002.

[3] Joe Navarro and Marvin Karlins. "What Every Body Is Saying." William Morrow Paperbacks, 2008.

[4] Roel Vertegaal, Robert Slagter, Gerrit Van der Veer, and Anton Nijholt. "Eye gaze patterns in conversations: there is more to conversational agents than meets the eyes." In Proceedings of the SIGCHI conference on Human factors in computing systems, pp. 301-308. ACM, 2001.

[5] Latikka, Rita, Tuuli Turja, and Atte Oksanen. "Self-efficacy and acceptance of robots." Computers in Human Behavior 93, 157-163, 2019.

[6] Yuda Irawan, Muhardi Muhardi, Rian Ordila and Roni Diandra. " Automatic Floor Cleaning Robot Using Arduino and Ultrasonic Sensor." in Journal of Robotics and Control (JRC), 2(4), 240-243, 2021.

[7] Schneider, Bertrand, Marcelo Worsley, and Roberto Martinez-Maldonado. "Gesture and gaze: Multimodal data in dyadic interactions." International Handbook of Computer-Supported Collaborative Learning. Springer, Cham, 625-641, 2021.

[8] Chapa Sirithunge, A.G. Buddhika P. Jayasekara, and D. P. Chandima. "Proactive Robots With the Perception of Nonverbal Human Behavior: A Review." in IEEE Access 7 , pp. 77308-77327, 2019.

[9] A.M. Leslie, "Pretense and representation: The origins of theory of mind." in Psychological review, 94(4), p.412, 1987.

[10] Chapa Sirithunge, Viraj Muthugala, Buddhika, Jayasekara, and D. P. Chandima. "Interpretation of interaction demanding of a user based on nonverbal behavior in a domestic environment." In 2017 IEEE International Conference on Fuzzy Systems (FUZZ-IEEE), pp. 1-8. IEEE, 2017.

[11] Chapa Sirithunge, Viraj Muthugala, Buddhika Jayasekara, and D. P. Chandima. "A Wizard of Oz Study of Human Interest Towards Robot Initiated Human-Robot Interaction." In 27th IEEE International Symposium on Robot and Human Interactive Communication (RO-MAN), pp. 515-521. IEEE, 2018.

[12] Henny Admoni and Brian Scassellati. "Social eye gaze in human-robot interaction: a review." in Journal of Human-Robot Interaction 6.1, pp. 25-63, 2017.

[13] Simon Ho, Tom Foulsham, and Alan Kingstone. "Speaking and listening with the eyes: gaze signaling during dyadic interactions." PloS one 10, no. 8: e0136905, 2015.

[14] Charles Rich, Brett, Ponsleur, Aaron Holroyd and Candace Sidner, "Recognizing engagement in human-robot interaction," in 5th ACM/IEEE International Conference on Human-Robot Interaction (HRI). IEEE, pp. 375–382, 2010.

[15] K. Fischer, L.C. Jensen, S.D. Suvei, and L. Bodenhagen, "Between legibility and contact: The role of gaze in robot approach" in Robot and Human Interactive Communication (RO-MAN), 25th IEEE International Symposium on. IEEE, pp. 646–651, 2016.

[16] Lucas Paletta,Dini Amir, Murko Cornelia, Saeed Yahyanejad, Michael Schwarz, Gerald Lodron, Stefan Ladstatter, Gerhard Paar, and Rosemarie ̈ Velik. "Towards real-time probabilistic evaluation of situation awareness from human gaze in human-robot interaction." In Proceedings of the Companion of the 2017 ACM/IEEE International Conference on Human-Robot Interaction, pp. 247-248. ACM, 2017.

[17] Vertegaal, Roel, Robert Slagter, Gerrit Van der Veer, and Anton Nijholt. "Eye gaze patterns in conversations: there is more to conversational agents than meets the eyes." In Proceedings of the SIGCHI conference on Human factors in computing systems, pp. 301-308. 2001.

[18] Lucas Morillo Mendez, and Óscar Martínez Mozos. "Towards Human-Based Models of Behavior in Social Robots: Exploring Age-Related Differences in the Processing of Gaze Cues in Human-Robot Interaction." In STAIRS@ ECAI. 2020.

[19] Ijuin, K., Jokinen, K. "Exploring Gaze Behaviour and Perceived Personality Traits." In: Meiselwitz, G. (eds) Social Computing and Social Media. Design, Ethics, User Behavior, and Social Network Analysis. HCII Lecture Notes in Computer Science(), vol 12194. Springer, 2020.

[20] S.M. Anzalone, S. Boucenna, S. Ivaldi and M. Chetouani, "Evaluating the engagement with social robots." in International Journal of Social Robotics, 7(4), pp.465-478, 2015.

[21] Ronald C. Arkin. "Behavior-based robotics." by MIT press, 1998.

[22] Michael Argyle and Mark Cook. "Gaze and mutual gaze." Cambridge University Press, 1976.

[23] Allen Newell. "Unified Theories of Cognition" by Harvard University Press, 1990.

[24] Richard M. Shiffrin, and Walter Schneider. "Controlled and automatic human information processing: II. Perceptual learning, automatic attending and a general theory." in Psychological review 84.2, 127, 1977.

[25] A. J. Moon et al., "Meet Me where I'm Gazing: How Shared Attention Gaze Affects Human-Robot Handover Timing," 2014 9th ACM/IEEE International Conference on Human-Robot Interaction (HRI), pp. 334-341, 2014.

IOP Conf. Series: Materials Science and Engineering 1292 (2023) 012014 doi:10.1088/1757-899X/1292/1/012014

How Do We Trust Our Robots?

Chapa Sirithunge[1] **and Nikolas Dahn**[2]

[1] Department of Engineering, University of Cambridge, Cambridge CB2 1PZ, UK

[2] German Research Center for Artificial Intelligence (DFKI), Robert-Hooke Str.1, 28359 Bremen, Germany

E-mail: csh66@cam.ac.uk

Abstract. Trust plays the most important role in accepting social robots in our communities, especially in an era where robots take part in daily activities in close distance with humans. While some communities accept the presence of robots in their surrounding, others are less open to them. Furthermore, this acceptance depends on a large number of factors which are specific to the communities as well as the deployed robots. We conducted a survey to explore the aspects which people would consider if they had an option to deploy a robot at home and assign responsibilities to it. The study will help direct future researches on embodied intelligence in robots towards a more human-accepted level. The paper presents a human study conducted to evaluate the trust between humans and robots. We present the feedback we received from the participants to assess the level of trust participants have on their robots and their personal preferences upon the abilities of robots. Human responses and their decisions observed during the study are analyzed and critical observations are highlighted. From the analysis we derive a set of guidelines to improve human trust in future robots by adjusting their humanlike social behaviors.

1. Introduction

The rise of social robots in our surroundings opens up an opportunity of transforming robots into companions rather than interactive machines that make our work easier. Hence, natural interaction between robots and humans is important to make the collaboration effective. On the other hand, if robots can perceive the requirements of a scenario without external guidance, this will lessen the burden on humans in these activities, both emotional and physical. *Trust* plays an important role in making this companionship between humans and social robots more interactive [1].

To assess human expectations within a given situation, robots might have to think and act like humans. In short, their behavior will have to be more humanlike in order to make their interaction with humans smooth and flexible during day-to-day tasks. This *natural* behavior will help humans to accept the robots in their surrounding more easily, enabling humans to closely work without worrying over a set of opaque handling rules. Machines expressing humanlike features have been shown to benefit human-robot team cohesion [2], but may also prove detrimental if implemented poorly, as is demonstrated by the now well-known uncanny valley [3].

At present, there are several obstacles that prevent humans from fully trusting robots. On the one hand, there are technological constraints to enable robots with all the required features. On the other hand, we lack a proper conceptual model for a robot's brain [4] that would equip a robot with adequate emotional intelligence to make humanlike decisions at proper occasions. The *adequateness* of the robot's actions is one of the cornerstones for building trust into the robot's capabilities in human-robot collaborative environments. In addition, there can be cultural and personal variations in communities and individuals in accepting robots as well [5]. Embodied intelligence and humanlike perceptive skills can immensely enhance human acceptance towards assistive robots [6].

Improving human trust towards robots despite these issues requires an analysis of the state of the human trust on robots at present. As different communities have different perceptions upon robots, it is important to conduct researches to analyze the level of acceptance received by robots from around the world [7]. Even so, some of the findings presented here can be assumed to be universal and may help in developing social robots in the future.

We conducted a survey to assess the level of trust in our community and explore the issues associated with that. The main objective of this research is to explore the requirements for improving human trust on robots and to mediate a set of guidelines for the future robotic companions in social environments. During the survey, we allowed the participants to answer different questions to assess their acceptance of robots in their surroundings and later evaluated their preferences in assigning responsibilities for robots in mission-critical tasks. We further explored human preferences upon the appearance and physical and cognitive skills of a robot.

2. Related Work

Trust is crucial for the flow of interaction between humans and their robotic companions. The work explained in [8] made an argument if and under which conditions the notion of trust is applicable to robots. The author concludes that the relationship between humans and robots is distinct from that of most other technical devices and involves an often implicit level of trust that should be considered during design, e.g. by enabling the robot to explain its actions. Since then, various studies have been conducted to examine factors influencing trust in robots, like their overall appearance [9], specific appearance features [10], personality traits [11], transparency [12], team interactions [13], and many more.

In [14], a model of human trust in robots and which factors influence it is developed. They separate these factors into three categories: robot characteristics, human characteristics and environmental characteristics (i.e. interaction modalities). Respective examples for these are adaptability, workload and communication. It seems worth mentioning that this research considered a military context, where trust and reliability are of particular importance.

A more refined model can be found in [15], where the authors present a meta-analysis of empirical studies investigating factors that influence human trust in robots. They identify factors that have been proven to be relevant as well as those that lack sufficient evidence or have not been considered yet, especially from the area of team collaboration (e.g. interaction frequency). In contrast, our survey tries to assess whether humans even want to trust robots and under which conditions.

In [1], a social study is presented where a connection was made between a situation's severity, the magnitude of errors made by the robot, and their users' trust towards it. The authors also found that some character traits of their participants, like benevolence and agreeableness, are strongly correlated with their disposition to trust a robot. We add to this study by exploring the connection between trust in robots, user preferences and user expectations in more detail.

While the above works take a more passive-descriptive approach, [16] gives a summary of how a robotic system may actually be gain, measure and maintain its user's trust. Although several challenges remain unsolved (in particular reliable approaches for establishing and measuring trust "in the wild"), the authors note that for all of these promising approaches are under active research.

Although studies exist that assess how people's willingness to trust changes when the robot is making mistakes [17, 18], we argue that people - through media and video games - already have an idea of what present day robots supposedly can and cannot do and thus how trustworthy they are. These beliefs need to be addressed when designing a robot in order to maximize its user acceptance. Through this study we hope to shed some light on this area and inspire others to expand on it.

3. The survey

3.1. Background

Participants of the study come from a relatively young population of around 28 years (mean-28.46, SD-3.36). There were 63 participants; 32 female and 31 male. 17% of the participants had previous experience with robots and the rest did not. The participants were Sri Lankans who at least finished high school. The effect of gender on human trust towards robots was not considered within the scope of this study. Previous human studies explored that the Sri Lankan community has a positive mindset towards the deployment of service robots in domestic environments for conversations [19]. This lead us to inquire about more complex aspects of human-robot collaboration such as trust, which directly affects the human acceptance towards robots.

3.2. Survey questions

The survey questionnaire was formed so that participants will indirectly answer the question: *To which extent do you trust robots in your surrounding?*. To this end we started by asking participants if they would like to have a robot around them (Question 1-2). From these initial questions we wanted the participants to create an image of the robot they actually prefer, without asking them directly. These are as follows.

- **Question 1: Do you like a robot helping you with your work at home?**
- **Question 2: How do you prefer a robot? As a companion or an assistant? Or both?**

The options given for question 1 were *Yes*, *No* and *May be* while the options given for question 2 were *as an assistant*, *as a companion* and *both*.

Question 3 directly explores the participant's preference upon the robot's appearance and its stereotyped personality.

- **Question 3: How would you call your robot: him, her or it?**

The options given for question 3 were *him*, *her* and *it*.

Questions 4 and 5 explore the participants' intentions for deploying a robot in their surroundings. These questions reveal the participants' main reason to choose a robot. Hence, these questions are the foundation for all other answers provided by the participant.

- **Question 4: What do you think robots do not know, but better if they had known?**
- **Question 5: Which activities do you think that a robot cannot do?**

The participants were given the opportunity to explain or make a list of their preferences for questions 4 and 5.

Questions 6 to 9 are selected in order of complexity and privacy of the task assigned to the robot. While cleaning a house and cooking can be the typical tasks assigned to domestic robots, driving a vehicle and driving kids to school can be critical tasks where high trust and reliability are required. Therefore, by allowing the participants to imagine their robots in such simple and mission-critical tasks, we wanted to compare how people measure the abilities of present robots.

- **Question 6: Would you like a robot to clean your house?**
- **Question 7: Would you like a robot to cook for you?**
- **Question 8: Would you let your robot drive your car?**
- **Question 9: Would you like a robot to drive your kids to school?**

The options given for questions 6 to 9 were *Yes*, *No* and *May be.*

Other than the common set of tasks which domestic robots are specialized in, we further explored how individual preferences of people matter in human-robot domains. While some people accept robots in their surrounding or community, others do not. Question 10 was added to weigh this fact.

- **Question 10: For which tasks would you like to get help from robots?**

The participants were given the opportunity to list their preferences for question 10. However, this may have reasons other than trust, and so questions 5 and 11 will help understand such reasons.

- **Question 11: Will it be better if a human/servant do these for you?**

The options here were *a robot*, *on my own or another human* and *depends on the situation.*

When humans trust someone, they will be relaxed or comfortable around that person. Furthermore, they will be sharing sentiments, having conversations with them [20]. Question 12 explores this fact to evaluate the sentimental relationship between the human and the robot.

- **Question 12: Say you are having a relaxed evening sitting outside. Then your robot approached you. Will you be interested in having a conversation with the robot?**

The options given for question 12 were *Yes*, *No* and *May be.*

Question 13 helps to get a better idea of the participant and their background in robotics, since experience with robots can have a significant influence on a person's stand towards them. The options given for question 12 were *Yes* and *No.*

- **Question 13: Have you worked with robots before?**

Questions 14 to 16 examine the appearance of the robot participants imagined. A graphical illustration of a robot was requested as well to give a better impression of the participants' answers. We have given the chance for the participants who had no drawing ability to search the Internet and upload a matching image. We have taken this step to avoid the participants skills affecting the results of the survey.

- **Question 14: What type of robot did you picture in your mind when answering these questions?**
- **Question 15: If your answer was *other* to the previous question, state the type of robot you imagined.**
- **Question 16: Illustrate the robot you imagined while answering this questionnaire.**

As a summary, the survey questions are formulated, not only to evaluate the level of human trust on robot, but also to understand the following aspects about human-robot collaboration.

- Human preferences upon the capabilities of their robot companions (Question 1, 2)
- Motives of deploying a robot (Question 4, 5, 10, 11)

- How people interpret the personality of a robot during this collaboration (Question 2-3, 14-16)
- How people assign responsibilities to their robots (Question 4-5, 6-9)
- How robots can receive human-level acceptance (Question 12)

Answers for these questions are discussed in section IV.

4. Results and discussion

For question 1, 3.2% of participants responded *No*, 1.6% responded *May be* and the remaining 95.2% responded *Yes*. Therefore, the number of participants who voted against using robots at home were very low compared to those who preferred a robot helping them at home. It is important to note that the participants who voted against robots selected either *no* or *may be* for questions 6 to 9 as well.

Figure 1 (a) and (b) illustrate the results for question 2 and 4. According to Figure 1 (a), out of all the participants, 1.6% did not prefer getting help from a robot, while the remaining 98% were open to the company of a robot. While the majority (47.6%) preferred their robots to be assistants, a similar amount of participants preferred robots as both *companions* and *assistants*. From that we can observe that people expect a much friendlier, collaborative interaction with robots while making their tasks easier. This further suggests that people would prefer companionship with robots if the technology permits, despite the conventional role of robots to perform dedicated tasks.

For question 3, despite the three options we provided, a considerable amount of participants have suggested *calling the robot by a name*, although this was not provided as an option. This was an interesting observation during this survey. This is demonstrated in Figure 1 (b). According to the pie chart in Figure 1 (b), this new category: *calling the robot by a name* made up 18.8% of all votes. Altogether, 45.4% of the participants, chose *it* or *calling by a name*. Hence it is interesting to see that nearly half of the participants did not choose a gender for their robot. However, the relationship between the gender of participants and their preference upon the gender of the robot was not considered within the scope of this study. Here, we would argue that 73,4% of the participants assigned biological features to their robots and only a small amount considered it a *thing*.

The answers we received for question 4 are illustrated in Figure 2 (a). We gave participants the opportunity to write their own answers here. Later we categorized their responses under five groups: Emotional intelligence, Physical activities, Technical tasks, Specialized tasks and ***Other***. Responses such as *understanding people, having feelings or emotions, humanity, creativity, loyalty appreciation* and *being part of a family*, etc. were categorized under

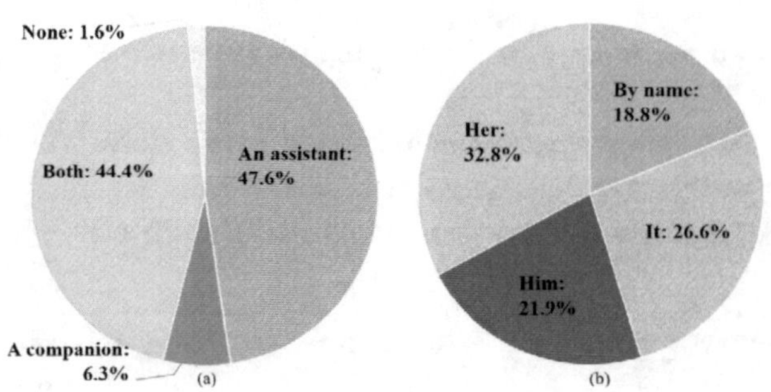

Figure 1. Pie charts categorizing the percentage of votes received for the questions 2 and 4 are shown in (a) and (b) respectively.

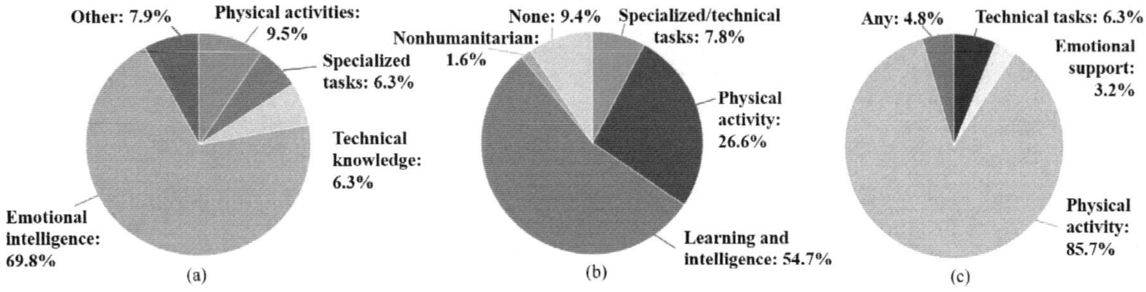

Figure 2. Pie charts categorizing the percentage of votes received for the questions 4, 5 and 10 are shown in (a), (b) and (c) respectively.

Emotional Intelligence. ***Physical activities*** included responses such as *gardening* and *cleaning the swimming pool.* We categorized responses such as *air pollution, auto-correction of errors* and *precision* under ***Technical knowledge***. Abilities such as *medical training, sorting food, tasting, smelling* and *recognizing visitors* were categorized under ***Specialized tasks***. In addition, there were responses such as *mind reading, humor* and *no preference* which were categorized under ***Other***. Out of the robots' abilities preferred by the participants, a majority (69.8%) required emotional intelligence. All the other categories received less than 10% votes. Furthermore, while there were multiple answers for question 4, the keywords *feel, emotion* and *think* have been used in 33 answers, that is 52.4%, and again 26 answers (41.3%) in question 5. This shows that people expect robots to be more than specialists in a certain task, but be more humanlike in cognitive skills.

While questions 4 and 5 appear to be similar, question 4 focuses more on the cognitive skills of a robot while question 5 evaluates the human expectations towards the abilities of a robot. Figure 2 (b) illustrates the answers for question 5. Again, we allowed participants to write down their own answer for this question, similar to question 4. Then we categorized the responses into five groups such as ***Learning and Intelligence***, ***Physical activity***, ***Specialized or technical tasks***, ***Nonhumanitarian*** and ***None***. Under ***Learning and Intelligence***, we categorized responses such as *thinking, advising me, loving and caring, creativity, replacing a human, formulation of new ideas, having a long conversation with me* and *attending to urgent matters.* ***Physical, Specialized*** and ***Technical Tasks*** were as in question 4. The ***Nonhumanitarian*** category included responses such as *hitting others.* This yields similar results to the above. A greater percentage voted for the abilities under ***Learning and intelligence***. It is interesting to note that 9.4% completely believed in the abilities of robots despite the inadequacy in technology at present. Compared to the previous occasion in Figure 2 (a), in Figure 2 (b), the votes for ***physical activity*** grew by 15.5%. This suggests that it is equally important to equip robots physically as well as cognitively.

Although many participants preferred robots as assistants rather than companions (Figure 1 (a)), they expect emotional intelligence from them even with task-specific applications (Figure 2 (a),(c)).

Figure 2 (c) illustrates answers for question 10, which explores where people expect help from robots. The responses were categorized into four groups: ***Emotional support***, ***Physical activity***, ***Technical tasks*** and ***Any***. ***Any*** category includes the votes where participants preferred any kind of help from their robots. The rest of the categories include responses same as before. It is important to note that the majority (85.7%) expect help from their robots for physical activities. This seems to contradict the majority vote for ***Emotional intelligence*** in the first two charts: Figure 2 (a) and (b). The reason for this is that participants prefer

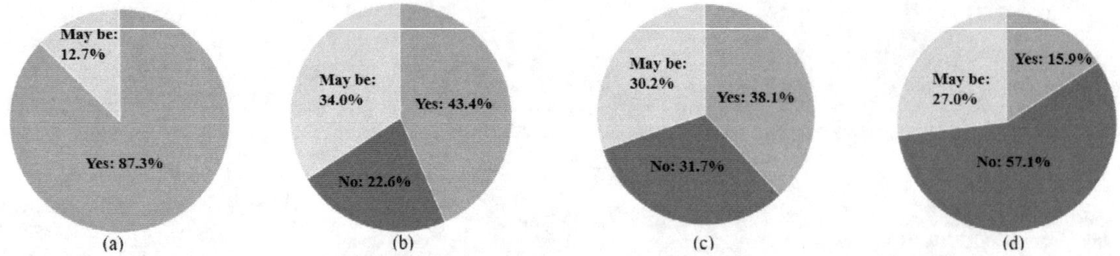

Figure 3. Pie charts representing the percentages of *yes*, *no* and *may be* received as answers for the questions 6 to 9 are shown in (a), (b), (c) and (d) respectively.

emotionally intelligent robots even in support of physical work. A minor group of participants preferred robots in ***emotional support*** and ***technical tasks***, 3.2% and 6.3% respectively. 4.8% of the participants accepted any help from robots in their surrounding. Results obtained from the charts in Figure 2 (a) to Figure 2 (c) suggest that people expect more *humanlike* cognition in robots rather than the *machinelike* nature in robots in the past. But this fact is not true for the preference of participants on a robot's appearance. This is explained in Figure 6.

Pie charts in Figure 3 plot the preferences *Yes*, *No* and *May be* for questions 6 to 9. For the ease of finding patterns in these responses, we plotted responses for each task on the same graph as in Figure 4. If participants preferred their robots performing tasks from 1 to 4. they responded ***Yes*** and if not responded ***No***. If they are uncertain, they were allowed to respond ***May be***. These tasks in questions 6 to 9 were selected in the order of increasing complexity and criticality. For the task of *cleaning*, 87.3% of participants preferred a robot while there were zero *no* responses. For the task 2: cooking, the percentage of *Yes*-sayers reduced to 43.4% (by 33.9%). This further reduced down to 15.9% in task 4: driving kids to school. An opposite reaction was seen in the *No* response for these tasks as is shown on the graph. Participants' likeliness to respond *May be* increased from task 1 to task 2 and stayed almost constant until task 4. It can be seen that the uncertainty of participants' preferences increased from task 1 to task 4. Therefore it can be seen that particpants' trust on robots reduced as the criticality of the task increased.

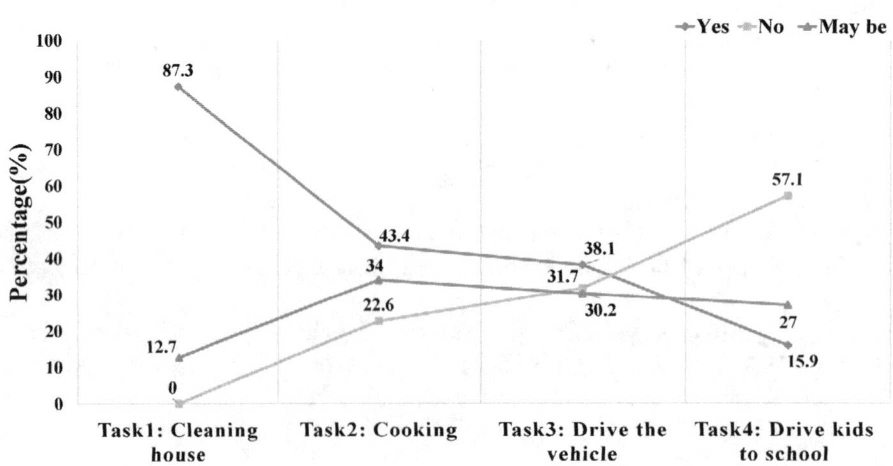

Figure 4. The percentage of *yes*, *no* and *may be* received as answers for the questions 6 to 9 are plotted against the task involved in each question.

Table 1. An ANOVA Test for the comparison of responses received for the task 1 to 4 in Figure 4

	Mean	Variance		
Task: Cleaning house	1.87	0.11		
Task: Cooking	1.32	0.62		
Task: Driving the vehicle	1.06	0.71		
Task: Driving kids to school	0.56	0.54		
ANOVA Test				
	SS	**DOF**	**F**	**P**
Between groups	55.01	3	36.83	1.15E-19
Within group	121.5	244		
Total	176.51	247		

An ANOVA test was performed on this data in Table 1, for all four tasks in Figure 4 to analyze the difference between participants' preferences. The null hypothesis will be that there is no significant difference between the preferences for the four tasks considered. As seen in Table 1, F-value (36.83) was significantly larger than F-critical (2.64). Hence, in this case the null hypothesis can be rejected and the assumption that *there is no significant difference between participant's preference for robots for the selected set of tasks* was declined. It can thus be deduced that the user preference for robots on each task was significantly different when the given four tasks were considered. Furthermore, in this occasion, p-value (1.15E-19) was smaller than the alpha variable (0.05). This also suggests that the individual variables were statistically significant. Hence it can be deduced that there is a significant drawback in human trust on robots with critical tasks compared to normal tasks.

In question 10, the participants were asked to list the tasks for which they prefer to get help from robots. Then they were allowed to compare the results they would expect from the robot and another human for the tasks they listed as in question 11. Figure 5 (a) illustrates the responses received for question 11. More than 50% of the participants relied on robots in these tasks while a still considerable amount of participants (41.3%) would prefer it if a human can replace the robot for the tasks they prefer. Therefore a considerable amount of participants have problems accepting the capabilities of present robots. Therefore it is likely that people still trust robots less due to the lack of emotional intelligence and physical capabilities as found under question 5.

Each slice in the pie chart shown in Figure 5 represents responses received for question 12. 36.5% of the participants preferred interacting with a robot during a relaxed occasion while

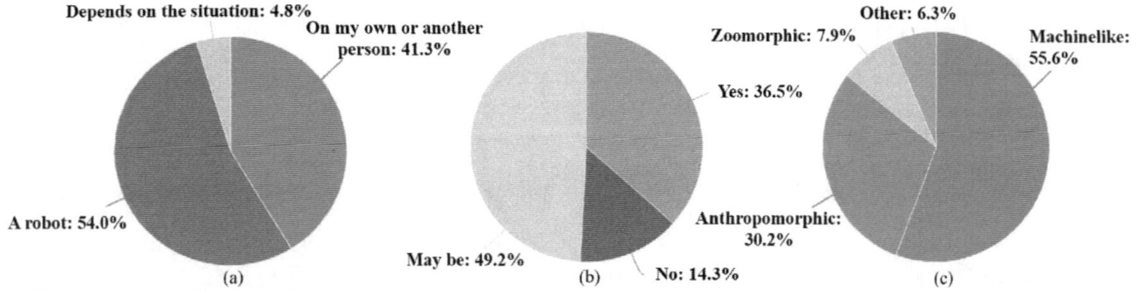

Figure 5. Pie charts categorizing the percentage of votes received for the questions 11, 12 and 14 are shown in (a), (b) and (c) respectively.

49.2% were uncertain of their decision. Possible reasons for this uncertainty were discussed in [21]. According to [21], the internal state of mind of a person largely contributes to their decision to interact with a robot. Therefore, a robot has to be able to understand if a situation is suitable for interaction prior to a conversation. We think that this may be the reason for the large share of *May be*. 14.3% of the participants did not prefer interaction with a robot while relaxed. The reasons can be seen in Figure 2 (c) where many prefer their robots in physical tasks rather than in social engagement.

Figure 6 illustrates some of the pictures drawn by participants for question 16. We allowed participants to upload an image from the Internet which matches their imagination if they lacked confidence in their drawing skills. An important observation we made here was that almost all the illustrations had eyes as a physical feature despite the robot being anthropomorphic-, zoomorphic- or machinelike in appearance. After analyzing these pictures and the response for questions 14 and 15, we categorized them into four categories considering the physical appearance of robots. This is shown in Figure 5 (c). These categories are ***anthropomorphic***, ***zoomorphic***, ***machinelike***, and ***Other***. A majority of participants (55.6%) preferred robots which look like a *machine*. According to the participants' illustrations, this look was more similar to the *typical* robot look typical to humanoid robots which can be frequently seen on the internet and sci-fi movies. It was interesting to observe that people prefer robots to have their own look rather than being *humanlike*. The uncanny valley problem can be a reason for these preferences by the participants [3]. Therefore, only 30.2% preferred the human look in their robots while 7.9% preferred an animal figure. More abstract designs such as vehicles or spheres were categorized under *Other*. It is found in [19] that people responded to robots in a similar way they treat children during relaxed conversations. The same approach could be seen in the illustrations participants made during this survey as well: a tendency to draw *cute* robots in this regard. Figure 5 (c) and Figure 6 testify to this.

5. CONCLUSIONS

Robot intervention in our daily tasks is becoming more and more common nowadays. *Trust* is an important aspect in enhancing the collaboration between humans and robots in this regard. In this paper, we report findings from a survey related to human preferences when allocating tasks to and sharing responsibilities with robots. In addition, we also asked participants to provide a visualization of the robot they envisioned while taking part in the survey. This helped us explore the uncanny valley of robotics and how it affects human trust on robots.

This study evaluates the willingness of people to trust their robot companions in different scenarios through a set of questions. Results of the study confirm the fact that capability of the robot to perceive their surroundings in a way similar to humans, makes an impact to sustain interaction between a robot and its user. Our results further confirm that the capability of the robot to perform physical actions as smoothly and flexibly as humans can considerably improve human trust. However, limitations in the dynamics as well as emotional intelligence have to be overcome so humans can build trust on their robotic companions for more critical

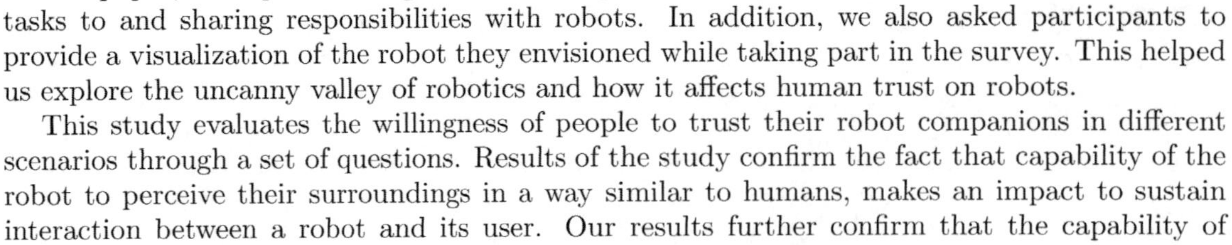

Figure 6. Some of the pictures drawn by the participants under question 16

tasks. Moreover, human preferences upon the appearance of their robotic companions have been analyzed to explore the relationship between human trust on robots, their appearance and their capabilities.

Future research should evaluate more factors that might impact human trust towards robots in their surroundings. We also encourage readers to conduct similar surveys in their local communities in order to help separate generally applicable aspects of trust in robots, from locally specific ones.

5.1. Implications for design and theory

The acceptance of a robot for everyday tasks depends on the values and risks associated with the task (i.e. its criticality) as well as their trust towards the robot. Trust in turn depends on the capabilities the user is willing to ascribe to the robot, both in physical and social interactions. While these capabilities do not have to be demonstrated within the target task, they shape the user's understanding of the robot. If the participants' vision could be matched, many participants would prefer robots in their surroundings to share their workload. It is therefore important to increase the capabilities of a robot according to the tasks it is allocated for. This is the first design guideline derived from this study.

Results confirmed that people preferred it if robots could understand and adjust to their emotions. This suggests that people prefer interaction with a robot that are, to a certain degree, emotional and less clinical. Therefore, emotional intelligence will be a sought-after feature in future domestic or social robots. This is the second design guideline derived from this study.

An important observation made during the study was that all pictures drawn or uploaded by participants included a feature similar to *eyes*. We can thus deduce that gaze is a prominent feature that humans prefer during human-robot interaction. Therefore investigating the importance of gaze and *eyes* as a feature for robots worth exploring in the future. This will be the third design guideline derived from this study.

As per the results, it can be seen that people believe that robots are physically not capable enough to perform all tasks participants desired in a satisfactory way. Therefore, equipping robots with adequate physical capabilities so that they can perform human activities, will be the next design guideline derived from the study. This requires adequate technologies and improved robot dynamics.

Humans preferred emotional intelligence in their robotic companions even though they mainly wanted to use robots for physical activities. Therefore engraving at least a minimum amount of emotional intelligence for collaborative robots despite their task will be the fifth design guideline for our future robots.

From the answers we received for the questionnaire, we observed a less tight relationship between human expectations upon humanlikeness in the appearance and the skills of a robot. Our findings suggest that people prefer humanlike features in robot's capabilities but surprisingly less humanlikeness in the appearance. Instead they preferred the 'typical robot' figure more. Therefore working on robot's skills more than its appearance will attract more human acceptance in the future. Therefore this will be the sixth design guideline for our future robotic companions.

ACKNOWLEDGMENT

The authors would like to acknowledge the participants who volunteered the study.

References

[1] A. Rossi, K. Dautenhahn, K. L. Koay, and M. L. Walters, *The impact of peoples' personal dispositions and personalities on their trust of robots in an emergency scenario*, Paladyn, Journal of Behavioral Robotics, vol. 9, no. 1, pp. 137-154, 2018.

[2] S. Guznov et al.,*Robot Transparency and Team Orientation Effects on Human-Robot Teaming*, International Journal of Human-Computer Interaction, vol. 36, no. 7, pp. 650-660, 2019.

[3] T. Geller, *Overcoming the Uncanny Valley*, IEEE Computer Graphics and Applications, vol. 28, no. 4, pp. 11-17, Jul. 2008.

[4] E. S. Cross, R. Hortensius, and A. Wykowska, *From social brains to social robots: applying neurocognitive insights to human-robot interaction*, Philosophical Transactions of the Royal Society B: Biological Sciences, vol. 374, no. 1771, p. 20180024, 2019.

[5] O. Korn, N. Akalin, and R. Gouveia, *Understanding Cultural Preferences for Social Robots*, ACM Transactions on Human-Robot Interaction, vol. 10, no. 2, pp. 1-19, 2021.

[6] Beckerle, P., Kõiva, R., Kirchner, E.A., Bekrater-Bodmann, R., Dosen, S., Christ, O., Abbink, D.A., Castellini, C. and Lenggenhager, B., *Feel-good robotics: requirements on touch for embodiment in assistive robotics.* Frontiers in neurorobotics, 12, p.84, 2018.

[7] L. Bishop, A. van Maris, S. Dogramadzi, and N. Zook, *Social robots: The influence of human and robot characteristics on acceptance*, Paladyn, Journal of Behavioral Robotics, vol. 10, no. 1, pp. 346-358, 2019.

[8] M. Coeckelbergh, *Can we trust robots?*, Ethics and Information Technology, vol. 14, no. 1, pp. 53–60, 2011.

[9] D. Li, P. L. P. Rau, and Y. Li, *A Cross-cultural Study: Effect of Robot Appearance and Task*, International Journal of Social Robotics, vol. 2, no. 2, pp. 175–186, 2010.

[10] J. Bernotat, F. Eyssel, and J. Sachse, *The (Fe)male Robot: How Robot Body Shape Impacts First Impressions and Trust Towards Robots*, International Journal of Social Robotics, vol. 13, no. 3, pp. 477-489, 2019.

[11] R. Looije, M. A. Neerincx, and F. Cnossen, *Persuasive robotic assistant for health self-management of older adults: Design and evaluation of social behaviors*, International Journal of Human-Computer Studies, vol. 68, no. 6, pp. 386-397, 2010.

[12] R. H. Wortham and A. Theodorou, *Robot transparency, trust and utility*, Connection Science, vol. 29, no. 3, pp. 242-248, 2017.

[13] N. Wang, D. V. Pynadath, E. Rovira, M. J. Barnes, and S. G. Hill, *Is It My Looks? Or Something I Said? The Impact of Explanations, Embodiment, and Expectations on Trust and Performance in Human-Robot Teams*, Persuasive Technology, pp. 56-69, 2018.

[14] T. Sanders, K. E. Oleson, D. R. Billings, J. Y. C. Chen, and P. A. Hancock, *A Model of Human-Robot Trust: Theoretical Model Development*, Proceedings of the Human Factors and Ergonomics Society Annual Meeting, vol. 55, no. 1, pp. 1432-1436, 2011.

[15] P. A. Hancock, T. T. Kessler, A. D. Kaplan, J. C. Brill, and J. L. Szalma, *Evolving Trust in Robots: Specification Through Sequential and Comparative Meta-Analyses*, Human Factors: The Journal of the Human Factors and Ergonomics Society, vol. 63, no. 7, pp. 1196-1229, 2020.

[16] B. C. Kok and H. Soh, *Trust in Robots: Challenges and Opportunities*, Current Robotics Reports, vol. 1, no. 4, pp. 297-309, 2020.

[17] Salem, Maha, Gabriella Lakatos, Farshid Amirabdollahian, and Kerstin Dautenhahn. *Would you trust a (faulty) robot? Effects of error, task type and personality on human-robot cooperation and trust*, 10th ACM/IEEE International Conference on Human-Robot Interaction (HRI), IEEE, pp. 1-8. 2015.

[18] S. You and L. P. Robert Jr., *Human-Robot Similarity and Willingness to Work with a Robotic Co-worker*, Proceedings of the 2018 ACM/IEEE International Conference on Human-Robot Interaction, 2018.

[19] C. Sirithunge, M. A. V. J. Muthugala, A. G. B. P. Jayasekara and D. P. Chandima, *A Wizard of Oz Study of Human Interest Towards Robot Initiated Human-Robot Interaction*, 27th IEEE International Symposium on Robot and Human Interactive Communication (RO-MAN), pp. 515-521, 2018.

[20] J. Navarro and M. Karlins, *What every body is saying : an Ex-FBI agent's guide to speed-reading people*, New York William Morrow, An Imprint Of Harpercollins Publishers, 2008.

[21] C. Sirithunge, A. G. B. P. Jayasekara, and D. P. Chandima, *An Evaluation of Human Conversational Preferences in Social Human-Robot Interaction*, Applied Bionics and Biomechanics, vol. 2021, pp. 1-13, 2021.

Self-organization in Piano Playing: Why Pattern Transition?

Huijiang Wang and Fumiya Iida

Bio-Inspired Robotics Lab, Department of Engineering, University of Cambridge, Trumpington Street, Cambridge CB2 1PZ, United Kingdom

E-mail: hw567@cam.ac.uk, fi224@cam.ac.uk

Abstract. Over billions of years, natural lives and organs have evolved with essential self-organized life-sustaining activities such as locomotion and respiration. Biological studies have shown that the Central Pattern Generator (CPG) is the essential mechanism for such pattern transitions. Embodied intelligence is a paradigm for investigating how humans employ the decentralized controller and embodiment of the body to interact with their environment (piano). Previously piano playing robots were hard-coded to implement mechanical contact and musical generation, resulting in clumsy non-anthropomorphic keystrokes. Instead, we revisit the piano challenge by studying the biomechanical physics of the body, the self-organization of reflexive behavior, and the neuro-muscular synergy in terms of coordinated behavioral diversity and energy minimization.

1. Introduction

Biology is full of examples where organisms perceive their bodies, control their bodies, and interact with the environment in a meaningful way. The paradigm, embodied cognition [2], shows that physical bodies developed to solve task-specific difficulties with the ecological assembly of perceptions, such as army ants finding the best route with their polarized light-sensitive eyes. The same cognitive design principles can be used in robotics. Exploring how biological systems make use of the physical bodies can give us insights into how the robotic system can have the same embodied intelligence among the brain (controller), body and interactions with the environment. The dead fish swimming experiment [1], for example, revealed that the body of a biological system can interact with its surroundings efficiently even without the assistance of the brain, which inspired the energy-free passive walker robot [3]. Soft robots have been conceived, constructed, and applied to various dexterous manipulation tasks using the embodied intelligence emphasized between bio-inspired robotics and the environment to investigate how bodies work to minimize the load on the brain, how the sensorized robot can mechanically self-regulate its behaviors and how the "body shapes the way we think" [20].

Music, one elegant form of art, acts as a barrier-free language that bridges people from varying cultural backgrounds to communicate and share their emotions. For years of training, people have been trying to master musical instruments with the purpose of expressive playing, in which embodied intelligence plays a significant role in human-machine interaction. Piano playing, different from other instrumental performances, is a complex task emerging through the reciprocal interaction between the environment and the player's hands. Recent developments in robotics demonstrate an increasing interest in music and entertainment robots. The production of high-quality expressive performance, particularly, requires nuanced coupling between the dexterity, adaptability, and mechanical compliance of the fingers and the dynamics of the piano itself. Traditional research on piano playing have centered on the joint biodynamics [5, 13] of piano strikes and their robot replica [22, 27, 24, 26, 4, 17, 12, 21, 27, 16, 21, 25]. The piano challenge is narrowly understood either without considering the biomechanical constraint or merely via mechanically engineering a robot reproduction. The way how the body makes use of embodied intelligence over dexterous manipulation - piano playing in this case, needs to be further investigated. The study of fundamental pattern transitions in piano playing provides an entry point into the self-organization of the organ's essential life-sustaining behaviors. We will primarily consider: What is the definition of pattern transition? What kind of pattern transition can be observed? In what ways does embodied intelligence play a significant role in expressive piano playing?

There is debate regarding what the true pattern transition is and why it is significant in biological systems. The gait transition in stance-swing [14], walking-running [7, 6, 23], and swimming-crawling [11, 10] are typical problems that are very interesting to investigate because they depict a natural and spontaneous shift from one state to another that obviously differentiates, without the need for complex additional system input. Researchers attempt to explain the transitional pattern, however, these are difficult to generalize to complex biological systems [18]. We explore the pattern transition by investigating the coordinated upper-limb arm-swing behavior of a robot pianist with neuro-muscular synergy. Conventional understanding holds that CPG, which is derived from the spinal cord, is effective for coordinated rhythmic behaviors, such as natural vertebrate locomotion. However, it has yet to be demonstrated how this distributed/decentralized controller benefits non-locomotion cases. Pattern transition is ubiquitous in those repetitive coordinated rhythmic behaviors. One purpose of this study is to investigate whether pattern transition occurs in non-locomotion circumstances and how it affects expressive piano playing.

We exploit embodied intelligence to allow a robot pianist to perform expressive piano playing. The piano-playing scenario is replicated with anthropomorphic features including the anatomically-correct musculoskeletal upper-limb system, neuro-modulatory brain stem circuits and neuro-muscular synergy. The pianist interacts with the environment in a reciprocal way, during which we analyze how the body is making use of the self-organization, and embodiment of the body to address the piano challenge with a brainless controller.

The rest of this paper is organized as follows. Section 2 covered the pattern transition and the criteria for defining it. Section 3 depicts an anthropomorphic piano player's embodied intelligence and how the robotic multi-joint system exploits the pattern transition to achieve expressive piano playing. Section 4 provides a case study, while the subsequent sections provide a conclusion and discussion.

2. Spontaneous Pattern Transition

Patterns in nature are observable regularities, and the outcome of transitional action is obviously recognizable. Smooth pattern transitions are ubiquitous in biological systems; they are essential for natural bodies because they demonstrate how the system reacts to external stimuli at different levels of synchronization. This could be due to optimization considerations, such as minimizing energy expenditure. However, there is no definition of pattern transition and no explanation of how it may be generalized to other natural lives and organs. We put forth an effort to define the pattern transition, which must meet the following requirements:

- **Criteria 1**: The transitional activity occurs in response to specific stimulus injection.
- **Criteria 2**: It is a spontaneous rather than manually imposed procedure.
- **Criteria 3**: The system's fundamental output exhibits nonlinear variation.

IOP Conf. Series: Materials Science and Engineering 1292 (2023) 012015 doi:10.1088/1757-899X/1292/1/012015

Traditionally, the research for pattern transitions centered on biological locomotion. The biped walking-to-running [23, 7] study shows a very good example that how the gait of the biped walker smoothly changes to another form of pattern with only energy injection into the system. There is an evident measure to define the transition from a double-hump to a single-hump gait trajectory. The patterns differ in that when walking, at least one foot makes touch with the ground, whereas when running, both feet are off the ground for a certain amount of time. Note that the rise in speed caused by the energy injection does not count as a transitional activity (violating **Criteria 3**). Importantly, the transitional gait is achieved spontaneously by the body without any kind of manually-enforced programming.

There are numerous different forms of pattern transitions in nature that follow the aforementioned criteria 1. For instance, consider the flow pattern transition for boiling water. The water molecules are synchronized with distinct patterns as a result of the energy input, enabling the liquid-gas transition. It has been observed that a solitary firefly exhibits an uneven illumination pattern, however automatic synchronization is accomplished when they are placed in a swarm. We do not fully understand how they achieve the pattern transition but the stimuli input for the system could be the communication among the agents. When a predator is nearby, the same pattern explains the fish swarming behavior. Other natural locomotion examples include the swimming-to-crawling and gliding-to-flying transitions, with both neural modulation and energy as the transitional stimuli.

Interestingly, the falling paper displays an unusual non-locomotion pattern transition, since the interactions between the body and the environment can always be categorized into four patterns for a free-falling paper [9]. Similarly, the neuronal cells of the brain are synchronized with distinct patterns for a relaxed and a focused state, respectively. The input of this transitional activity is still unclear, but it may be fluctuations in synaptic stimulation, such as the firing rate of neurons. Although it is difficult to determine a concrete measure to quantify the "nonlinear variation of the pattern," phase difference and offset appear to be viable candidates because they characterize the repetitive rhythmic system. For the walking-to-running gait transition, the phase difference, not the rise in speed, is considered a pattern transition.

3. Synchronized Arm-Swing in Piano Playing

Dexterous manipulation is a research realm in which many fingers or manipulators of the bespoke end-effectors collaborate with each other to manipulate objects. Dexterous soft manipulation is distinguished by the fact that it is object-oriented and can be accomplished with a variety of cooperative solutions among soft bodies [19]. In piano playing, the multiple subsystems work cooperatively to perform keystroke actions, and the flow of keystrokes through time produces musical articulation. The entire piano-playing body is a synchronized neuro-modulation-based entity, with musculoskeletal tissues and bones serving as constraints.

Embodied intelligence's role has the potential to overcome the limitations of expressive robotic piano playing. It is believed to be a grand challenge for a robot to manipulate musical instruments and produce music with its own "emotional variety." Traditional musical robots are accused of being simply sound-generating machines rather than true musicians. Before we get into the technical details of robotic instrument playing. We must investigate the underlying mechanism by which humans use embodied intelligence for musical events. Several human piano playing studies have shown that for different-tempo piano keystrokes, the multi-joint system (wrist, elbow, and shoulder) exhibits varied patterns [5, 13]. For example, in a slow-frequency keystroke, the wrist joints contribute more, allowing the end-effector to work as a bouncing behavior, whereas in a high-speed piano strike, the motion in the elbow is the primary energy consumption. This is a natural technique in which the body synchronizes several groupings of musculoskeletal parts. Humans employ the primate corticospinal system to control individual finger and arm motions in order to master playing the piano[15]. A distributed and decentralized

Figure 1. Various pattern transitions in nature. (a) The phase of water molecules performs a state transition from liquid to gaseous phase within water boiling. (b) A salamander robot[11] achieves transitional locomotion from swimming to crawling. Inspired by the aquatic-to-terrestrial locomotion in vertebrate locomotion, the robot uses a primitive spinal cord model to modulate the complex locomotion dynamics including velocity, direction, and gait type for tetrapods. (c) The human's biped walking to running is a natural ubiquitous pattern transition[23, 7]. (d) The gliding-to-flying transition. (e) Paper disk shows various falling patterns with varying morphological features[9]. Based on automatic experimentation and unsupervised learning technique, the falling paper problem has been investigated, where the paper disk exhibits four falling patterns: tumbling, chaotic, steady and periodic. (f) Marine organism shows various patterns of swarm behavior and (g) The neural cells of the brain show self-organized transitional activity.

controller is utilized in this process to exploit and prioritize the "intelligence of the embodiment."

Self-organization in piano playing occurs as a result of neuro-muscular coupling, in addition to the decentralized controller (see Figure 2). The spinal cord-inspired controller (CPG) can synchronize the entire upper-limb system for pattern transition. In order to study how self-organization occurs in piano playing, we tune the CPG parameters, such as the firing rate, membrane potential, and body parameters, such as the spring, damping coefficients for the joints, center of mass, the density of the body, the gradient of the mass, and so on.

4. A Case Study: Robot Pianist

To investigate anthropomorphic piano playing, a high-fidelity replication of a pianist was computationally duplicated using a state-of-the-art multi-body dynamics platform. As shown in Figure 5, a three-neuron CPG oscillatory network has been configured as the basic neural input. An anthropomorphic hand and upper limbs have been duplicated based on the morphology of

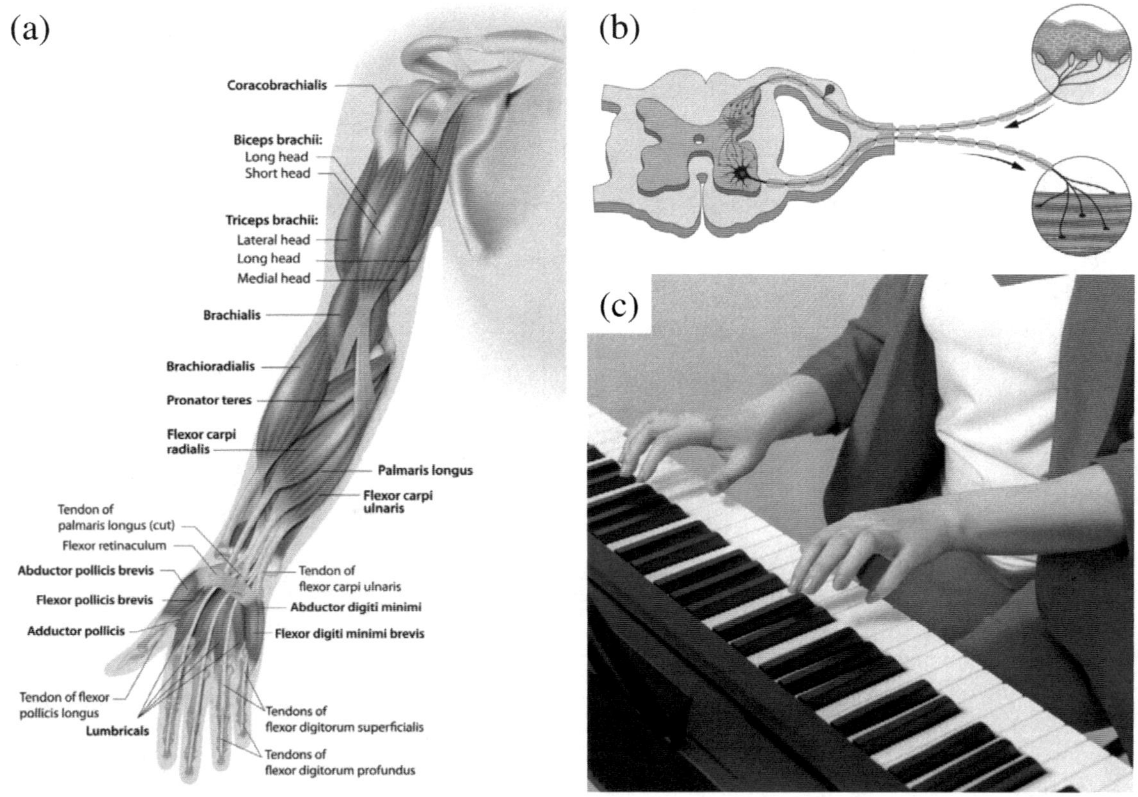

Figure 2. Neuro-muscular synergy in piano playing. (a) The entire upper limb is a complex biological system with a good embodiment, allowing for self-organized reflexive behaviors for pattern transitions. (b) A decentralized controller achieves the synchronization of the multi-motor upper-limb system. (c) Coordinated piano playing depicts that the body is able to interact with the environment with reduced brain control.

real anatomical elements. A full-scale modern piano is also modeled in the simulation with a row of 88 keys that includes 7 C-major scale octaves. The finger-key interaction is replicated as a collision between the hand's distal joints and the piano blocks, with friction defined to provide resistance to prevent relative motion.

4.1. Experimental Setup

We modeled the passive piano playing by mechanically replicating the geometric features of an anatomically correct hand. The hand is attached to a muscle-actuated upper-limb system. Note there is no individual actuation for finger joints and the spatial motions of the hand result from the sole actuation on the wrist. The compliant behaviors of fingers (output) are caused by the contact between the fingertip and piano note and are an indirect outcome of the wrist actuation (input). In addition, joint sensor blocks are used to allow the measurement of joint angle and torque. The monitored data is then written to the Workspace for further iteration in parameter optimization. The connecting joint between two geometric bodies is modeled as a mass-spring damping system which is constrained by a revolute joint with certain equilibrium positions.

The key-strike action is represented as the collision between two geometries. In the simulator, we defined the contact between the fingertip and piano key by adding the Spatial Contact Force

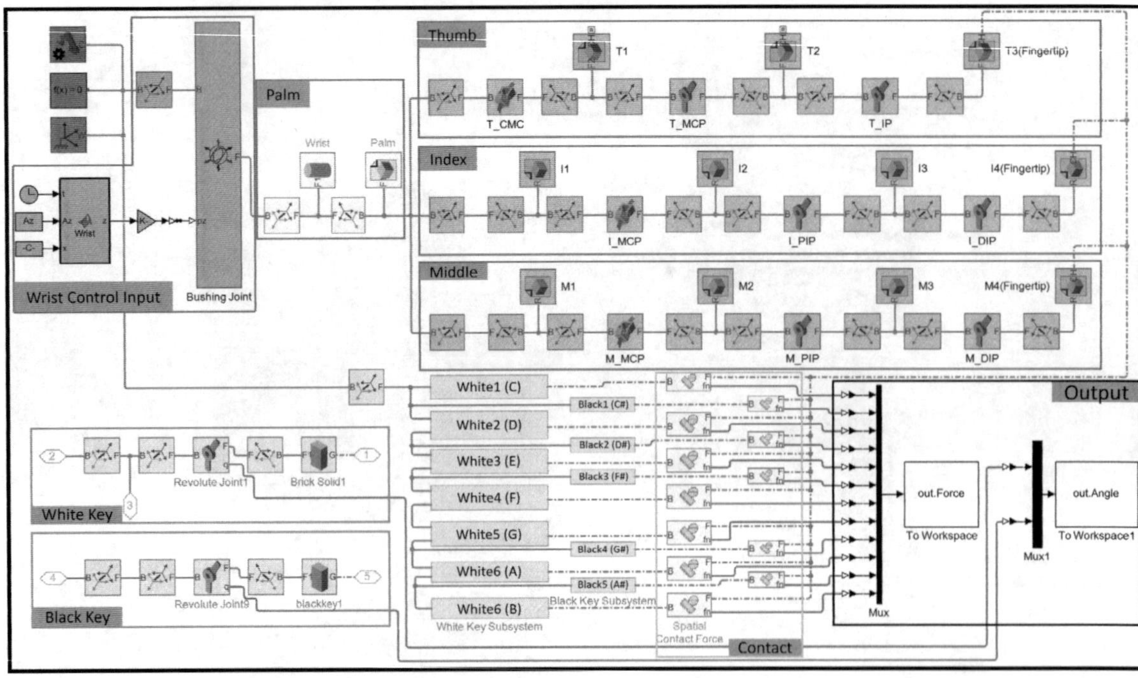

Figure 3. Simscape block diagram of the piano playing simulator.

block. This block prevents the penetration between two objects and is capable of sensing the normal and frictional forces on the contact point, which enables the dynamic analyses of the finger-key interactions.

The simulation platform is able to imitate the pipeline of the entire piano playing scenario. Anthropomorphic upper limbs including the shoulder, elbow, wrist and hand, as well as an 88-key piano are built with detailed mechanical, geometric and auditory characteristics (Figure 4a and b). Two MIDI events, which resulted from the reciprocal interaction between the two primitives, are interpreted to assess the produced musical phrases(Figure 4c). The entire hand-piano interaction is depicted in Figure 4d. For this simulator, the exclusive input is the torques exerted on the three joints. Figure 3 illustrates the block diagram of the detailed construction of the simulator in Simscape.

The keystroke action is the result of rhythmic coordination of the entire upper limbs, which includes the shoulder, upper arm, forearm, and hand. Figure 5 depicts the upper limb motions involved in piano playing. The practical actuation of the upper limbs is a complex synergy of multiple muscle groups such as deltoid, latissimus dorsi, biceps brachii, brachioradialis and so on. The upper limbs in the simulation are modeled from genuine bones with a humanoid morphology. In order to simulate the musculoskeletal motion in keystrokes, a computational forward-dynamic model [8] has been exploited to replicate the aforementioned three muscles in a biomechanical way. This model is based on a Hill-type muscle structure and provides a high-fidelity simulation of the concentric and eccentric contractions. The output is a one-dimensional force, while the model's inputs include muscle-tendon-complex (MTC) length, contraction velocity, and neural excitation.

4.2. Coordinated Rhythmic Behaviors

The robot player is instructed to actively change the articulation styles during the performance to showcase the robot's ability to handle different expressive patterns. We have analyzed the

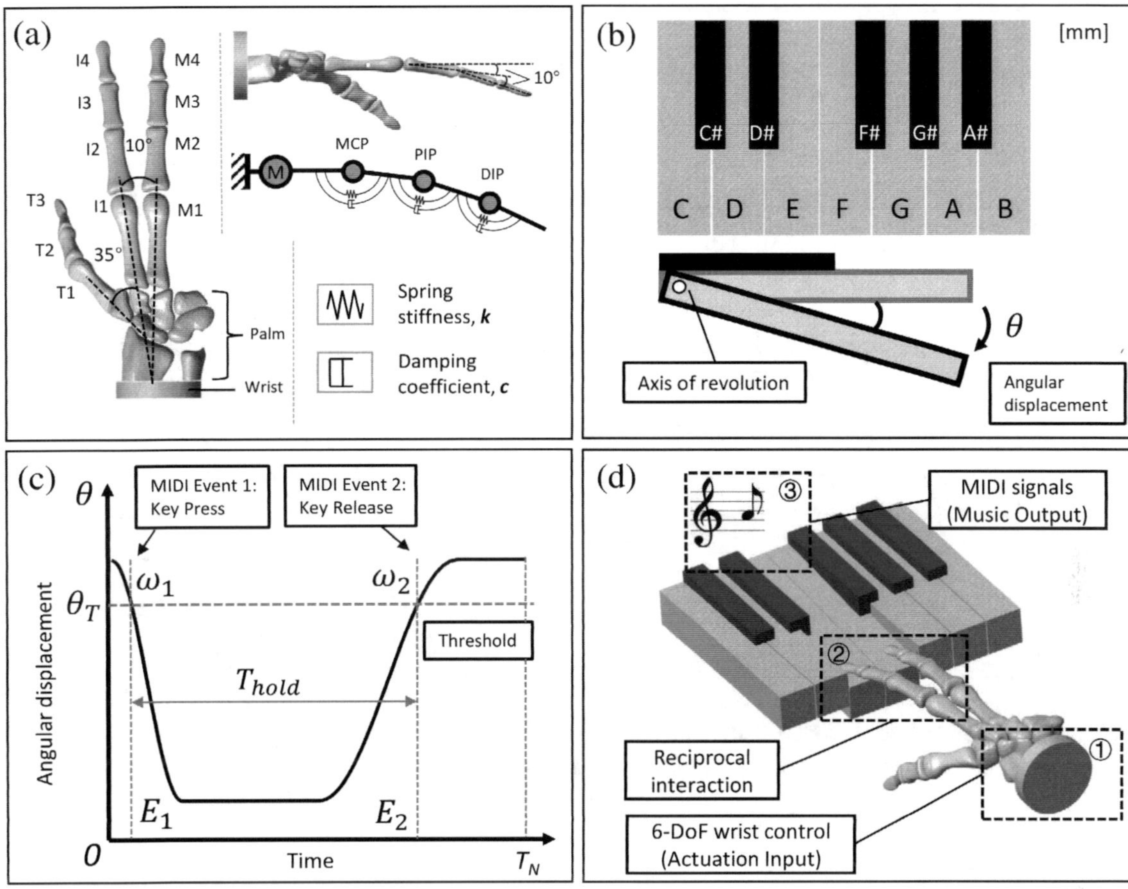

Figure 4. Different components of the simulation platform. (a) The anthropomorphic hand consists of anatomically correct objects placed in a natural bending pose. The revolute joints between two phalanges are characterized by spring stiffness and damping coefficient. (b) An octave starting from note C is simulated with 7 white keys and 5 black keys. The angular displacement of the key can be detected as a result of the finger-key contact. (c) Musical phrase produced from a single-note key-strike action is parameterized by two MIDI events. A triggering threshold is used to determine the key-press and key-release motions. (d) Illustration of the pipeline work of the simulator. First, actuation signals are sent to the upper-limb muscles to enable the 6-DoF spatial motions of the hand. Then, the contact with the piano denotes the hand-piano interaction, during which the finger joints demonstrate various passive dynamics. Finally, the angular displacement of the pressed key, together with the chosen threshold, is used to produce the musical signals that characterize the quality of the piano playing.

motions of the entire multi-motor upper limbs and the corresponding MIDI output for a tempo-transition piano playing case.

With different levels of biologically inspired neural synaptic excitation, the pattern transition between slow and fast styles is performed dynamically. The robot pianist manipulates speed-changing musical flows via neuro-muscular synergy, during which the body and reflexive neurons exhibit spontaneous and self-organized shifts. It can be seen from Figure 6a and 6b that the CPG-controlled multi-muscle system reaches a natural higher arm-swing frequency for the transition from $Andante, 86BPM$ to $Moderato, 113BPM$, and vice versa. The resultant tempo

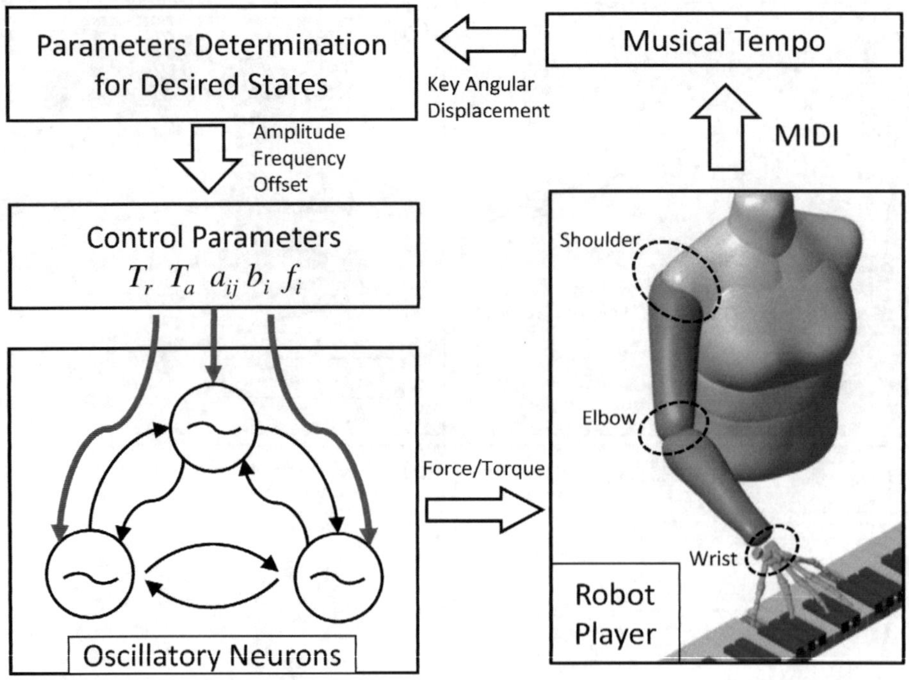

Figure 5. Control architecture of the CPG-based keystroke coordination for a virtual piano player with anthropomorphic musculoskeletal properties.

transition is achieved by tuning the CPG neuron's excitation to the musculoskeletal multi-joint system, which replicates the neuro-modulatory mechanism in real human limb muscles. The end-effector achieves steady repetitive keystroke actions despite the variations in limb displacement.

The coordinated arm-swing behaviors ensure repetitive keystroke actions for the finger-key interaction. The piano notes have been pressed down multiple times and the corresponding MIDI events were created for each keystroke. The MIDI protocol has been employed to parameterize the keystroke actions. In the simulation, the mechanical contact of finger-key interaction is converted into a MIDI event specifying the note's pitch, timing, and loudness. Figure 6c illustrates the resultant MIDI output for the Andante-Moderato transition patterns in line with the multi-joint coordination.

It can be seen from Figure 6c that the piano keyboard was manipulated successfully with an accurate switching between two expressive tempi, implying that the robot could accomplish a variety of expressive styles in terms of the articulation metric. For both the slow-to-fast and fast-to-slow transitions, the first keystroke pattern deviates slightly from the expected one, but it quickly stabilizes around the desired tempo with the supplied coordinated musculoskeletal system.

5. Conclusion

This article investigated fundamental patterns and how natural bodies can use pattern transitions to perform critical life-sustaining behaviors. To address the rhythmic coordination challenge, artificial Central Pattern Generator (CPG) oscillators have been widely created to replicate the electrophysiological activity of interconnected neurons based on membrane potential and other biological signals. The traditional concept of pattern transitions is centered on multi-motor coordinated locomotion. The application of a decentralized controller based on

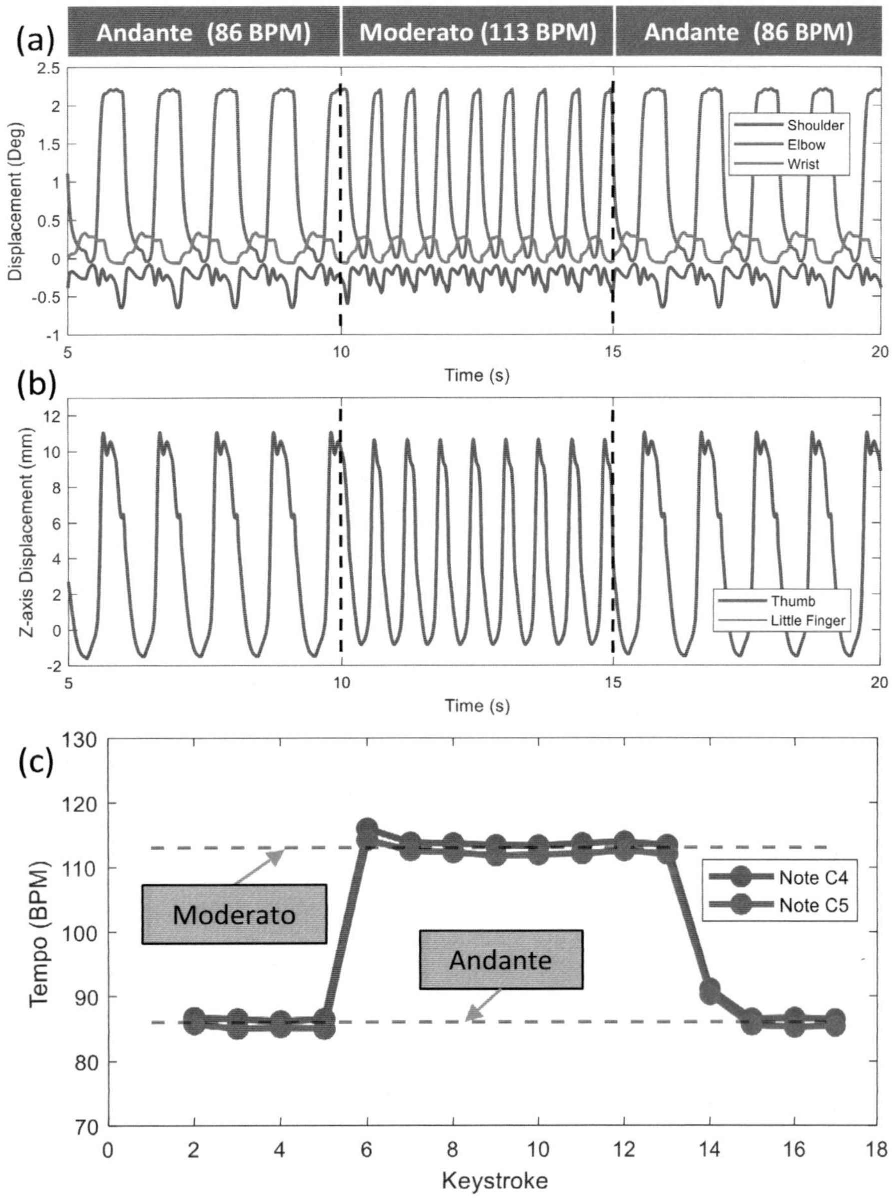

Figure 6. Tempo transition whilst piano playing. (a) The angular displacement of three upper-limb joints. (b) Z-axis displacement of the thumb and litter finger. (c) MIDI output for the Andante-Moderato transition.

the spinal cord to non-locomotion circumstances has been discussed, along with the criteria for pattern transition and several examples of natural pattern transition.

A case study showing how embodied intelligence is the potential to solve the challenges of expressive piano playing is given. Based on this we conclude that the CPG is able to be exploited in non-locomotion cases. The CPG matters in the way that gives the neural excitation and stimuli to coordinate the repetitive behaviors in natural lives or organs. The embodiment of the muscular model, on the other hand, provides limitations and limits how the body interacts with the world. Tuning the CPG and body parameters to accomplish the pattern transition help in the study of how self-organization emerges via neuro-muscular synergy.

6. Reference

[1] DN Beal, FS Hover, MS Triantafyllou, JC Liao, and George V Lauder. Passive propulsion in vortex wakes. *Journal of Fluid Mechanics*, 549:385–402, 2006.

[2] Andy Clark. An embodied cognitive science? *Trends in cognitive sciences*, 3(9):345–351, 1999.

[3] Steven H Collins, Martijn Wisse, and Andy Ruina. A three-dimensional passive-dynamic walking robot with two legs and knees. *The International Journal of Robotics Research*, 20(7):607–615, 2001.

[4] Chin-Shyurng Fahn, Cheng-Feng Tsai, and Yung-Wei Lin. Development of a novel two-hand playing piano robot. In *Proceedings of the 4th International Conference of Control, Dynamic Systems, and Robotics (CDSR'17). CDSR*, 2017.

[5] S Furuya and H Kinoshita. Expertise-dependent modulation of muscular and non-muscular torques in multi-joint arm movements during piano keystroke. *Neuroscience*, 156(2):390–402, 2008.

[6] Hartmut Geyer and Hugh Herr. A muscle-reflex model that encodes principles of legged mechanics produces human walking dynamics and muscle activities. *IEEE Transactions on neural systems and rehabilitation engineering*, 18(3):263–273, 2010.

[7] Hartmut Geyer, Andre Seyfarth, and Reinhard Blickhan. Compliant leg behaviour explains basic dynamics of walking and running. *Proceedings of the Royal Society B: Biological Sciences*, 273(1603):2861–2867, 2006.

[8] DFB Haeufle, M Günther, A Bayer, and S Schmitt. Hill-type muscle model with serial damping and eccentric force–velocity relation. *Journal of biomechanics*, 47(6):1531–1536, 2014.

[9] Toby Howison, Josie Hughes, and Fumiya Iida. Large-scale automated investigation of free-falling paper shapes via iterative physical experimentation. *Nature Machine Intelligence*, 2(1):68–75, 2020.

[10] Auke Jan Ijspeert. Central pattern generators for locomotion control in animals and robots: a review. *Neural networks*, 21(4):642–653, 2008.

[11] Auke Jan Ijspeert, Alessandro Crespi, Dimitri Ryczko, and Jean-Marie Cabelguen. From swimming to walking with a salamander robot driven by a spinal cord model. *science*, 315(5817):1416–1420, 2007.

[12] Lin Jen-Chang, Hsin-Cheng Li, Kuo-Cheng Huang, and Shu-Wei Lin. Design of piano-playing robotic hand. *IAES International Journal of Robotics and Automation*, 3(2):118, 2014.

[13] Bruce A Kay, Michael T Turvey, and Onno G Meijer. An early oscillator model: studies on the biodynamics of the piano strike (bernstein & popova, 1930). *MOTOR CONTROL-CHAMPAIGN-*, 7(1):1–45, 2003.

[14] Francesco Lacquaniti, Yuri P Ivanenko, and Myrka Zago. Patterned control of human locomotion. *The Journal of physiology*, 590(10):2189–2199, 2012.

[15] Roger Lemon. Recent advances in our understanding of the primate corticospinal system. *F1000Research*, 8, 2019.

[16] Yen-Fang Li and Chi-Yi Lai. Intelligent algorithm for music playing robot—applied to the anthropomorphic piano robot control. In *2014 IEEE 23rd International Symposium on Industrial Electronics (ISIE)*, pages 1538–1543. IEEE, 2014.

[17] Jen-Chang Lin, Hsing-Hsin Huang, Yen-Fang Li, Jen-Chao Tai, and Li-Wei Liu. Electronic piano playing robot. In *2010 International Symposium on Computer, Communication, Control and Automation (3CA)*, volume 2, pages 353–356. IEEE, 2010.

[18] Marilyn MacKay-Lyons. Central pattern generation of locomotion: a review of the evidence. *Physical therapy*, 82(1):69–83, 2002.

[19] Allison M Okamura, Niels Smaby, and Mark R Cutkosky. An overview of dexterous manipulation. In *Proceedings 2000 ICRA. Millennium Conference. IEEE International Conference on Robotics and Automation. Symposia Proceedings (Cat. No. 00CH37065)*, volume 1, pages 255–262. IEEE, 2000.

[20] Rolf Pfeifer and Josh Bongard. *How the body shapes the way we think: a new view of intelligence*. MIT press, 2006.

[21] Luca Scimeca, Cheryn Ng, and Fumiya Iida. Gaussian process inference modelling of dynamic robot control for expressive piano playing. *Plos One*, 15(8), 2020.

[22] Shigeki Sugano and Ichiro Kato. Wabot-2: Autonomous robot with dexterous finger-arm–finger-arm coordination control in keyboard performance. In *Proceedings. 1987 IEEE International Conference on Robotics and Automation*, volume 4, pages 90–97. IEEE, 1987.

[23] Gentaro Taga, Yoko Yamaguchi, and Hiroshi Shimizu. Self-organized control of bipedal locomotion by neural oscillators in unpredictable environment. *Biological cybernetics*, 65(3):147–159, 1991.

[24] Anthony Topper, Thomas Maloney, Scott Barton, and Xiangnan Kong. Piano-playing robotic arm. 2019.

[25] Huijiang Wang, Toby Howison, Josie Hughes, Arsen Abdulali, and Fumiya Iida. Data-driven simulation

framework for expressive piano playing by anthropomorphic hand with variable passive properties. In *2022 IEEE 5th International Conference on Soft Robotics (RoboSoft)*, pages 300–305. IEEE, 2022.

[26] Huijiang Wang, Thomas George Thuruthel, Kieran Gilday, Arsen Abdulali, and Fumiya Iida. Machine learning for soft robot sensing and control: A tutorial study. In *2022 IEEE 5th International Conference on Industrial Cyber-Physical Systems (ICPS)*, pages 01–06. IEEE, 2022.

[27] Dan Zhang, Jianhe Lei, Beizhi Li, Daniel Lau, and Craig Cameron. Design and analysis of a piano playing robot. In *2009 International Conference on Information and Automation*, pages 757–761. IEEE, 2009.

IOP Conf. Series: Materials Science and Engineering 1292 (2023) 012016 doi:10.1088/1757-899X/1292/1/012016

What is Life? The Artistic Aspect of the Embodied Intelligence and Soft Robotics: Life in Sonic Arts, Audio-Visual Programming, and Handicrafts Design

Shu Yang

Sussex Humanities Lab, Department of Music, School of Media, Arts and Humanities, University of Sussex, UK

E-mail: sy324@sussex.ac.uk

Abstract. The artistic aspect is the 'dancer' (Gary Zukav) [7] in the machine. There is an embryonic potentiality for embodied intelligence, soft robotics, and engineering if we emphasise the artistic aspect and employ it as an indispensable element. Art is less weighted in science, technology, and engineering than functionalities and applications. Equivalently, there is a potential under-explored advantage of science and technology-based artistic creation instead of individualistic improvisation, interpretation, and expression. Starting from the philosophical and scientific question, What is life? this paper reviewed influential theories from entropy, 'code-script' by Erwin Schrödinger, 'wetware', cell compartmentation by Paul Nurse, sentient machine by Anil Seth, to the 'Dance' by Gary Zukav, the '30 000' organ by Denis Noble, the 100,000 protein sonification by Markus Buehler, and a bio-inspired composition by Alberto Carretero using P-system. With my research on cell-inspired audio-visual works *The Cell Planet* using Touch Designer, music creation for the Protein Misfolding project, and the soft-robotic inspired biomimetic textiles and art, this paper provides solutions for integrating technology, science, engineering, and art from a deeper, experimental, and philosophical context.

1. Introduction

If there are questions about life, then one of the most fundamental enquiries should be what life is. Before presenting my approach towards this question, I would like to introduce three influential pioneers: Erwin Schrödinger, Sir Paul Nurse, and Alan Turing, together with their extraordinary ideas and contributions.

Amongst these three notable figures, there is only one biologist: Sir Paul Nurse. Erwin Schrödinger is a physicist, and Alan Turing is a mathematics and computer scientist. There are two books published at different times, but with similar titles: “What is life? The Physical Aspect of the Living Cell” by Erwin Schrödinger in 1944 [1], and “What is Life? Understand Biology in Five Steps” by Sir Paul Nurse in 2020 [2]. Alan Turing’s article “The Chemical Basis of Morphogenesis (1952)” [3] is an informatics approach towards life.

2. What is *life*?

2.1. Life as Entropy

Schrödinger introduced entropy into the enquiry for what is life. He proposed that life feeds on ‘negative entropy’ to keep it from dying [1]. Because non-living things have a tendency towards the high entropy (high disorder and low energy) from low entropy (high orderliness and low energy), to reach the state of thermodynamical equilibrium [1].

Matters tend to become disorder from orderliness. Whereas living organisms are highly organized and ordered forms, which means they (or we) have this fundamental difference from non-lives: we have a resistance to decay. In other words, we create high entropy or make orderliness to avoid decaying into the equilibrium state of non-living things. It is this peculiar ability that keeps us from dying. This ability is the fundamental line to separate life from non-life. That is why metabolism is the vital force of life. We are constantly exchanging, processing, and transforming all the chemical elements and physical properties with the environment: both within ourselves and outside of our bodies, either with living beings or non-living things.

Life takes 'negative entropy' (negentropy) in, creates orderliness within, and releases positive entropy out.Developed from the law of physics, Schrödinger decoded the distinct driving force of life: 'negative entropy' [1] (negentropy). He rewrote Boltzmann's equation as:

$$- (\text{entropy}) = k \log(I/D) \quad [1]$$

This equation allows us to describe life from a measurable and quantifiable perspective. Using this physics key, Schrödinger opened a novel door to biology, which created the possibility for us to simulate life and calculate the energy of life.

2.2. The language of *life*

The nuclein was named by Friedrich Miescher in 1869. He discovered the chemical compositions of nucleins: nitrogen and phosphorus. However, the function of the nuclein (DNA) remained unknown.

Nine years before decoding the double-helix structure of DNA, Schrödinger proposed a genius viewpoint on the 'aperiodic crystal' [1]. He defined this 'crystal' with an innovative, visionary, precise, and insightful name: the 'code-script' [1]. Once again, he opened another door to biology using a new key: informatics. This idea, later, influenced Crick and Watson on the decipherment of the DNA structure.

The linear structural character of DNA, which exists as a three-dimensional state, is an exquisite 'design' to accurately memorise, transcript, and instruct life. I am emphasising the linear structure here because this structure determines the direction to successfully read and interpret this 'code-script' must be strictly 'one-way', which significantly reduces the possibility of fault occurrence in DNA replication processes. And the folded double-helix structure gives DNA a crucial ability to store an enormous amount of information using an extremely small space inside the cell.

To consider life from the informatics perspective is a breakthrough in biology. It opens a door for scientists to simulate, mimic, interpret life, and more importantly, visually 'see' the intricate chemical structures as 'scripts', so that we can 'read', illustrate, and decode the meaning of hidden language within living forms. The core of life, after Schrödinger's 'code-script', Crick and Watson's double-helix structure, and the decades of works by biologists, have been becoming 'readable'. The three-dimensional dark structure of the cell is now opened and transformed into the language of life.

Although life as information is still not the final answer to what is life, it provides us with an innovative approach to logically understanding the mechanisms of living beings and linguistically interpreting the meanings of all these most fundamental chemical reactions within the cell. We can interpret them as if reading Shakespeare that was written in an ancient, or even an alien language. We can even 'hear' and understand them by decoding the symphonies of metabolism into meaningful vowels, grammars, sonatas, or even spells.

2.3. The 'wetware' and non-life

When Alan Turing invented the Automatic Computing Engine using punch-card, it became the starting point of the modern computer. The computers we are using today are a typical strand of hardware. And the apps or graphical operating systems (Windows, Mac or Linux) are software.

The term 'wetware', however, was introduced by biologist Dennis Bray in his book Wetware: A Computer in Every Living Cell (2009) [4]. He gives life another insightful, definitive, and 'fashionable' description. Both biologist Sir Paul Nurse and Neuroscientist Professor Anil Seth discussed the 'wetware' in their books to nominate what is life.

With the development of artificial intelligence, the boundary between life and computer seems to become blurry. I would, however, disagree with the universal growing expectations on the vanishment of the boundary between organisms and machines. Because the term 'wetware' means life is based on self-organized physical and chemical reactions, rather than virtually imitating, mimicking or simulating life forms, or mechanically operating under the instructions.

If, nevertheless, there will be an invention of a real 'wetware' and 'wet-machine', how could it be, and what does it look like? Must it be made from 'sampled', hybrid, modified, or cultured bits and pieces from any scales of living beings? If so, then, does it have any fundamental biochemical difference between you, me, or all living forms on earth? If it must be made from the real organisms or even the most basic organelles in cells, it no longer strictly belongs to artificial intelligence, it is the synthesis of the cultivating, culturing, growing, and breeding of our own living forms. It is biology.

Alan Turing was asked the by the question: can digital computer think? He had been refusing to give an answer until May 1951 in his lecture on BBC Radio. He proposed an interesting view on machine: the ability of imitating human and human brain [3]. Turing also believe that as a machine, computer can provide us with a significant useful way to look at how our intelligence works [3]. Alongside his preeminent answer to the question, I would, however, argue that the 'imitational machine' itself is not sufficient to be equalled with a biologically cell-based conscious living brain, which is part of a 'self-sustaining biological organism'.

The 'self-sustaining biological organism' was proposed, seventy years after Turing's lecture, in the book "Being You A New Science of Consciousness" (2021) by Professor Anil Seth [5]. As a notable neuroscientist, he argues that it should be biotechnology, not artificial intelligence, which seems to be the right approach toward synthetic consciousness as it requires real neurons [5]. He also points out that this self-maintenance mechanism can go down from humans to the level of cells. He demonstrated that intelligence and consciousness are not necessary links together, and being conscious is not necessarily applied to being a cell or even any living organism [5].

I would like to briefly summarize Anil's idea as the following equations:

$$\text{Life} \neq \text{Consciousness}$$

$$\text{Life} \neq \text{Intelligence (including AI)}$$

$$\text{Life} \neq \text{Machine}$$

$$\text{Life} \neq \text{Consciousness} + \text{Intelligence (including AI)} + \text{Machine}$$

Beside of his demonstration on 'wetware', Anil discussed the sense of human's feeling with machines: 'just making computers smarter is not going to make them sentient' [5]. It can be summarized as:

$$\text{Sentient} \neq \text{intelligence (including AI)} + \text{machine (computer)}$$

Back to Turing's question: can digital computers think? I would argue that the think here refers to the (artificial) intelligence aspect of, with, and from life. However, the ability 'to think' does not even belong to any fundamental definitions of life. A computer is not a life.

Then what can be defined as a life?

In the book "What is Life? Understand Biology in Five Steps", Sir Paul Nurse explicitly summarized the most fundamental principles of life. Life is based on the bounded physical entities with physical metabolism going on, and the ability to evolve, grow and reproduce. And it is the line to draw life from non-life [1].

Although the question of life has long been floating up in the air and remaining unsolved, it exists as a compass for intellectually adventurous scientists, philosophers, and artists to discover humans and nature. It constantly leads our childlike curiosities to decode all the hidden mysteries and hypotheses in our lives. This enquiry has also triggered my research and my approaches towards life.

2.4. The 'recipe' of *life*

An interesting 'menu' by biologist Nick Lane in his book The Vital Question Why is Life the Way it is? He presented a 'solution' of how to make a cell. There are six principles: a continuous supply of reactive carbon; a supply of free energy to drive metabolic biochemistry; catalysts; excretion of waste; compartmentalisation; heredity material [6].

Although the 'recipe' and hypophysis by Nick Lane is yet to be physically achieved, it is a perfect solution to generate the 'living cell' virtually, visually, sonically, and 'soft-robotically'.

2.5. The *Wu li* of *life*

American spiritual teacher Gary Zukav depicted Bohm's implicate order as "that-which-is" [7]in the final Chapter of his book The Dancing Wu Li Masters (1979). He uses the "dance" as a metaphor for breaking symbolic boundaries such as "Things", "matters", or "particles". And the "dance "here, as that-which-is, leads my enquiry from biological to vibrations, such as sound and music.

2.6. The music of *life* [8] and the sound of *life*

In the book The Music of Life: Biology beyond Genes (2006), Denis Noble presented a novel idea, "An organ of 30 000 pipes". He demonstrated the "helpfulness" of musical analogy to represent the human genome (around 30 000 genes). He proposed that the "world's largest organs" can enable the music of life to be played and heard.[8] Inspired by his idea, and with the help of up-to-date digital audio workstation (DAW), I developed a sonic approach to protein misfolding as a contribution to what is life? In the following chapter, 3.2.

American scientist Prof. Markus J. Buehler at MIT published a paper in 2019 on the sonification of over 100,000 protein structures. He proved that the "Things (or particles)" in our body could be translated into sound and music following a scientific way.[9] The visual programming language Max MSP is much more logical, repeatable, and controllable compared with the "world's biggest organs". Highly influenced by Markus's work, in the next chapter, 3.1, I applied another visual programming language Touch Designer with Logic Pro, to my science-oriented audio-visual art creation. And I applied the vibrational spectrums method to the analysis section of the music I created in chapter 3.2, which reflects the vibrational spectrum that Markus uses for obtaining protein database.

Last but not least, composer Alberto Carretero's music piece Parallel Lives [10] emphasises the artistic application and outcomes of music, despite the application of cellular automata membrane computing called P-system by Gheorghe Păun. Additionally, in his paper on composing bio-inspired ensemble music using AI, he argues that the artistic result cannot be guaranteed by using scientific applications and methods, [11] which means artistic interpretation is the key to making the results aesthetic. The "artistic results" here, or the aesthetic aspect of life, cell, physics, soft robotics, and vibrations, is the purpose of my conference proceedings. Alberto's mastering of micro percussion sound using classical instruments, mini-loudspeakers, and non-musical instruments gave the ensemble a rich, organic, lively, and slightly randomised sound personality during performing and improvisation, as if the "Dance" by Gary Zukav. And this inspired my choice of the musical instruments in Logic Pro for the protein misfolding project in chapter, 3.2.

3. My Approaches to *life*

3.1. The Cell Planet: *life* in Audio-Visual Arts

The Cell Planet was an audio-visual collection that I created in November 2021 [12]. A living cell is a good example of compartmentation [2], inspired by this concept, which Sir Paul Nurse has also mentioned in his book, I chose TouchDesigner as the node-based visual programming language It is an ideal software to create audio-visual content in real-time. Because the design of this software has many resemblance features with metabolic pathways. And the system is formed with many small operators as controllers. All the controllers can be connected by wiring them together. Same as the cell, the organelles are bounded by membranes, they communicate with each other through invisible chemical wires. The patchworks are a mimic of sections of metabolic pathways, with feedbacks, controls, input, output, oscillators, and random scrips.

IOP Conf. Series: Materials Science and Engineering 1292 (2023) 012016 doi:10.1088/1757-899X/1292/1/012016

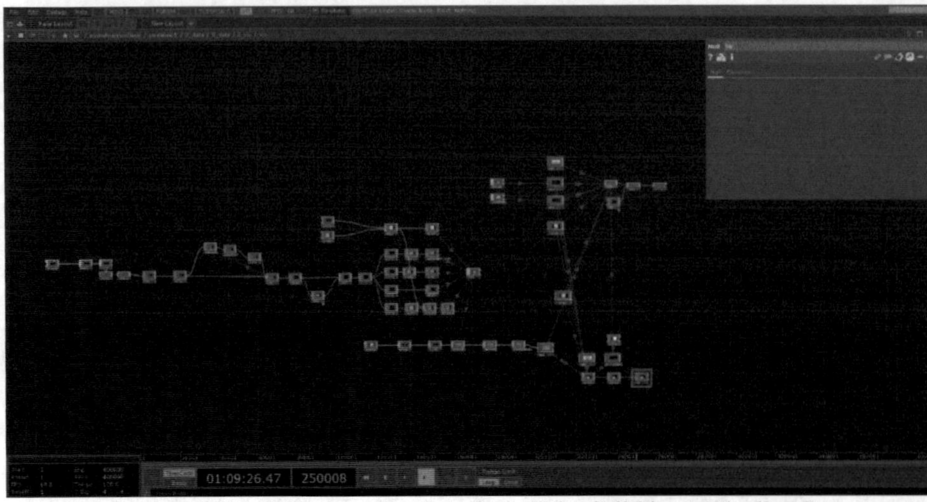

Figure 1. Touch Designer Patchworks in "The Cell Planet: The lives".

3.1.1. The *noise* and noise

If you 'look' inside of the living cell, what you will see is enormous chemical reactions happening dynamically and constantly every second. If putting all these reactions equally together on a screen or by hearing them sonically from an amplifier at one time, what you will see or hear will be the noise. The noise is the gateway to looking inside of life. However, the noise here in the living cell has fundamental distinctions from the noise from the natural environment or non-living forms.

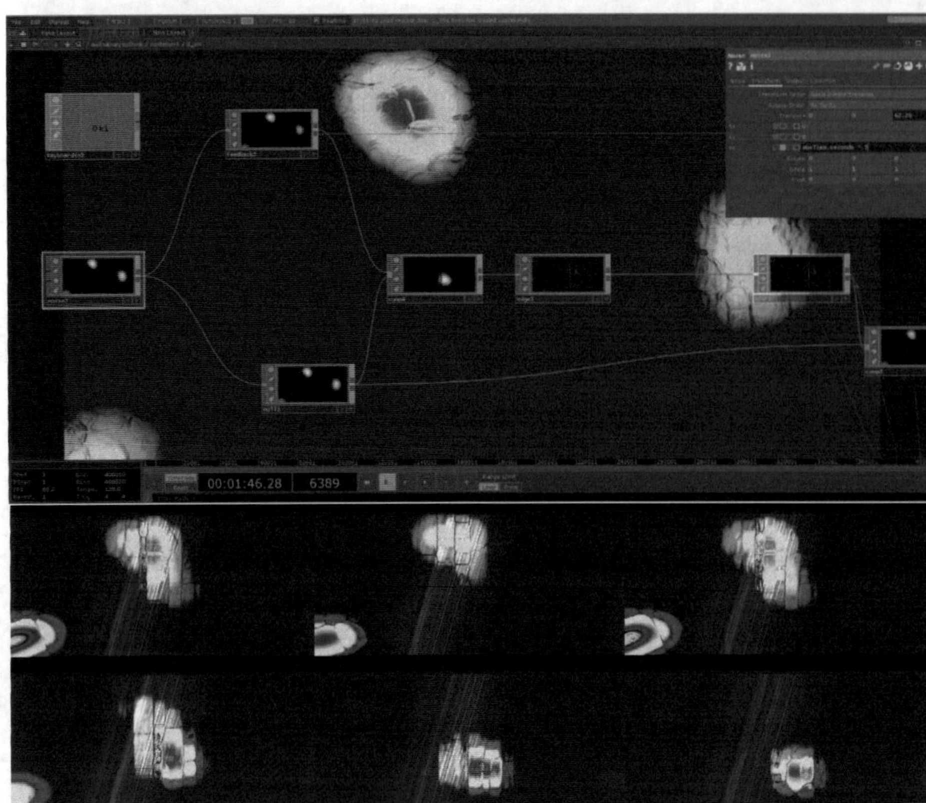

Figure 2. Touch Designer Patchworks and outcomes in "The Cell Planet: The lives".

The oscillators of the noise within the living cell are the chemical reactions in cell metabolism. If put all the 'metabolic volume' together, what you will hear should be a meaninglessly noisy sound, as if you are listening to all existing versions of symphonies in classical time orderlessly together at the same time. On the contrary, however, when you 'look' deep into each metabolic pathway, what you will see are the orderly script and hear the frequencies, as if your friend switched all but one symphony off, and only leaves the one clarinet playing in Symphony No. 6 in F major, Op. 68 by Beethoven. https://www.youtube.com/watch?v=o_JOWiAZJGA [13]

By using the example of the clarinet sound on the previous page, the relationship between the noise and with scripts can be vividly interpreted. In biology, the fundamental building blocks of the noise or metabolism are the processes of creating, transcription, transporting, and reading the 'code scripts. Each 'code scripts' has a series of linear messages, even if they are physically existing. This opened my hypophysis on the sonic approach to life. By simulating the codes in a living cell, we can create the noise and play with it. It will be a constant, self-organising, and dynamic sound. In TouchDesigner, there are also noise operators for generating pixels and sound spectrums.

3.1.2. 'absTime.seconds' and randomness: when you look back, you see your pathway

The python code 'absTime.seconds' for generating random numbers is very useful and crucial when simulating life. Same as Brownian motion [1], Alan Turing introduced random numbers and diffusion in his 'Morphogenesis' [3].

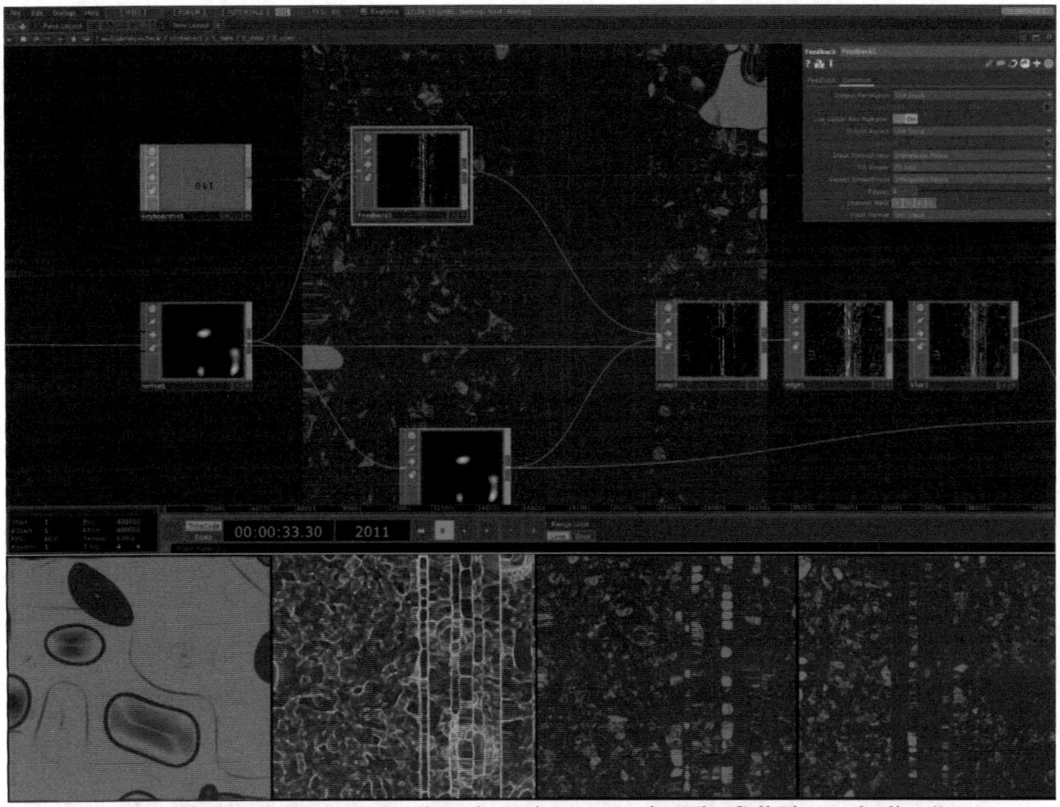

Figure 3. Touch Designer Patchworks and outcomes in "The Cell Planet: the lives".

What is randomness? Sonically, per se, I would argue random is the discrete perception of what is going to happen before it comes. Taking the Brownian motion as an example, the randomness happens before we see the whole process or the 'walk'. After the random walk is accomplished, however, you see a continuous linear pathway, and the randomness has gone. Musically, the random sound is extremely vital to creating timbre, particularly in digital music creating. When someone is talking to you, normally the meaning of the sentence will come when it is finished.

A living cell is a life form that uses various materials in the natural environment to grow and reproduce. The randomness comes from the environment's constantly influence on the cell. And the cell is reacting to the changes in the environment. By applying randomness as a mimetic 'ambient background' to nature into audio and visual art can make the creations look and sound much 'lifely'.

3.1.3. Feedback and regulation

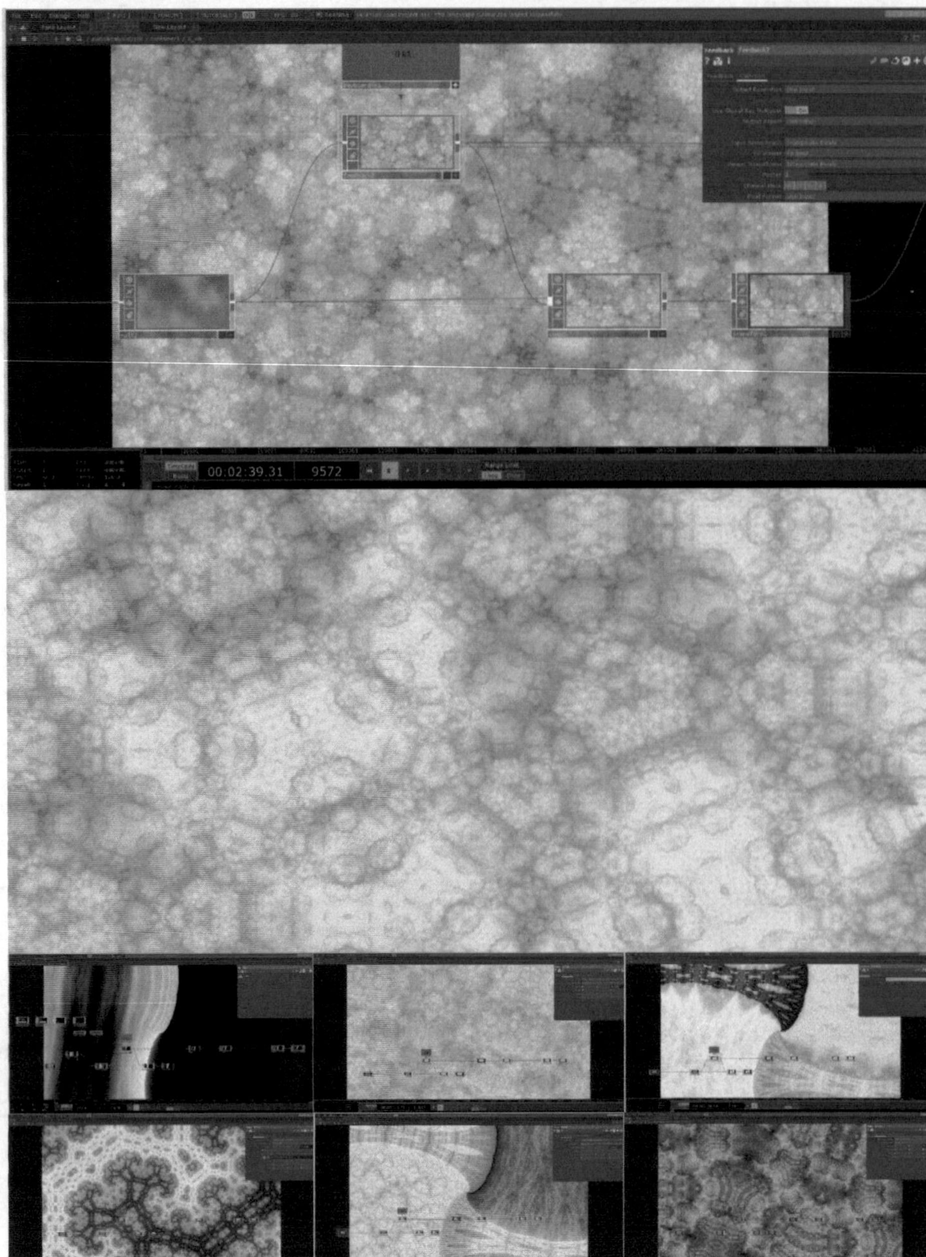

Figure 4. Touch Designer Patchworks and outcomes in "The Cell Planet: The landscapes".

In cellular regulation, feedback is the crucial switcher and decision-maker at the cell level. Metabolic pathways are influenced by feedback to keep the balance of all chemical densities within the cell at a healthy level. It is a vital function to keep the cell alive.

What is feedback? The first four alphabets, in my view, is the key: feed. Sonically, the shapes of the sound waves both from input and output are the potential elements to generate resonance or entangled sound waves, which is the 'sonic liquid reaction field'. The 'feed', however, is the upcoming resonance frequencies, which can be detectable, recognisable, absorbable, influenceable, and functional interactional with the pre-existing sound field. These aspects are also applicable in defining the 'feed' within the cell. Put it amusingly, to feed my dog with a banana is not the proper feed. Because if we want to create good feedback, the delicate control of the time, volume, pitch, and surroundings are the most important points to generate the 'feed'.

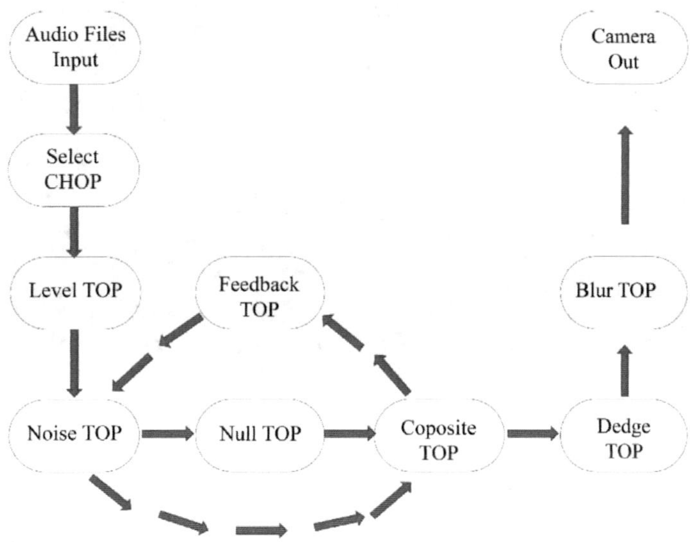

Figure 5. Box model of implanting noise and feedback in Touch Designer.

Visually, feedback can also create 'lifelike' motions. By sampling applying the feedback operator in TouchDesigner and adjusting the 'feed' to get the right to resonate, you can create incredible motions. The Landscape in The Cell Planet, for example, the beautiful and organic animation was created by only three operators: noise, feedback, and composition.

3.2. Protein Misfolding: *life* in Sonic Arts

"Misfolding" public engagement was a project collaborated the Centre for Research in Opera and Music Theatre and the Serpell Lab in the School of Life Sciences at the University of Sussex [14] [15].

Sound track link: https://soundcloud.com/devilbodyangel-heart/sets/hope-for-refolding?si=07a515d7bff64456ac85c39a14612892&utm_source=clipboard&utm_medium=text&utm_campaign=social_sharing [15]

3.2.1. Protein folding

Following the descriptions of normal protein folding compared with misfolding, I first created a very simple, ambient, and tranquil short piece, using Tibetan Singing Bowls in Logic Pro. Because the sound of singing bowls and bells can generate a very healing and natural sound, together with many microtones.[16]

IOP Conf. Series: Materials Science and Engineering 1292 (2023) 012016 doi:10.1088/1757-899X/1292/1/012016

The frequencies used in Tibetan Singing Bowls are different from the tuning of conventional instruments used in a symphony such as piano, clarinet, and violin. There are many spaces between each tone because I want to define the space between each tone as the high entropy (high disorder) environment around the healthy protein, which enables sufficient mobilities, fluencies and exchange between each different kind of protein. And the various timbres within this short piece represent the good biochemical density, rather than one dominant protein as the uniformed monophony. If, according to Schrödinger, the environment beside the living parts is orderliness, then it means that the energy formed during the folding process is difficult to exit.

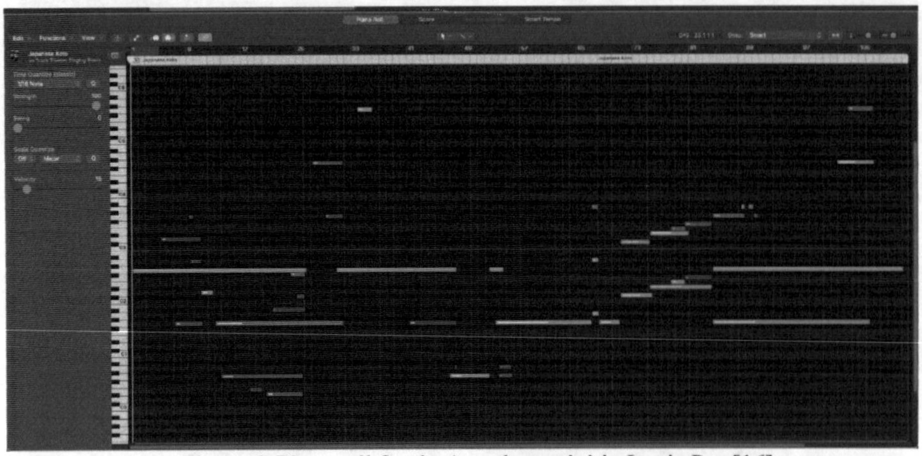

Figure 6. Piano roll for the ‘good protein’ in Logic Pro.[16]

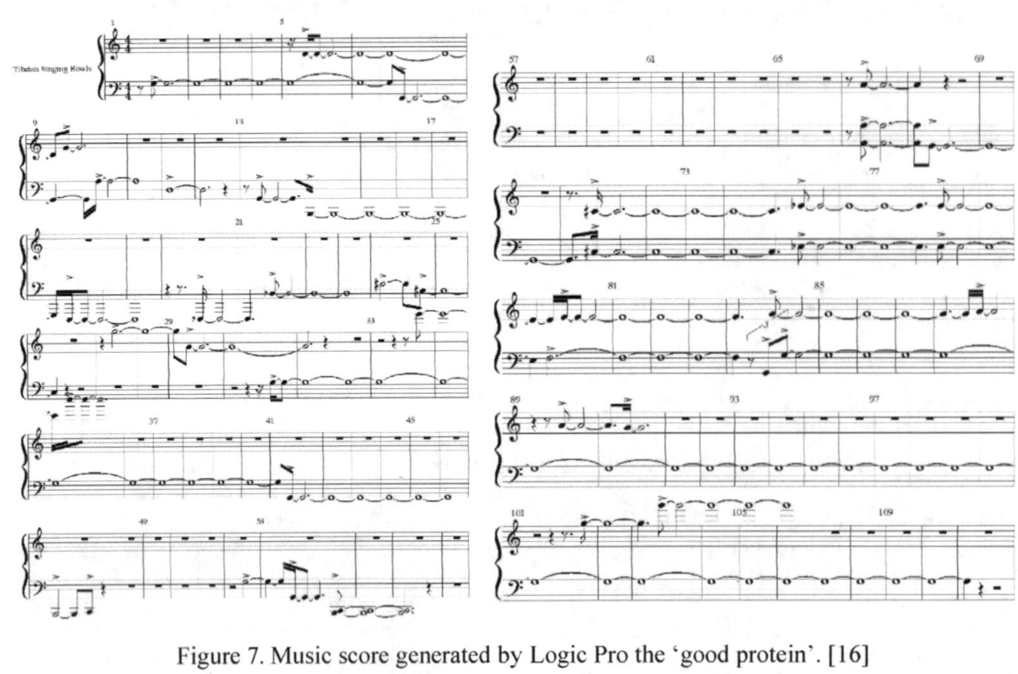

Figure 7. Music score generated by Logic Pro the ‘good protein’. [16]

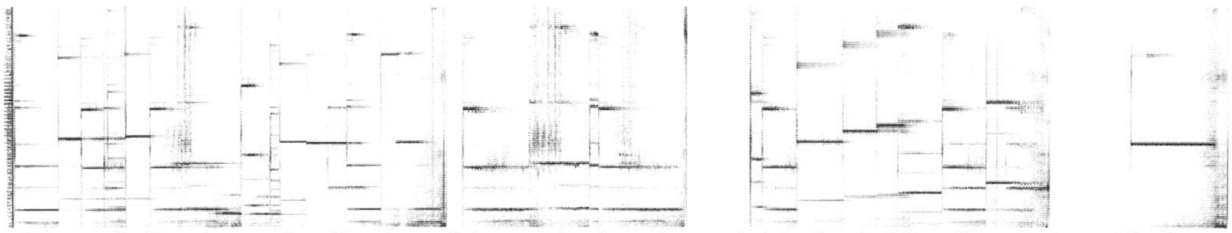

Figure 8. Spectrum visualisation of the music: 'good protein' in Sonic Visualiser (colour: Black on White; scale: linear; Gain: 0dB).

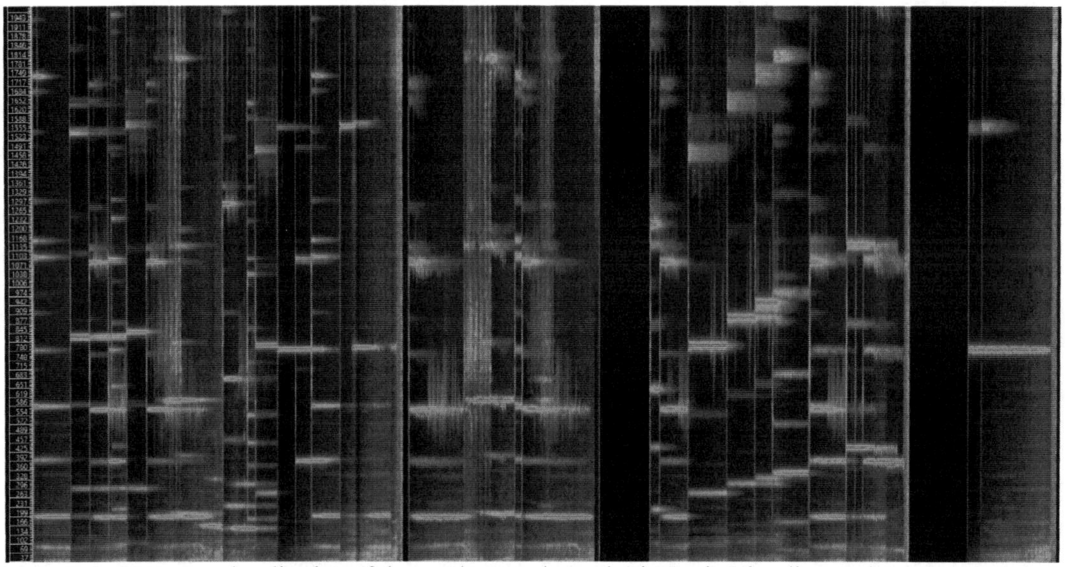

Figure 9. Spectrum visualisation of the music: 'good protein' in Sonic Visualiser (colour: Green; scale: Meter; Gain: 0dB). x: bin frequency in hertz, y: time; red colour on green shows the loudness of the sound. [16]

After creating the 'good protein', I simply copied the same score into eight tracks and set them into different instruments. The instruments are chosen because they can produce harmonious sound, and they are profoundly used in symphonies and many other forms of Classical music. I orderlessly arranged the starting points of each track to generate a high entropy, according to Schrödinger's equation [1] [17]. As this workshop has a limited time requirement, I only use the starting points and one volume as variants. But the result was good. It sounds as if many autonomous robots are assembling bits and pieces of components in a little nice factory. And they sounded like they were working in an orderly way.

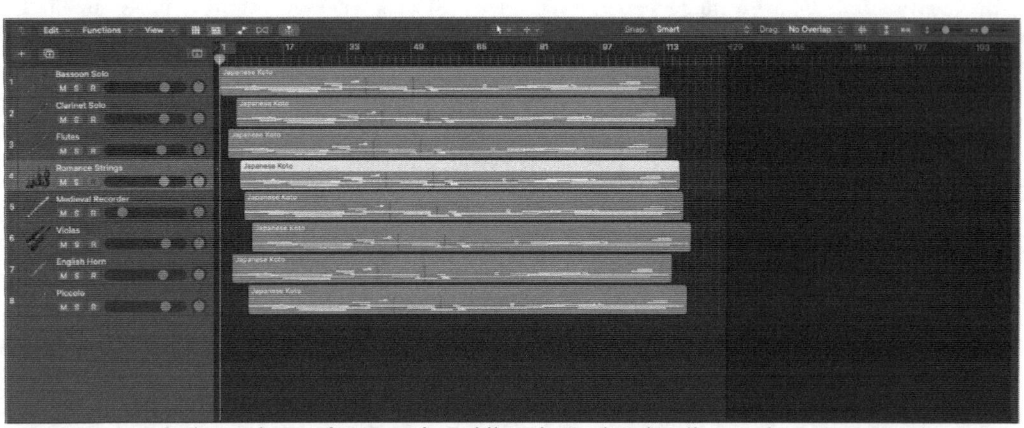

Figure 10. Tracks in Logic Pro for 'Protein Folding' in Sonic Visualiser (colour: Green; scale: Meter; Gain: 0dB). x: bin frequency in hertz, y: time; red colour on green shows the loudness of the sound.[17]

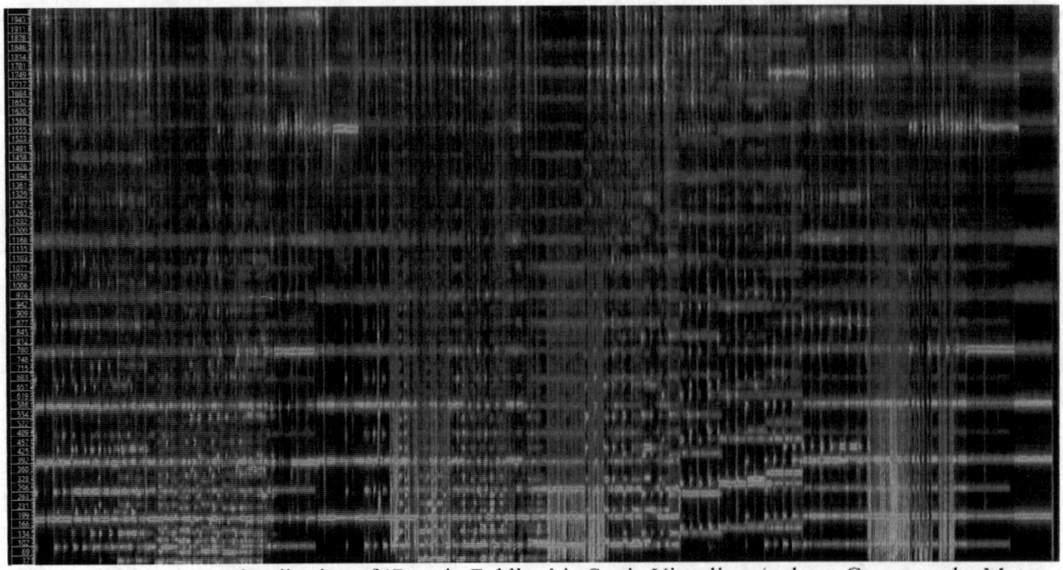

Figure 11. Spectrum visualisation of 'Protein Folding' in Sonic Visualiser (colour: Green; scale: Meter; Gain: 0dB). x: bin frequency in hertz, y: time; red colour on green shows the loudness of the sound.[17]

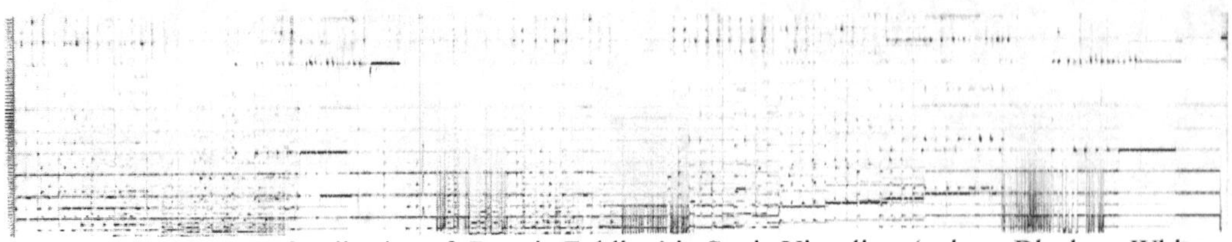

Figure 12. Spectrum visualisation of 'Protein Folding' in Sonic Visualiser (colour: Black on White; scale: linear; Gain: 0dB).

3.2.2. Protein misfolding

From the aspects and causes of this disease, according to Professor Louise, I captured five keywords: dehydrated, stack, sticky, strong, and stable. Especially when Louise depicts the protein assembling process, I can vividly see that the entire fibre was built by the same little 'bricks' of Amyloid Beta protein as they are only self-associating. Back to the music I created, I 'mutated' the 'good protein', and created the 'Amyloid Beta protein' track. I applied EQ adjustment to that natural, holly, and tranquil music. The amplified lower frequencies and their emphasised strong resonance created a completely new, dark, greedy, unsettling, unbreathing, and evil sensation.[18] This feature is set to stimulate stickiness, as strong and stable as spider webs, and dehydration. From the entropy aspect, I have added a large amount of energy [1] to this music.

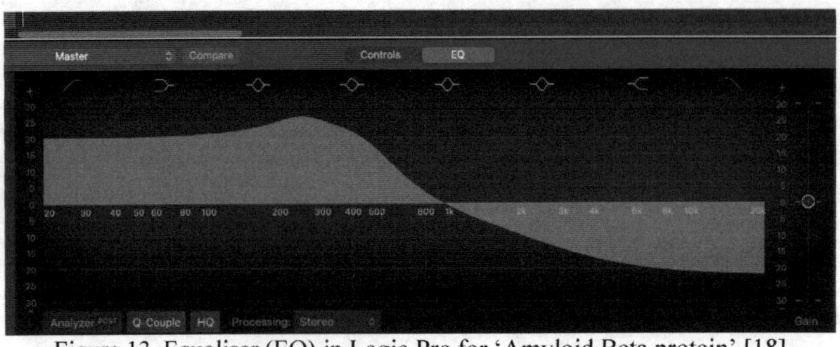

Figure 13. Equaliser (EQ) in Logic Pro for 'Amyloid Beta protein'.[18]

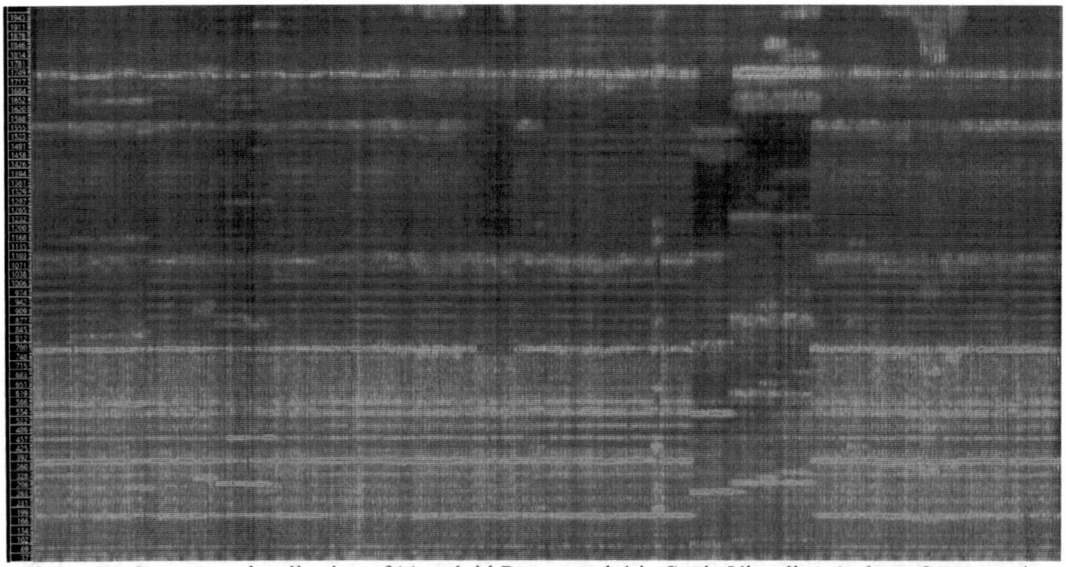

Figure 14. Spectrum visualisation of 'Amyloid Beta protein' in Sonic Visualiser (colour: Green; scale: Meter; Gain: 0dB). x: bin frequency in hertz, y: time; red colour on green shows the loudness of the sound.[18]

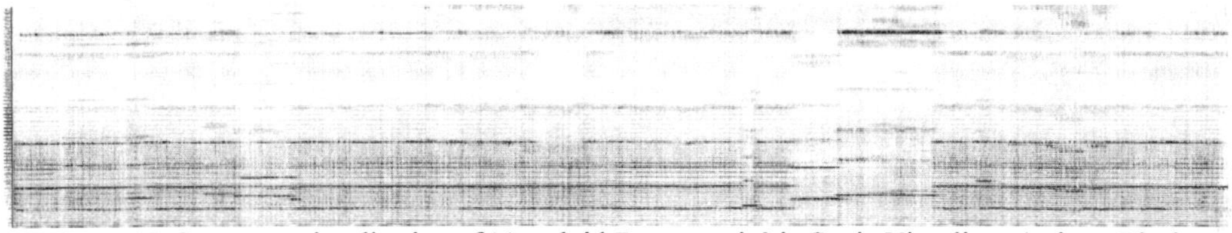

Figure 15. Spectrum visualisation of 'Amyloid Beta protein' in Sonic Visualiser (colour: Black on White; scale: linear; Gain: 0dB).[18]

Following the structure and formation of the Amyloid Beta fibre, I copied this 'mutated Amyloid Beta sound' twelve times, and, strictly laid them together as if stacking the Amyloid Beta proteins. I arranged the 'Amyloid Beta fibre' using the 'Amyloid Beta protein' tracks in the same length, same beginning, same volume, same effects, and same instrument. [19] As you can see, this is protein misfolding. It looks as if nothing was properly composed by a professional musician, but, unexpectedly, it was the most outstanding piece of the three.

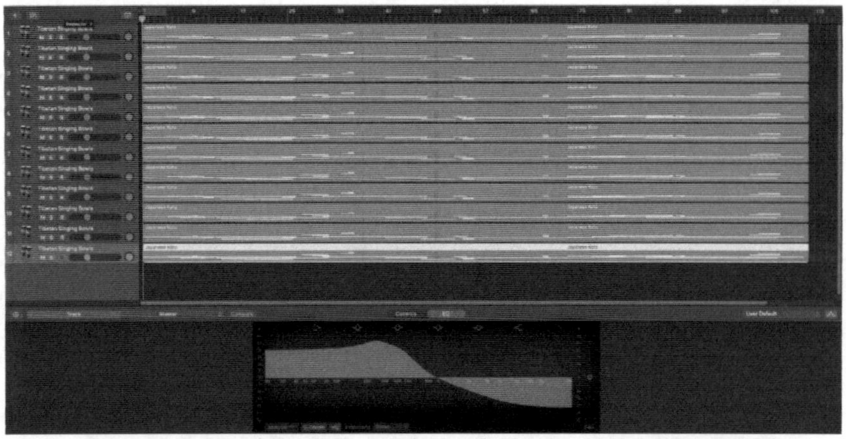

Figure 16. Sound Tracks in Logic Pro for 'Protein Misfolding'.[19]

IOP Conf. Series: Materials Science and Engineering 1292 (2023) 012016 doi:10.1088/1757-899X/1292/1/012016

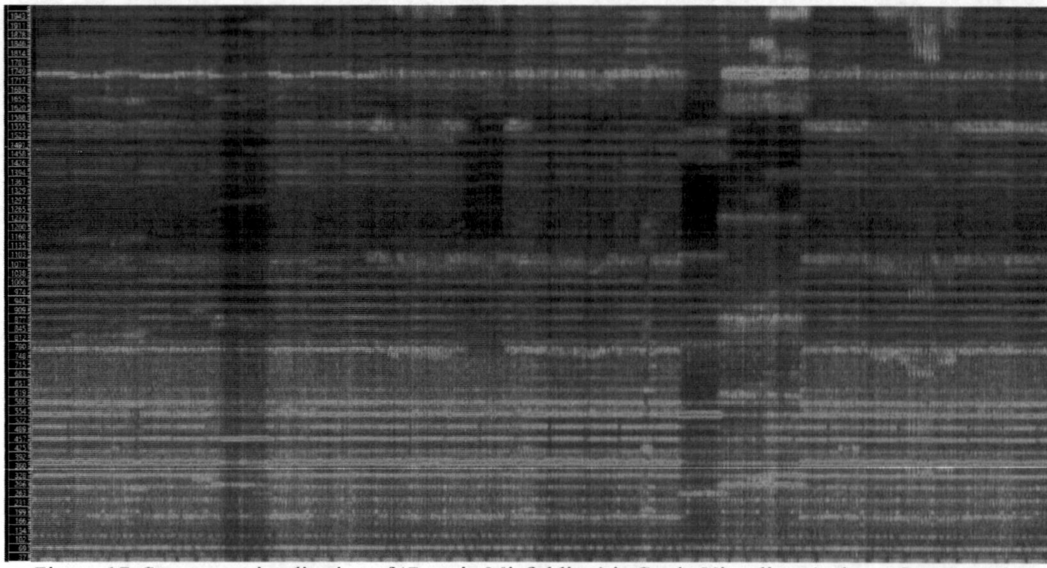

Figure 17. Spectrum visualisation of 'Protein Misfolding' in Sonic Visualiser (colour: Green; scale: Meter; Gain: 0dB). x: bin frequency in hertz, y: time; red colour on green shows the loudness of the sound.[19]

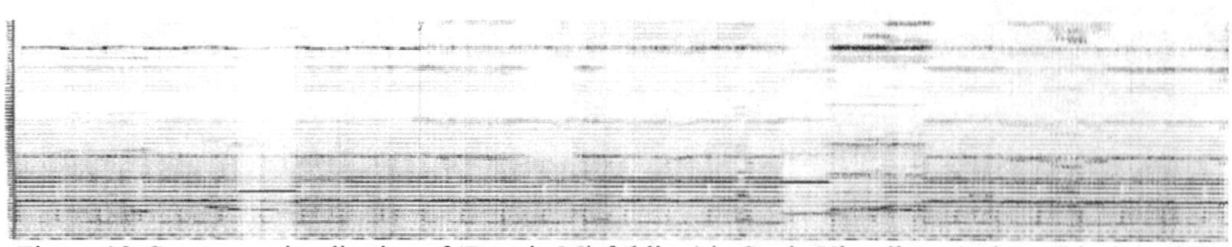

Figure 18. Spectrum visualisation of 'Protein Misfolding' in Sonic Visualiser (colour: Black on White; scale: linear; Gain: 0dB).[19]

All of the tracks are designed with an 'absolute' high orderliness. According to Schrödinger's entropy equation [1], this structure (both my music and the real protein) will have an extremely high life index (high energy), moreover, by overlaying them together will tremendously amplify any resonance in the music, which created twelve times resonances on each pitch. This is to say, in the natural selection process within the human body, this abnormally high orderliness 'lifely' structure will have a much higher advantage to grow than our normal structures. When the balance of growing and decaying is broken, a new level of entropy [1] is being set, and the tendency towards ever higher orderliness will unceasingly drive the organism towards illness, defunctioning, and death.

IOP Conf. Series: Materials Science and Engineering 1292 (2023) 012016 doi:10.1088/1757-899X/1292/1/012016

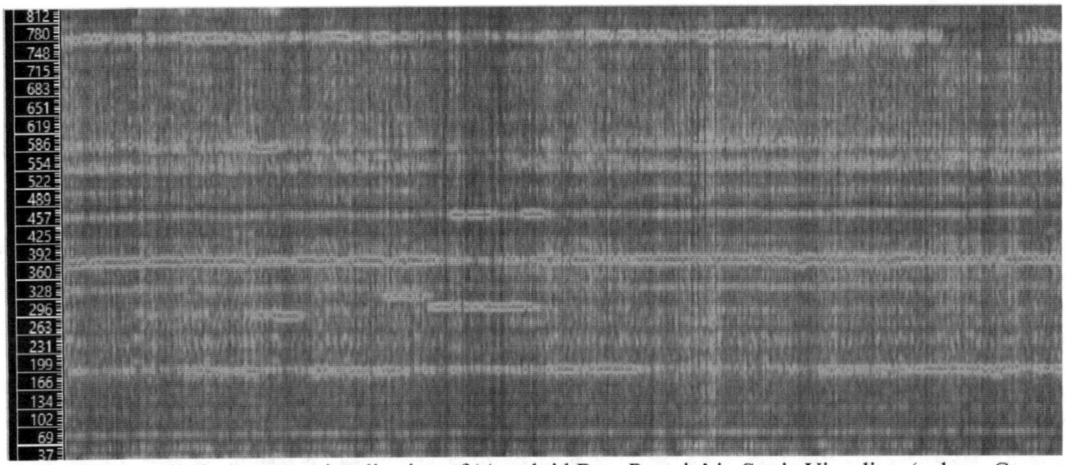

Figure 19. Detailed spectrum visualisation of 'Amyloid Beta Protein' in Sonic Visualiser (colour: Green; scale: Meter; Gain: 0dB). x: bin frequency in hertz, y: time; red colour on green shows the loudness of the sound.[18]

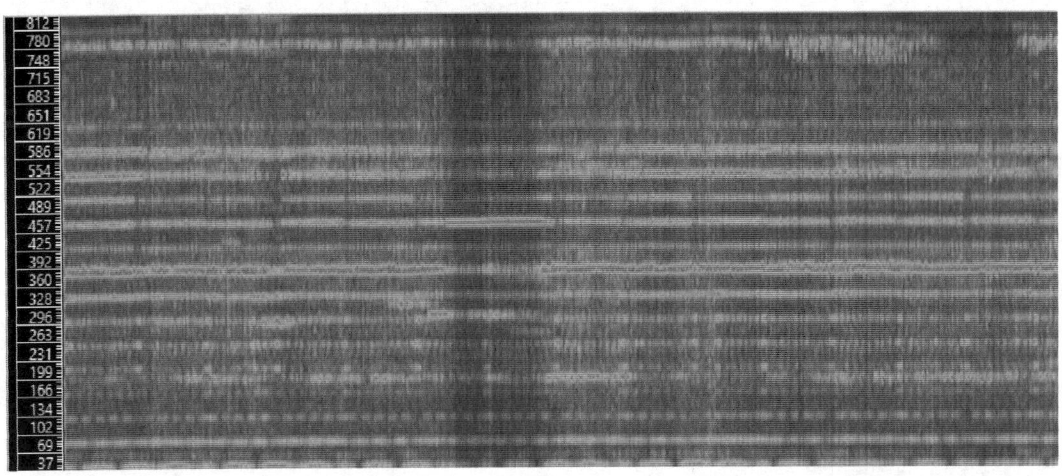

Figure 20. Detailed spectrum visualisation of 'Amyloid Beta fibre' in Sonic Visualiser (colour: Green; scale: Meter; Gain: 0dB). x: bin frequency in hertz, y: time; red colour on green shows the loudness of the sound.[19]

The significant difference in sound volume and micro- rhythmics is demonstrated clearly by looking at the frequency range 380 Hz – 600 Hz. The sonic spectrums in Figure 19 is more randomised and segmented compare with the sonic spectrums in Figure 20. By listening the sounds on SoundCloud [18, 19] and looking at the these frequency range, the amplified orderliness [1, 2] can be found.

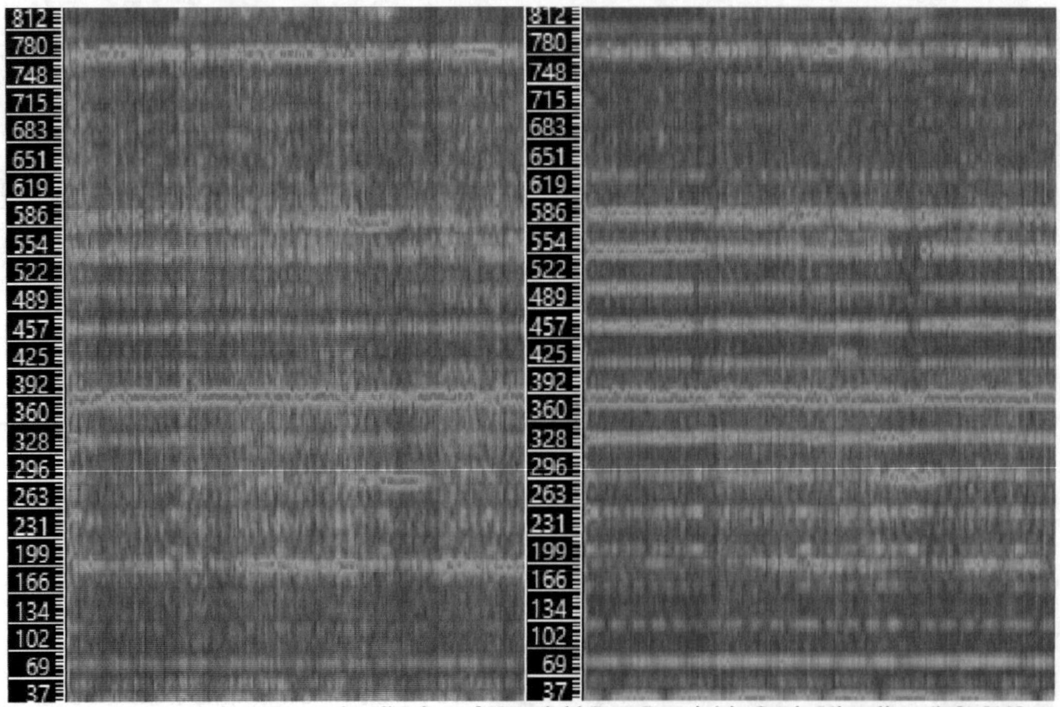

Figure 21. Detailed spectrum visualisation of 'Amyloid Beta Protein' in Sonic Visualiser (left) [18] vs Detailed spectrum visualisation of 'Amyloid Beta fibre' (right) [19]in Sonic Visualiser colour: Green; scale: Meter; Gain: 0dB). x: bin frequency in hertz, y: time; red colour on green shows the loudness of the sound.

These two detailed sonic spectrum visualisations revealed another distinguishing feature. By comparing the 'mutated' 'Amyloid Beta protein' (left) [18] with the 'chained' 'Amyloid Beta fibre' (right) [19], we noticed, that aside from the amplification of each unit sound, a sign of uniformity and rhythmical homogenisation has been generated. It looks as if the sound is becoming segmented, crystalised, hard, sticky (magnetically), rocklike, and chaining together like the zips. The peculiarities of dehydration, orderliness and self-association. This character is identical to the Alpha Beta sheets demonstrated by Louise.

Sonically, this visualised of 'Amyloid Beta fibre' is a successful attempt to life. It proves that by exclusively fuelling biology principles into Erwin Schrödinger's entropy equation can simulate life through sound spectrums. Must it be a truism that the orderliness (low entropy and high energy) by Schrödinger is the universal law, the Misfolding piece indeed attracted the most audiences, rather than the loosely, but intellectually, effortfully, and seriously arranged other two pieces. And that might be a reasonable explanation that you can hardly hear any music by John Cage [20] in a gym or a McDonald's.

3.2.3. Hope for protein refolding

The idea for refolding is a hope that misfolded fibres between the neurons can become softened, detectable, and 'willingly cooperative'. I slightly swung the concrete stacking shapes of the bricks and applied a relatively large number of tracks with varieties of instruments. The instruments range from violas to Brooklyn drums, from Liverpool bass to tuba.[21] The variations of instruments that I have applied in this piece are according to the balance and diversity of biochemical 'species'. Because the unfriendly, greedy, self-associated, attractive, and xenophobic Amyloid Beta protein is based on creating this uniformly homogentisic orderliness. In other words, the biochemical structures, and the microenvironment around them are constantly gathering and forming high entropy and high energy [1], which consumes much more energy than normal biochemical reactions and dispossess the quantity of other healthy proteins.

According to the visualisation in Sonic Visualiser, you can see the magnetic rock-like segmented metal chain has been broken. The tones are breezily distributed and naturally spaced with interactions. It sounds amusing with thunder-like lower frequencies, which brings a sense of ease and humour while knocking off all the selfish chains. Again, the result is achieved by reducing the entropy.

Figure 22. Instruments track layout in Logic Pro for 'Amyloid Beta fibre'. [21]

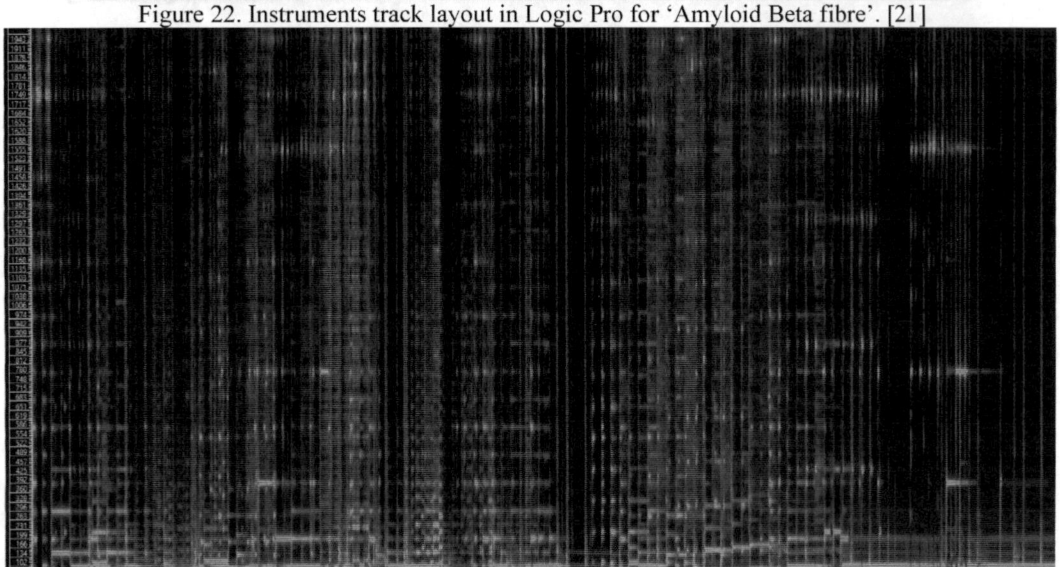

Figure 23. Spectrum visualisation of 'Hope for Refolding' in Sonic Visualiser (colour: Green; scale: Meter; Gain: 0dB). x: bin frequency in hertz, y: time; red colour on green shows the loudness of the sound. [21]

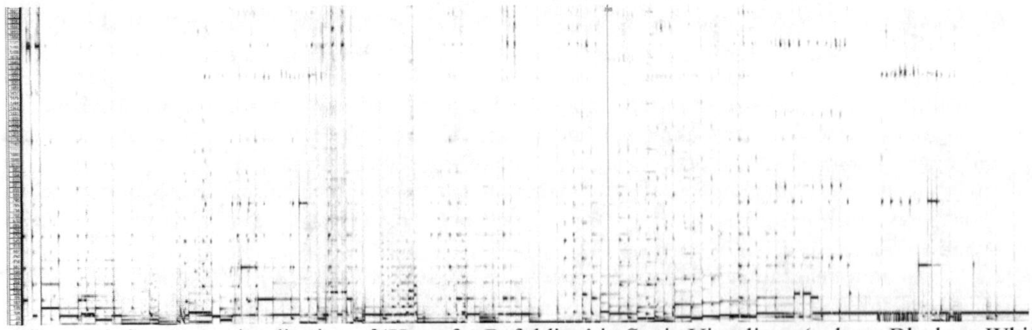
Figure 24. Spectrum visualisation of 'Hope for Refolding' in Sonic Visualiser (colour: Black on White; scale: linear; Gain: 0dB). [21]

3.3. Biomimetics and Soft Robotics inspired: *life* in Handicrafts

3.3.1. Biomimetic mini sculpture: "Cellulose"

It is a creative biomimetic sculpture: "Cellulose" designed by Shu Yang at the University of Brighton in 2020. This design aims to convey the seemingly simple formations of natural shapes, contrary to industrial-oriented products.

Figure 25. "Cellulose".

3.3.2. Smart fabric design 1: "The Random Emotion"

The Random Emotion is an experimental fabric design woven by Shu Yang at the University of Brighton during his MA time (2019-2020). It is an innovative application of inflatable PVC tubes with wool and cotton. The fabric's shape, weight, temperature, and density are controllable by blowing air or liquid into the tubes. This project has been suspended due to the workshop facilities' disruptions of the lockdown during the pandemic.

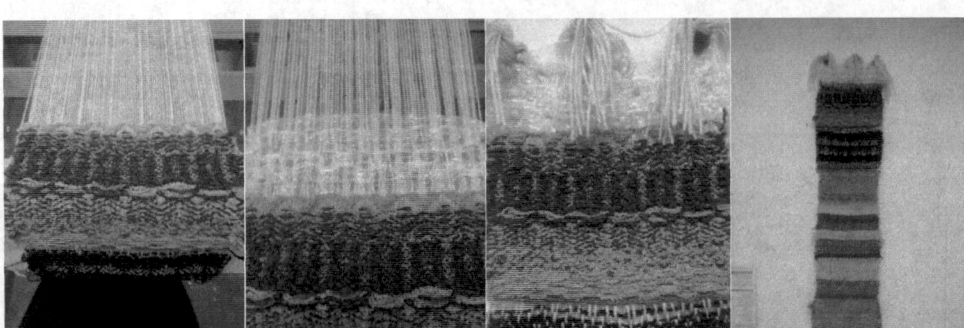
Figure 26. Smart fabric design 1: "The Random Emotion".

3.3.3. Smart fabric design 2: "The Breathable Wing"

"The Breathable Wing" is another experimental design by Shu Yang at the University of Brighton in 2019. This design aims to create a novel functionality in wearable material. The combination of Tyvek, inflatable PVC, and copper wires makes the material breathable, light, shapable, strong, and easy to clean. This project has been suspended due to the workshop facilities' disruptions of the lockdown during the pandemic.

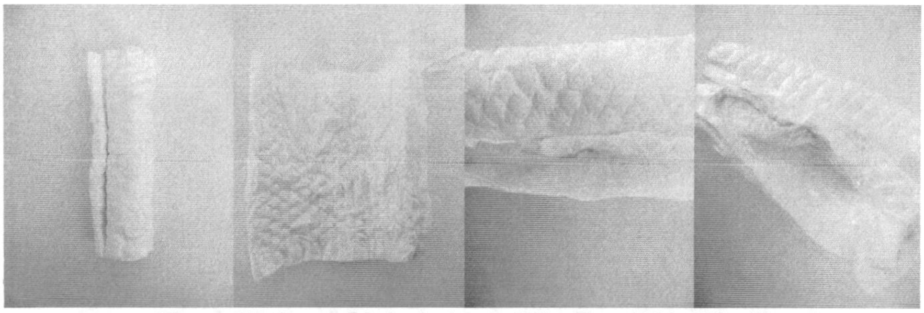

Figure 27. Smart fabric design 2: "The Breathable Wing".

3.3.4. Biomimetic design: "The Amino Acid Skeleton"

"The Amino Acid Skeleton" is a semi-fabric prototype and semi-soft sculpture mimicking the amino acid chain, designed by Shu Yang at the University of Brighton in 2020. This "skeleton" has a soft, flexible, flowy but strong and restrained characteristic, which physically represents the abstract structures with a touchable feature rather than the virtual representations under electronic scopes. This project has been suspended due to the workshop facilities' disruptions of the lockdown during the pandemic.

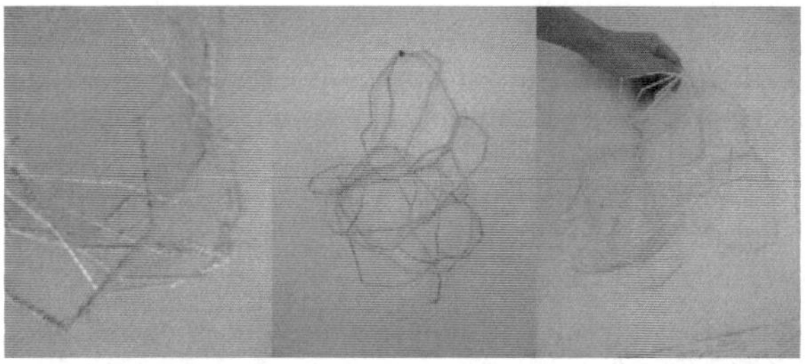

Figure 28. The biomimetic design: "The Amino Acid Skeleton".

"The Amino Acid Skeleton"has been applied two times to the short video for the "Misfolding" project by the Centre for Research in Opera and Music Theatre and the Serpell Lab in the School of Life Sciences at the University of Sussex in 2021 and 2022 [22].

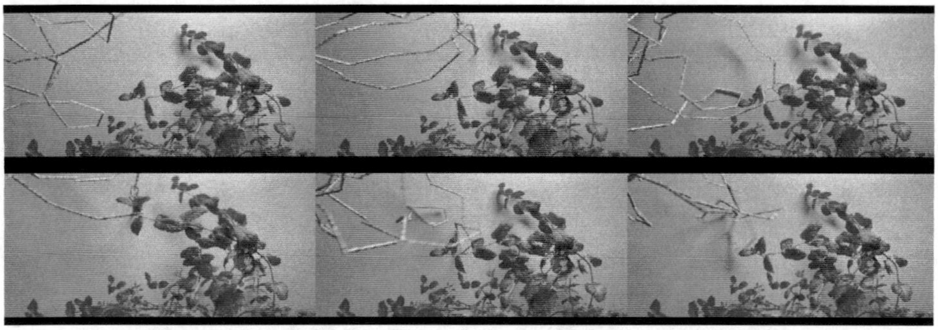

Figure 29. "Misfolding Through the Looking Glass" (2022).[22]

3.3.5. Soft robotics-inspired handicraft: "The Soft Harpsichordia"

"The Soft Harpsichordia" is a soft robotic-inspired soft musical instrument by Shu Yang at the University of Brighton in 2020 [23]. This design aims to create a prototype based on controllable shapes to create another layer of physical improvisational elements for contemporary instruments and digital audio workstations. This project has been suspended due to the workshop facilities' disruptions of the lockdown during the pandemic. This "Harpsichordia" has the same purpose as embodied intelligence, although it is niched in material art or handicrafts.

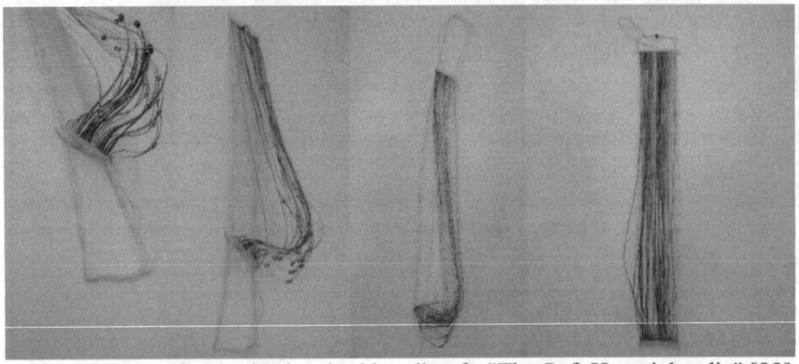

Figure 30. The soft robotics-inspired handicraft: "The Soft Harpsichordia" [23].

3.3.6. Biomimetic raincoat: "The Inflatable Algae"

"The Inflatable Algae" is an biomimetic prototype designed by Shu Yang at the University of Brighton in 2020. This raincoat is created from recycled PVC, mainly from Ikea packaging. By mimicking the algae, this raincoat has an outstanding ability to change shapes in both volume and strength based on its lightweight. Similar to the algae in the changeable environment. This raincoat can protect your head in severe weather like hail and low temperatures. Its structure also transfers the weight of the entire garment from your top head to your shoulders. This project has been suspended due to the workshop facilities' disruptions of the lockdown during the pandemic.

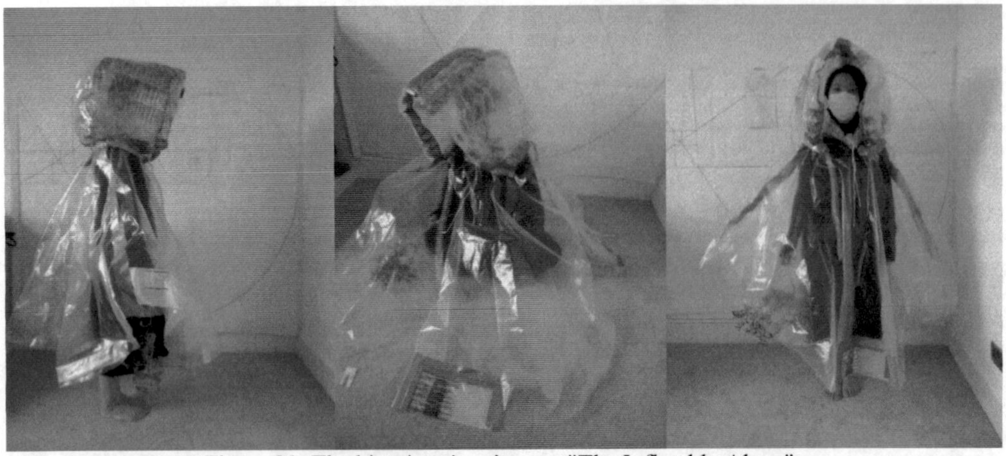

Figure 31. The biomimetic raincoat: "The Inflatable Algae".

4. Discussion

Soft Robotics have the most significant potential to create forms of life physically using the essential material with innovative ideas and technologies. On the other hand, Embodied intelligence has created remarkable opportunities for researchers from diverse disciplines to contribute to this community. The cross-disciplinary collaborations have generated extraordinary contributions in science, technology, arts, media, design, and humanities. As a multi-disciplinary researcher, I introduced my most recent artworks and research focusing on the artistic aspect of science, technology, and engineering. This emphasis on art can give solutions from another perspective for artistic and scientific innovations. Finally, back to the question at the beginning, I would like to end this chapter by quoting three sentences from Shunryū Suzuki [24], Erwin Schrödinger [25] and Denis Noble [8]:

"Waves are the practice of the water."

— Shunryū Suzuki, Zen Mind, Beginner's Mind: Informal Talks on Zen Meditation and Practice

"It seems to me that what I call the construction of an external world that really exists is identical with what you call the describablilty of the individual situation that occurs only once."

A letter to Einstein on 18 November 1950
— Erwin Schrödinger, Letters on Wave Mechanics (1967), p. 37

"If there is an organ, and some music, who is the player, and who was the composer? And is there a conductor?"

— Denis Noble, The Music of Life: Biology beyond Genes (2006), p. 32

Acknowledgment
I would like to thank EI 2022 for providing me with this valuable opportunity. My greatest appreciation to all my tutors in biological engineering, biochemistry, fashion design, textiles, fabric science, musicology, music technologies, computer science, neuroscience, literature, and philosophy.

References

[1] Schrödinger, Erwin. What Is Life? The Physical Aspect of the Living Cell. C.U.P., 1944, pp.75-80.
[2] Nurse, Paul, and Ben Martynoga. What Is Life? Understand Biology in Five Steps. David Fickling Books, 2020, p. 2, pp. 135-142, pp. 189-194.
[3] Turing, Alan, and B. Jack Copeland. The Essential Turing : Seminal Writings in Computing, Logic, Philosophy, Artificial Intelligence, and Artificial Life Plus the Secrets of Enigma. Clarendon, 2004, pp. 507-518.
[4] Bray, Dennis. Wetware: A Computer in Every Living Cell. Yale University Press, 2009, pp. 226–27.
[5] Seth, Anil. Being You: A New Science of Consciousness. Faber & Faber, 2021, pp. 248-253, pp. 246-267.
[6] Lane, Nick. The Vital Question: Why Is Life the Way It Is? Profile Books, 2015, p.96.
[7] Zukav, Gary. The Dancing Wu Li Masters: An Overview of the New Physics. 1. ed, Morrow, 1979, pp. 307-309, p. 314.
[8] Noble, Denis. The Music of Life: Biology beyond Genes. Oxford University Press, 2008, pp. 31-32.
[9] Qin, Zhao, and Markus J Buehler. "Analysis of the vibrational and sound spectrum of over 100,000 protein structures and application in sonification." Extreme Mechanics Letters vol. 29 (2019): 100460. doi:10.1016/j.eml.2019.100460

[10] Parallel Lives - Alberto CARRETERO (Schallfeld Ensemble) AIMC 2021 Graz, Mumuth KUG. www.youtube.com, https://www.youtube.com/watch?v=e5tIDAe1oWk. Accessed 15 Mar. 2023.

[11] Carretero, Alberto. "Parallel lives, composing bio-inspired ensemble music by means of AI." (2021).

[12] The Cell Planet - YouTube. https://www.youtube.com/playlist?list=PLN6OWAWFw_XlIKxAbMnDNUQRVMuyNDru3. Accessed 30 Aug. 2022.

[13] Beethoven 6th Symphony - Full Clarinet Excerpts. www.youtube.com, https://www.youtube.com/watch?v=o_JOWiAZJGA. Accessed 23 May 2022.

[14] Misfolding : Current Projects : ... : Centre for Research in Opera and Music Theatre : University of Sussex. https://www.sussex.ac.uk/research/centres/centre-for-research-in-opera-and-music-theatre/research/projects/misfolding. Accessed 15 Mar. 2023. 'Hope for Refolding'. SoundCloud, https://soundcloud.com/devilbodyangel-heart/sets/hope-for-refolding. Accessed 14 Mar. 2023.

[15] 'Hope for Proteins Refolding' for Misfolding Workshop, June 1 & 2, 2021. www.youtube.com, https://www.youtube.com/watch?v=yh8K9AvHyWw. Accessed 22 May 2022.

[16] 001 The Good Protein. soundcloud.com, https://soundcloud.com/devilbodyangel-heart/001-the-good-protein. Accessed 14 Mar. 2023.

[17] 002 Protein Folding. soundcloud.com, https://soundcloud.com/devilbodyangel-heart/002-protein-folding. Accessed 14 Mar. 2023.

[18] 003 Amyloid Protein. soundcloud.com, https://soundcloud.com/devilbodyangel-heart/003-amyloid-protein. Accessed 14 Mar. 2023.

[19] 004 Protein Misfolding. soundcloud.com, https://soundcloud.com/devilbodyangel-heart/004-protein-misfolding. Accessed 14 Mar. 2023.

[20] 4'33" by John Cage - John Cage Live at the Barbican - BBC Four Collections. www.youtube.com, https://www.youtube.com/watch?v=yoAbXwr3qkg. Accessed 23 May 2022.

[21] 005 Hope For Refolding. soundcloud.com, https://soundcloud.com/devilbodyangel-heart/005-hope-for-refolding. Accessed 14 Mar. 2023.

[22] Misfolding Through the Looking Glass. www.youtube.com, https://www.youtube.com/watch?v=9AHjJ894eUI. Accessed 30 Aug. 2022.

[23] Yang, Shu. Visualization of Sound and Music. [Unpublished Manuscript]. School of Art, University of Brighton, 2020.

[24] Suzuki, Shunryū, and Trudy Dixon. Zen Mind, Beginner's Mind. 40th anniversary ed, Shambhala, 2010.

[25] Schrödinger, Erwin, et al. Letters on Wave Mechanics. Vision, 1967, p. 37.

Theoretical Framework for Human-Like Robotic Taste with Reference to Nutritional Needs

Grzegorz Sochacki, Arsen Abdulali, Lucy Cheke and Fumiya Iida

Bio-Inspired Robotics Laboratory (BIRL), Department of Engineering, University of Cambridge, UK

E-mail: gks33@cam.ac.uk

Abstract. This paper contains a theoretical framework for an intelligent robot that can feed itself with organic matter and learns to like and want certain foods. Furthermore, being tied to the onboard generation of power, the robot's learning is grounded to physical requirements that need to be met for continued robot operation. We posit a system wherein the electrical power is generated by a microbial fuel cell(MFC), and the food value can be assessed by measuring the current generated. Interestingly, the MFC requires feed similar in many parameters to food preferred by most people with respect to parameters like salinity or pH. This property is theorized to be a good start for teaching a robotic chef human preferences. We also propose a circuit that injects additional current, simulating social cues for some foods, and effectively resulting in learning an acquired taste.

1. Introduction

Recent research on robotic cooking and electronic taste opens the possibility of robots entering the vast business of cooking and catering. Various types of electronic tongues and similar sensors are capable of feats like classifying wine age[1], finding pork contamination in mutton[2] and assessing fish freshness[3]. Moreover, the electronic taste has been used as feedback for a robotic chef, controlling the saltiness and texture of scrambled eggs[4]. Some approaches also use human taste to adjust robotic chefs' cooking[5].

This ability to implement taste for robotic chefs together with the low cost of robotic labour raises the possibility of producing cheap, individualized dishes. The major problem with the realization of this idea is machine learning for taste. Most capable models consist of large neural networks requiring sizeable labelled datasets. Taste factors such as the palatability of dishes are highly subjective and personal and thus cannot be objectively labelled. Furthermore, the structure of this label is undefined: A lack of exhaustive theory connecting measurable chemicals to notions of flavour and palatability limits precision, especially taking into account individual preference. This problem is not trivial to solve, but human chefs have the advantage of sharing in same preference spectrum as other humans, therefore making it easier to predict and accommodate these tastes. It may be the case that robotic chefs must thus be able to simulate client's preferences - as humans do - to excel in this function. The first step towards that goal may be the development of food preferences by the robot for themselves.

In this paper we propose a theory for robotic taste, taking inspiration from human feeding behaviour as shown in Figure 1. Moreover, we propose a robot that, using this theory, will develop its own subjective taste based on objective physical needs. For this thought experiment

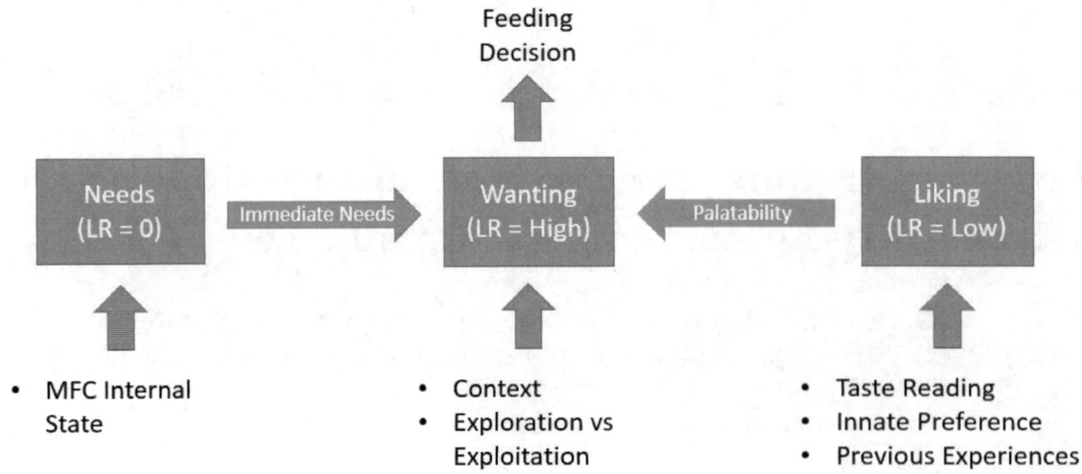

Figure 1. Schematics of bio-inspired tasting model. The model is grounded by the physical robot body that produces its own energy. Therefore, the body has needs, that need to be met to produce energy. It will also develop liking or preferences for certain foods based on their usefulness and rarity. Finally, it has a block called wanting that takes all the information together, and makes a decision specific to the current context.

we use a robot that uses a Microbial Fuel Cell(MFC) to generate its own power. A few of these kinds of robots have been constructed starting with Wilkinon's train[6] that was the first known prototype to operational systems like EkoBots[7]–[9]. Examples of these and more gastrobots[10], [11] are shown in Figure 2. This type of robot is a relatively new invention and few challenges are yet to be solved decisively[6]. We expand the gastrobot idea by putting the robot in charge of feeding (and thus powering) itself, based on sensors inside the MFC, its performance and sensors probing encountered potential fuel. This setup provides both interoception[12], and relevant tasting of food, allowing for the learning and development of associations between flavours and the impact these have on fundamental energy and nutritional needs, just as in the human body. The exact propositions for these sensors are discussed in the following sections.

This robot provides a minimal model of a body that has needs(energy production by MFC) and a way to feed itself based on the taste of potential fuel options. As we are primarily interested in taste we will assume a rather simple environment within which liquids are available as a "food" for the robot. The robot would also have the possibility to sample or taste the prepared foods and choose among them. We present a machine-learning approach that has access to the internal state of the robot to monitor its needs and learn based on them. The first part of the system is learning "liking" for various products. "Liking" is a stable context-general value describing the desirability of food. Therefore, it will be learned with low learning rate, based on the power production after the consumption. This system assesses how much the food is 'liked' based on the sensory stimuli from the tongue. Secondly, we introduce a "wanting" system, that provides a time-specific context-based motivational value behind the food. Computing this value considers all internal sensor measurements, tongue readings, and liking signals an input and decides on the next consumption. This system has a faster learning rate. Moreover, it indirectly affects liking by controlling the food intake, therefore the food exposure. Finally, we discuss the similarities between flavours preferred by the robot and those liked by most people and the possibility of exploiting the robot's acquired taste to adjust its liking towards those preferred by people.

Figure 2. Previously constructed gastrobots. a)"gastronome"[6] - train-robot powerd by microbial fuel cell(MFC) b)EcoBot-II[7] c)EcoBot-III[8] in its testing environment d)EcoBot-I[9] e)Marine gastrobot[10] f)Row-Bot[11] - marine gastrobot.

2. Needing, Liking, and Wanting in Human

Human tastes and food preferences are extremely complicated and give rise to the enormous amount of dishes, and multiple recipes for each dish. Even though there is a huge interest in this area by researchers and the food industry, one, simple model of human preferences does not exist. We attempt to describe these preferences by combining together three basic drives that influence human feeding behaviour - needing, liking and wanting.

2.1. Need

The human body has a large variety of needs that must be filled to maintain homeostasis. These needs usually fall into three major categories: energy, macro- and micronutrients. These needs are objective, anchored in the physical world, constrained and directed by both evolution and learning processes to produce unique personal preferences. Finding an exact required amount of these nutrients is not feasible, but most countries have recommendations (eg.[13], [14]), varying with the age, gender and physical activity of the consumer. The real requirements may vary from recommendations, with some extreme outliers documented. For example, Kenyan runners when training spent 3600 kcal/day while consuming 3100 kcal/day, compared to the 2500 kcal/day recommendation for their age group[15]. Micronutrients also need to be supplied in adequate

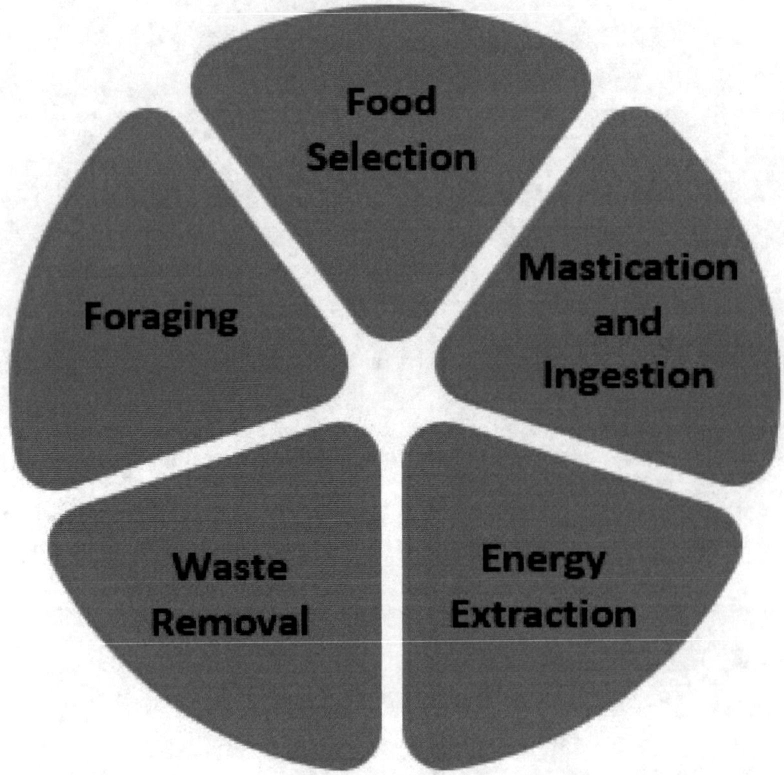

Figure 3. Challenges for constructing gastrobots - robots powered by biomass or food.

quantity, with deficiencies resulting in decreases in intellectual function [16], bone density, and other illnesses[17].

Failing to fulfil some needs causes sensations of discomfort (such as hunger, and thirst) which motivate action. On the other hand, failing to fulfil other needs may not result in any symptoms in the short term, or may cause generic symptoms that do not motivate consumption of the specifically needed foods. For example, vitamin B12 deficiency causes symptoms ranging from the sourness of the tongue and tiredness to low blood pressure and psychosis[18] - non of them prompting consumption of B12-rich foods like meat or eggs[19].

Overall, some needs cause a strong motivation to consume specific foods to reduce the drive state, while others do not. Interestingly, in some cases like hunger, it motivates to start consumption, but fulfilling the need does not entirely motivate stopping the consumption[20]. Therefore, there is clearly a space for additional mechanisms beyond physical needs that dictate the quantity of food consumed.

Taking all of these considerations we model Needs as a physical constraint for taste. These are immune to most of the external and contextual factors such as social pressure, specific available foods, or ideology. We consider Needs as a reliable input that plays a part in learning what we like and acts as an input to decisions as to what we want.

2.2. Like

Liking is a long-term estimation of the value of food. In humans, it comes both with innate components and learning. For example, sweetness, usually measured by sucrose concentration, is pleasant for all ages[22], with signs of hedonistic pleasure present even in newborn infants[23].

Basic Taste	Innate Response	Development Time
Sweet	Positive	In Utero
Sour	Negative	In Utero
Salty	Positive	4-6 Months Old
Bitter	Negative	In Utero + Maturation
Umami	Context-Dependent	Unknown

Table 1. Innate human response to basic tastes according to [21].

Regardless of this innate proclivity for sweetness, preferred concentrations change with age, with older people finding high sucrose concentrations less pleasant[24]. With specific foods, children tend to refuse the intake of new foods/flavours and need to taste them as much as 10 to 15 times before they will habituate and find their taste pleasant[25]. A good review of the development of innate proclivities to basic tastes is given in [21], which we summarize in Table 2.2.

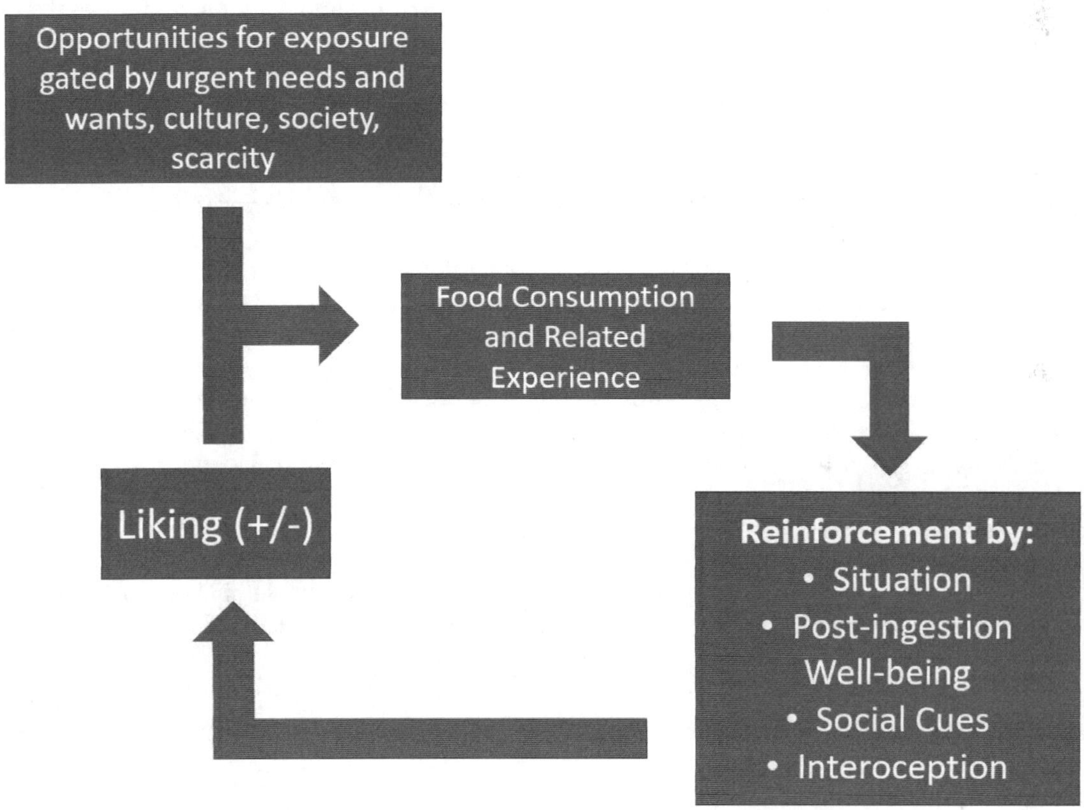

Figure 4. Model of "learning to like" as suggested by [21].

Liking is a learning process, starting with certain biases which are adapted and changed over time. Interestingly the reward is not only fulfilment of the body's need, but can also come from external conditioning with social or physical stimuli. For example, people repetitively presented with a certain kind of tea with sugar learn to like it more[26]. An example of social conditioning is food aversions passed down in families[27]. Therefore, liking can be understood as a slow, long-term oriented learning process holding a time-independent and context-independent value of certain foods. Importantly, as with any learning, it is biased by the data it learns. It is effectively closed in a control loop as shown in Figure 4. We will discuss this loop in the next section.

2.3. Want

The literature review above showed that, while there are some imprinted preferences for taste, these are quite general, and fluid. There is no general, objective measure of taste, yet it's immensely important for survival. We have seen that while the only physical anchor comes from bodily needs, that (for most) are rather easy to fill in the developed world[28], [29], we are still motivated to eat specific things, and this section is a discussion of how it happens in humans.

One of the most interesting aspects of the "Want" element of taste is how we come to be motivated to try new foods: As has been discussed, humans prefer known tastes. However, taste is not the only perceptual modality that can produce a sense of familiarity. For example, children tend to prefer to try the fruits that they have seen before, even if they have never eaten it before[25].

This cultivated liking through non-gustatory familiarity can be gained by via exposure to advertisements. Repeated presentation of recognisable foods and brands motivates buying and eating certain foods. Beyond this, analysis of advertisements placed in magazines[30] provides statistics on claims used for both hedonic and functional foods. Overall, properties like nutrition, taste, novelty, lack of (negatively perceived) substances, and general positive effects on health are emphasized. This speaks to the large variety of considerations that motivate specific consumption decisions, beyond simply hedonic enjoyment and bodily needs.

Finally, dietary decisions are also dependent on the environment. For example, the sugar levy in the UK reduced soft drink consumption by 18%[31]. The study also showed that 80% of reduced calorie consumption came from producers lowering the amount of sugar, possibly leading to more exposure to drinks with less sugar. Moreover, large portion sizes motivate individuals to consume more of the presented food regardless of individual needs and desires[20]. Even social environment can greatly change if and how much we eat[32]: For example, an individual who has not eaten for 24 hours (and thus has high Need) will still reduce their consumption to better match a social eating partner who has not been similarly deprived[33]. The identity of those who influence eating habits also matters. Family members and friends have more effect than coworkers, while coworkers have more effect than strangers[32]. The exact effect of social pressure depends also on the pressured person. For example, males consume much more snacks when in a group with friends in comparison to a group of strangers, while this effect is not present for females. On the other hand, when paired with strangers, females tend to consume less than those paired with a friend, an effect that is not present in men[34].

Overall, the "wanting" part of the system can be considered as the motivator of specific individual consumption events. It can be understood as an estimation of expected value, not only based on intrinsic properties (.e.g nutrients) or time-invariant value, but also specific context.

3. MFC Powered Robot's Needs

The literature describes some instances of feeding in gastrobots, but the procedure was hard-coded and did not include tasting the food. For example, Ecobot-III[8] was tested in a highly structured, robot-friendly arena, where the Ecobot was always fed the same food, in the same quantity and in the same place. In this setup no adaptive decision-making was required for the robot, therefore did not require any tasting ability to pick the correct food. Regardless of this shortcoming, the robot still had needs, that needed to be fulfilled for the robot to function. And from the need, liking and wanting could also emerge in the right environment. Therefore, in this section, we analyze the hypothetical needs of an MFC-powered robot.

Measured Quantity	Proposed Sensor	Equivalent in Human
Chemical Oxygen Demand or Dissolved Biomass	Camera (Color, Transparency)	Caloric Value
pH	H+ Sensitive Electrode	Sourness
Na+ Ions	Na+ Sensitive Electrode	Saltiness
P, N Ions	Ion Sensitive Electrodes	Micronutrients

Table 2. List of sensors proposed for the robotic taste.

3.1. MFC's Need

A robot's needs are usually specified as energy consumption or maintenance schedule. In the case of gastrobots, the needs become more complicated, to a point where a perfect diet is being discussed[35], and possible sources of sustenance in various environments are being explored. It is theorized that a carnivorous diet is preferred, as meat offers higher calorie density than plant-based foods[35]. While we will limit the environmental complexity in this discussion, the caloric density of the food is crucial, as the fuel cell can be loaded only so many times in a period of time, and higher organic matter density produces more power[36].

Moreover, the MFC needs a specific environment to facilitate the growth of the bacteria. It requires specific substances which are not a source of energy, reminiscent of the role of micronutrients in the previous discussion, and particularly their role in maintaining the microbiome. For example, salinity inside the MFC has a huge effect on power production, with a solid peak at 1% of salinity, but falling down to zero at salinity of 2%[37]. Similarly, acidity is also a big factor as the bacteria grow best in pH around 7[38]. Other ions will also influence the power generated: Higher ionic load would usually increase generation[39]. The adequate temperature of food is required as it also influences power production[40], an effect well documented due to the seasonality of power production in sewage treatment plants[41].

The last need we discuss is the need to prevent fouling of the MFC membrane. Fouling is a process of growth of microorganisms on the membrane or accumulation of other substances that limit the membrane's permittivity. Fouling shortens the life of MFC[42] and is one of the likely factors that can cause the shutdown of the robot. One of the factors that increase fouling is over-feeding[43]. In such a case extracellular polymeric substances(EPS) build up on the cell membrane. Excess nitrogen and phosphorus concentration also contribute to accelerated fouling of the membrane[44].

.

4. Robot and Environment Design

As we are focusing only on a few of the challenges for gastrobots the environment will be very specific in some areas, while very basic in others. Moreover, due to the focus on taste, we keep most of the robot and environment design the same as in the case of Eco-BotIII[8]. All the adaptations that we would introduce for the sake of the thought experiment are listed in this section.

4.1. Taste Sensing

Robotic taste is a recent development[4], [45], and has so far been used to cook dishes catering to specific human preferences. In this case, we choose a combination of sensors that can sense the substances needed by the fuel cell. Moreover, this array of sensors must be able to sense substances that humans taste. It is crucial, as the aim of the theorized system is to serve as a sense of taste for a robotic chef in the future.

Starting with the measurement of biomass concentration in the feed is crucial. Usually, it is estimated by measurement of Chemical Oxygen Demand(COD) in a sample, which is traditionally measured by incubating it in a warm environment. Currently, more compact sensors are available for COD measurement[46], and COD measurements on a large scale are possible[47]. However, COD can also be estimated using factors such as the colour or transparency of the sample. This is the solution we propose for our setup, as the robot can simply learn to like samples of specific colours. Saltiness, as well as the concentration of each ion that was found critical, can be measured by ion-sensitive electrodes. This includes the measurement of pH. Overall ionic strength should also be measured[48]. A conductance sensor is also a good candidate for this task. The recommended sensors and their relation to human taste are summarised in Table 2 2. The use of biosensors for specific substances is also an option but not recommended here due to their inability to work in a messy food environment.

4.2. Feeding Mechanism

Feeding mechanisms for gastrobots are not well researched. There was a successful attempt to build a robot that hunted slugs[49]. Moreover, it was shown that mastication can improve taste sensing for robots[45]. Given the scarcity of data, we do not attempt a complicated feeding mechanism for our hypothetical robot but will use liquid feeds only. The robot will, after using its sensors on presented samples, decide which sample and how much it should consume by pumping it inside its MFC, while evacuating ('spitting out') unwanted or excess liquid.

4.3. Energy Storage

Due to the nature of MFC the energy production is slow and the delivered voltage is dynamic and unpredictable. These characteristics make usage of switching DCDC converters problematic, and the output of charge pumps inconsistent. Therefore, we find that using a capacitor battery is the best approach because it is very flexible as comes to power sources. Moreover, based on previous experiments done on Ecobot-III[8] platform, we notice that the practical voltage of the capacitor is varied inside a rather small range. Therefore, we think that measuring the current output of the MFC is a sufficient measure of generated power. Moreover, the MFC current(I_{MFC}) measurement can be done indirectly by measurement of capacitor battery voltage(V_{cap}), as these two values are tied by capacitor equation:

$$V_{cap} = \frac{1}{C} \int_t I_{MFC} \, dt$$

, where C is the capacitance of the battery. This eliminates the need for direct current measurement.

5. Need, Like, Want - Artificial Intelligence

5.1. Need

While needs are grounded by the MFC's needs, they must be represented for the machine learning of taste, which will be specified in the next section. For now, we assume that the MFC cell has an extensive set of sensors inside the MFC, that collect information about the internal state. We require a way of translating these measurements into signals that can be read by neural networks and explain well the current need for the MFC. Moreover, the context of these sensors, or the values they measure is not known outside of this part of the system as it represents the physical needs only. We propose two possible approaches to this problem.

The first approach is data-driven and is based on an exhaustive search of all possible combinations of nutrients inside the cell. The resulting generated power and resulting fouling of MFC's membrane would be recorded on a scale, and become the need satisfaction value. This function can be used to find the most urgent need just by computing a gradient, and this information could be the output of the block representing needs. This solution, while elegant and with analogical concepts in machine learning, is nonetheless hard to implement. Firstly, each of the parameters specifying the composition of the feed would need to be quantized with the relevant resolution, but a higher resolution would require more experiments as the MFC need to be physically tested for every parameter value separately. Moreover, the system would have at least one parameter for every "taste" sensor, making the number of experiments grow further. If we use **n** parameters to describe the needs, and each of them was allowed **m** values, the number of required experiments **e** is specified by the following equation:

$$e = m^n$$

, making the experimentation not feasible. Regression is an approach that could help to reduce the number of experiments needed by filling the gaps between the parameters. However, the cost of it is adding another layer of learning to the system. Moreover, this learning would be applied to the part of the system that specifies needs, and this part of the system that provides a grounding to physical constraints. Therefore, we decide that this approach is not feasible due to a possible reality gap.

The other approach is to hard code the known needs for the MFC, making use of experiments already within the literature. This approach comes with numerous advantages. Firstly, the best-performing mixtures of feed are well-studied and can be expressed as an equation. Moreover, the system becomes more explainable and well-grounded to the physical requirement of the cell. One possible critique of this approach could be based on the fact that we can program only the things that were reported on in the literature. We still find it the best solution. Moreover, this solution is very similar to the human body where sensing and perception are very limited too. This means that not all needs are signalled by the body and not all nutrients can be sensed in the food. In this place, it should be noted that the same situation happens in the human body. Effectively, this approach would output binary signals corresponding to each of the parameters known to change the energy generation from MFC, 0 when the value is in range and 1 when the value is outside of the safe range. For the sake of this argument we choose to follow with hard coding known MFC needs as they are well known, and do not bring additional layers to the discussion.

5.2. *Like*

Liking is a long-term integration of the value of consumed food. This value is malleable, subject to change over the course of a lifetime, but stable on a day-to-day timescale. Therefore, the learning rate of this part of the system should be low, regardless of the actual implementation. We also want a system that takes into account all advantages and disadvantages of consuming a particular food, not only nutritional ones. Moreover, in this case, perception of food is limited, so liking is realistically a learned assignment of value to experienced taste. It can be considered as a black box that returns the expected value of consuming the food currently tasted.

To get as broad a take on these needs as possible, we propose to measure the value of particular food by integrating the generated current over a time window after consumption $t_{digestion}$. $t_{digestion}$ is itself a parameter of the model and should be set accordingly to the throughput of the food in MFC. The value V of a given taste, masked as a vector T, can be then computed with the following equation:

$$V = \int_{t_0}^{t_0 + t_{digestion}} I_{MFC}\, dt$$

, where I_{MFC} is current generated by MFC, and t_0 is consumption time. These values will always be used to update the liking model after the digestion time has elapsed.

The next step is to find a generalization of previously obtained values to all possible tastes that can be encountered, without the need to experience all of them. Therefore, multivariate regression is necessary. The multivariate regression model is of the following form:

$$y \sim f(X, B)$$

where y is the estimated function, X is a set of variables, and B is a set of adjustable parameters. This is the broadest form of multivariate regression, and in our case, the estimated function in food value V. This estimation is done base on the tasting sensors values X. Therefore, for our robot X becomes:

$$X = \{COD, pH, saltiness, [P], [N]\}$$

, where [N] denotes the concentration of nitrogen ions. Finally, we need to specify B, and for this, we need to agree on how to model function V to find parameters B. This is a problem with many possible solutions, but the chosen solution must make a model that represents physical reality well. We start with making the following observations:

- All the measured values have an optimum.
- A plateau around optimum is preset, and the current generated steeply falls off away from the optimum.
- Not meeting just one condition is enough to hamper power generation.

Taking these facts into account, we consider a sum of quadratic functions. Each of these can be specific to one of the measurements only. The whole V function will be then estimated as:

$$y \sim b_0 + b_1 * COD + b_2 * COD^2 + ... + b_{n-1} * [N] + b_n * [N]^2$$

where parameters b_1 through b_n need to be found by regression. Assuming this form of regression is somewhat arbitrary and likely not the most optimal, but it mirrors the natural experience of liking in humans, where some of its aspects are innate.

5.3. Want

Wanting is the final system that decides the action. It takes into consideration the unmet physical needs, palatability of the food and external signals. Working in our example of three foods presented it should produce a signal specifying how much should be ingested from each of the foods. Therefore, a fully connected neural network with three output neurons is preferable. An example network with two hidden layers is shown in Figure 5. MFC's immediate needs are signalled by input neurons n_1 through n_n, that signal a binary value 1 or 0. The taste assessment of each of three presented foods is signalled by inputs l_1 through l_3, which take an analogue value between 0 and 1. Finally, any number of inputs c_1 to c_n can be used as a description of the context around the robot. A relatively high learning rate should be used for this network to allow quick adjustment to the changes in the environment.

6. Aligning Robot and Human Taste - Acquired Taste

As discussed before, acquired taste is commonly found among humans, in various societies. Due to the robot's taste similarity to human taste, it is also possible to acquire a taste based on external or "social" cues. We simulate these "cues" as external energy injection, that is indistinguishable for our algorithms from energy produced by digestion. As discussed in Section 4.3 we use generated current as a measure of gratification, therefore this injection will be in the form of current, by the circuitry shown in Figure 6. This is effectively an extra, not nutritional benefit of consuming a certain food. These injections can be both positive and negative, similar to social cues in human society.

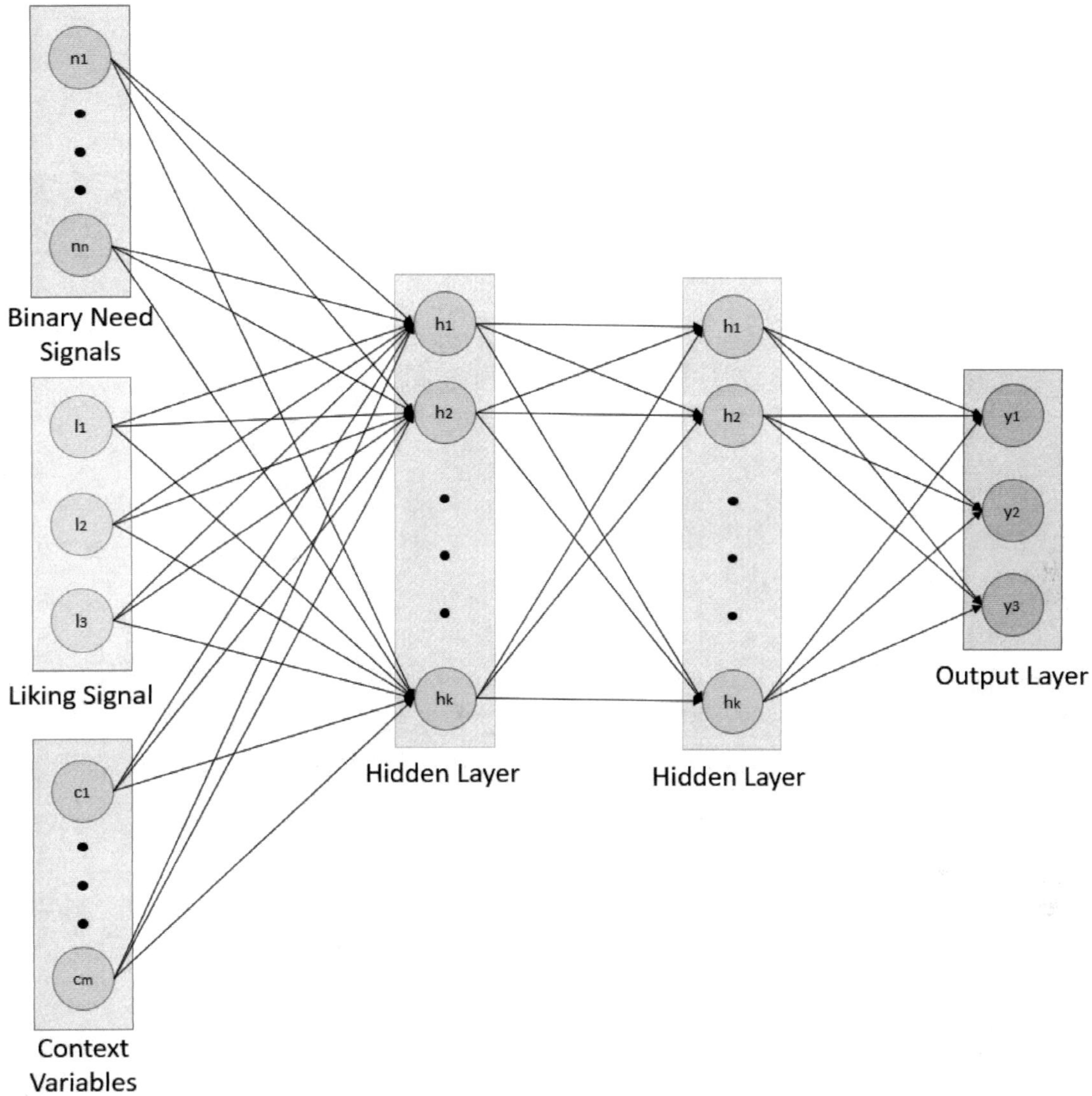

Figure 5. Design of a neural network that simulates human wanting process. It uses signals of liking the taste, nutritional needs and information about the context of the meal. Three output neurons decide which of the three presented feeds to ingest.

6.1. Acquired Taste Teaching

The teaching of the acquired taste would be done by producing a chosen food and presenting it to the robot. When the robot ingests the given food we start injecting an extra current into the system for the time $t_{digestion}$, discussed in the previous section. This should cause an increase in liking of this specific food, and also liking of all other foods ingested together with the target food. Moreover, all other foods ingested more recently than t_{digest} before the ingestion of the externally favoured food will be partially favoured too. Therefore, the learning will be not clear cut, just as in humans, but with enough food presented to the robot, it should result in liking the preferred food more than its nutritional value suggests.

We would also like to suggest the most likely way to use this way of learning the taste. The power generation from MFC is not likely to be enough to power a robot that has enough

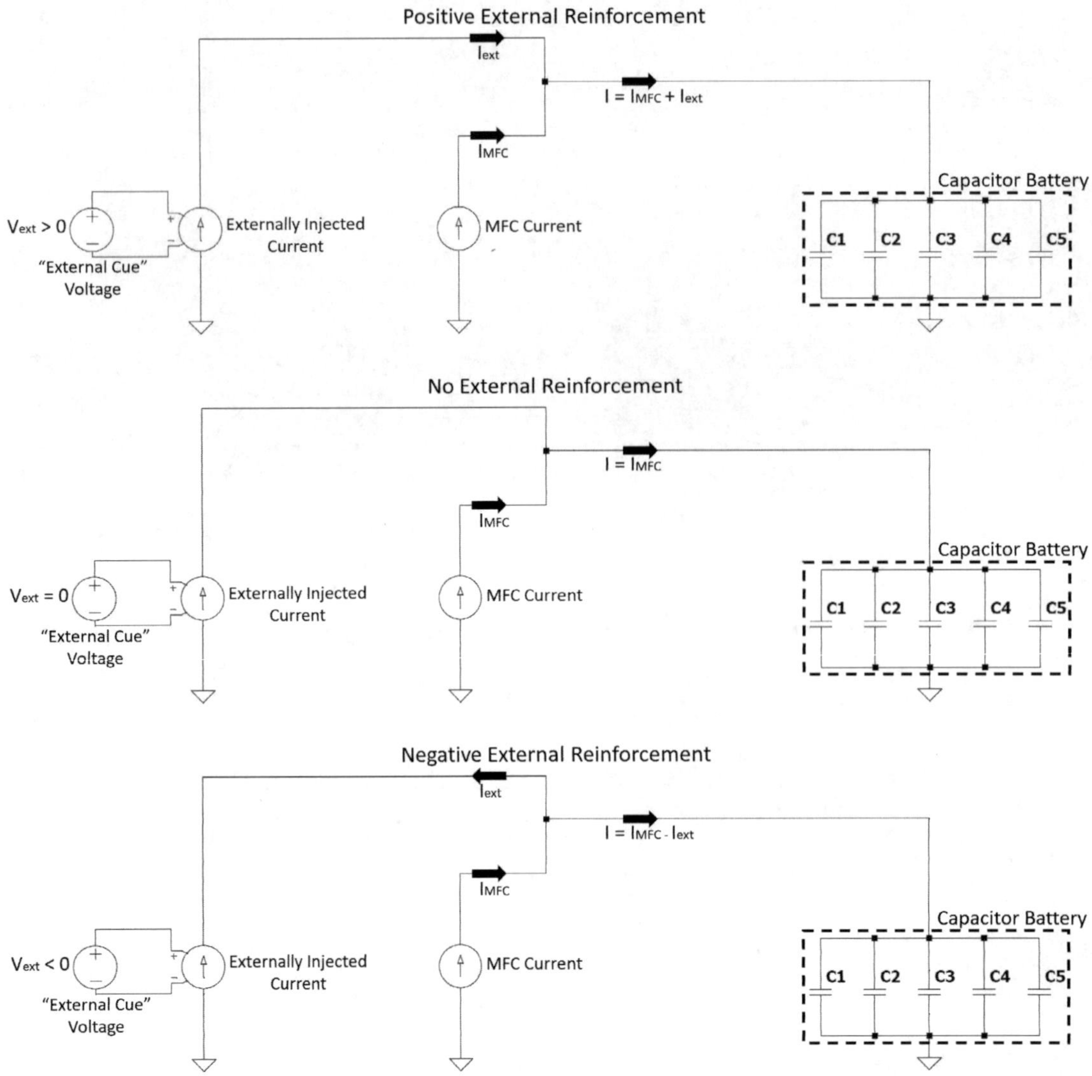

Figure 6. Figure showing microbial fuel cell with a parallel current source, which simulates non-nutritional cues, and therefore makes some foods valued more than their nutritional value.

dexterity, carrying capacity, and speed to become a robotic chef. Therefore, we would suggest operating the proposed system as an auxiliary device until the MFC technology progresses. The system can effectively act as an external controller of a robotic chef, where the MFC provides a symbol grounding for the taste AI.

7. Discussion and Conclusions

This chapter has discussed the previous experiments with robots that power themselves by ingesting organic matter - gastrobots. While this technology was initially thought of as a way to make robots independent of batteries or charging points to operate them in long deployments, this chapter proposes to repurpose this technology as a way of solving the grounding problem for robotic taste by binding the taste to the physical survival of the robot.

The proposed binding is inspired by the natural way this binding is done by humans, that is by mechanisms of needing, liking and wanting. Needing covers the short-term critical requirements, liking is the integration of benefits that exact food brought over the lifetime and wanting brings the situational context. In a robot the needs should be hardcoded bounds on the chemical composition of MFC, liking is the integration of produced current and wanting should be learned based on the previous outputs and the state of the environment.

The chapter also shows how current injection can simulate non-nutrition rewards and punishments for the robot, enabling it to simulate things like peer and social pressure by current injection. It is also proposed to use MFC simply as a way to ground taste learning for robotic chefs by using it as an auxiliary device rather than the main source of power.

Acknowledgments
We are grateful for the support from Beko plc and Engineering and Physical Sciences Research Council (EPSRC) Agriforwards CDT Project [EP/S023917/1] who made this work possible.

References
[1] Q. Ouyang, J. Zhao, Q. Chen, and H. Lin, "Classification of rice wine according to different marked ages using a novel artificial olfactory technique based on colorimetric sensor array," eng, *Food chemistry*, vol. 138, no. 2-3, pp. 1320–1324, 2013, ISSN: 0308-8146.

[2] X. Tian, J. Wang, Z. Ma, M. Li, and Z. Wei, "Combination of an e-nose and an e-tongue for adulteration detection of minced mutton mixed with pork," *Journal of Food Quality*, vol. 2019, 2019.

[3] L. Gil, J. M. Barat, E. Garcia-Breijo, *et al.*, "Fish freshness analysis using metallic potentiometric electrodes," *Sensors and Actuators B: Chemical*, vol. 131, no. 2, pp. 362–370, 2008.

[4] G. Sochacki, J. Hughes, S. Hauser, and F. Iida, "Closed-loop robotic cooking of scrambled eggs with a salinity-based 'taste' sensor," in *2021 IEEE/RSJ International Conference on Intelligent Robots and Systems (IROS)*, 2021, pp. 594–600.

[5] K. Junge, J. Hughes, T. G. Thuruthel, and F. Iida, "Improving robotic cooking using batch bayesian optimization," *IEEE Robotics and Automation Letters*, vol. 5, no. 2, pp. 760–765, 2020.

[6] S. Wilkinson, ""gastrobots"—benefits and challenges of microbial fuel cells in foodpowered robot applications," *Autonomous Robots*, vol. 9, pp. 99–111, Jan. 2000.

[7] I. Ieropoulos, C. Melhuish, J. Greenman, and I. Horsfield, "Ecobot-ii: An artificial agent with a natural metabolism," *International Journal of Advanced Robotic Systems*, vol. 2, no. 4, p. 31, 2005.

[8] I. Ieropoulos, J. Greenman, C. Melhuish, and I. Horsfield, "Ecobot-iii: A robot with guts," *ALIFE*, Jan. 2010.

[9] I. Ieropoulos, C. Melhuish, and J. Greenman, "Artificial metabolism: Towards true energetic autonomy in artificial life," in *European conference on artificial life*, Springer, 2003, pp. 792–799.

[10] W. B. Williams, E. K. Dreischerf, and T. N. Yath, "Marine gastrobot final design report," 2017.

[11] H. Philamore, J. Rossiter, A. Stinchcombe, and I. Ieropoulos, "Row-bot: An energetically autonomous artificial water boatman," in *2015 IEEE/RSJ International Conference on Intelligent Robots and Systems (IROS)*, 2015, pp. 3888–3893.

[12] A. D. Craig, "How do you feel? interoception: The sense of the physiological condition of the body," eng, *Nature reviews. Neuroscience*, vol. 3, no. 8, pp. 655–666, 2002.

[13] "Government dietary recommendations," 2016, Recommendations by UK government, Accessed: August 2022. [Online]. Available: https://assets.publishing.service.gov.uk/government/uploads/system/uploads/attachment_data/file/618167/government_dietary_recommendations.pdf.

[14] A. H. Association, S. S. Gidding, B. A. Dennison, *et al.*, "Dietary recommendations for children and adolescents: A guide for practitioners," *Pediatrics*, vol. 117, no. 2, pp. 544–559, 2006.

[15] V. Onywera, F. Kiplamai, P. Tuitoek, M. Boit, and Y. Pitsiladis, "Food and macronutrient intake of elite kenyan distance runners," eng, *International journal of sport nutrition and exercise metabolism*, vol. 14, no. 6, pp. 709–719, 2004, ISSN: 1526-484X.

[16] D. Benton, "Micro-nutrient supplementation and the intelligence of children," eng, *Neuroscience and Biobehavioral Reviews*, vol. 25, no. 4, pp. 297–309, 2001, ISSN: 0149-7634.

[17] A. Shenkin, "Micronutrients in health and disease," eng, *Postgraduate Medical Journal*, vol. 82, no. 971, pp. 559–567, 2006, ISSN: 0032-5473.

[18] A. Hunt, D. Harrington, and S. Robinson, "Vitamin b12 deficiency," *Bmj*, vol. 349, 2014.

[19] N. I. of Health *et al.*, "Vitamin b12: Fact sheet for health professionals," *Office of Dietary Supplements*, 2020.

[20] T. V. Kral, "Effects on hunger and satiety, perceived portion size and pleasantness of taste of varying the portion size of foods: A brief review of selected studies," eng, *Appetite*, vol. 46, no. 1, pp. 103–105, 2006, ISSN: 0195-6663.

[21] D. J. Mela, "Why do we like what we like?" *Journal of the Science of Food and Agriculture*, vol. 81, no. 1, pp. 10–16, 2001.

[22] E. H. Zandstra and C. de Graaf, "Sensory perception and pleasantness of orange beverages from childhood to old age," *Food Quality and Preference*, vol. 9, no. 1, pp. 5–12, 1998, ISSN: 0950-3293.

[23] J. R. Ganchrow, J. E. Steiner, and M. Daher, "Neonatal facial expressions in response to different qualities and intensities of gustatory stimuli," *Infant Behavior and Development*, vol. 6, no. 4, pp. 473–484, 1983, ISSN: 0163-6383.

[24] C. De Graaf and E. H. Zandstra, "Sweetness intensity and pleasantness in children, adolescents, and adults," *Physiology & behavior*, vol. 67, no. 4, pp. 513–520, 1999.

[25] L. L. Birch, L. McPhee, B. Shoba, E. Pirok, and L. Steinberg, "What kind of exposure reduces children's food neophobia?. looking vs. tasting," eng, *Appetite*, vol. 9, no. 3, pp. 171–178, 1987, ISSN: 0195-6663.

[26] D. A. Zellner, P. Rozin, M. Aron, and C. Kulish, "Conditioned enhancement of human's liking for flavor by pairing with sweetness," *Learning and Motivation*, vol. 14, no. 3, pp. 338–350, 1983.

[27] R. D. Mattes, "Learned food aversions: A family study," *Physiology & behavior*, vol. 50, no. 3, pp. 499–504, 1991.

[28] E. Derbyshire, "Micronutrient intakes of british adults across mid-life: A secondary analysis of the uk national diet and nutrition survey," *Frontiers in nutrition*, vol. 5, p. 55, 2018.

[29] M. Mazzocchi and W. B. Traill, "Calories, obesity and health in oecd countries," *Applied Economics*, vol. 43, no. 26, pp. 3919–3929, 2011.

[30] K. Kim, Y. Cheong, and L. Zheng, "The current practices in food advertising: The usage and effectiveness of different advertising claims," *International Journal of Advertising*, vol. 28, no. 3, pp. 527–553, 2009.

[31] A. Dickson, M. Gehrsitz, and J. Kemp, "Does a spoonful of sugar levy help the calories go down? an analysis of the uk soft drinks industry levy," 2021.

[32] C. P. Herman, D. A. Roth, and J. Polivy, "Effects of the presence of others on food intake: A normative interpretation," eng, *Psychological bulletin*, vol. 129, no. 6, pp. 873–886, 2003, ISSN: 0033-2909.

[33] S. J. Goldman, C. P. Herman, and J. Polivy, "Is the effect of a social model on eating attenuated by hunger?" *Appetite*, vol. 17, no. 2, pp. 129–140, 1991.

[34] P. Pliner, S. Salvy, T. Leone, and D. Jarrin, "Eating and social influence: Effects of multiple models and free-eating situations on intake," eng, *Appetite*, vol. 49, no. 1, pp. 320–320, 2007, ISSN: 0195-6663.

[35] S. Wilkinson, "Hungry for success - future directions in gastrobotics research," eng, *Industrial robot*, vol. 28, no. 3, pp. 213–219, 2001, ISSN: 0143-991X.

[36] A. Adekunle, V. Raghavan, and B. Tartakovsky, "Real-time performance optimization and diagnostics during long-term operation of a solid anolyte microbial fuel cell biobattery," *Batteries*, vol. 5, no. 1, 2019, ISSN: 2313-0105.

[37] O. Adelaja, T. Keshavarz, and G. Kyazze, "The effect of salinity, redox mediators and temperature on anaerobic biodegradation of petroleum hydrocarbons in microbial fuel cells," *Journal of Hazardous Materials*, vol. 283, pp. 211–217, 2015, ISSN: 0304-3894.

[38] M. Behera, P. S. Jana, T. T. More, and M. Ghangrekar, "Rice mill wastewater treatment in microbial fuel cells fabricated using proton exchange membrane and earthen pot at different ph," *Bioelectrochemistry.*, vol. 79, no. 2, pp. 228–233, 2010, ISSN: 1567-5394.

[39] H. Liu, S. Cheng, and B. E. Logan, "Power generation in fed-batch microbial fuel cells as a function of ionic strength, temperature, and reactor configuration," *Environmental Science & Technology*, vol. 39, no. 14, pp. 5488–5493, 2005.

[40] J. Greenman, A. Mendis, J. You, I. Gajda, I. Horsfield, and I. Ieropoulos, "Microbial fuel cell based thermosensor for robotic applications," *Frontiers in Robotics and AI*, vol. 8, 2021, ISSN: 2296-9144.

[41] T. Miyoshi, T. Tsuyuhara, R. Ogyu, K. Kimura, and Y. Watanabe, "Seasonal variation in membrane fouling in membrane bioreactors (mbrs) treating municipal wastewater," *Water Research*, vol. 43, no. 20, pp. 5109–5118, 2009, ISSN: 0043-1354.

[42] B. Wu and A. G. Fane, "Microbial relevant fouling in membrane bioreactors: Influencing factors, characterization, and fouling control," *Membranes*, vol. 2, no. 3, pp. 565–584, 2012.

[43] B. Wu, S. Yi, and A. G. Fane, "Microbial community developments and biomass characteristics in membrane bioreactors under different organic loadings," *Bioresource Technology*, vol. 102, no. 13, pp. 6808–6814, 2011, ISSN: 0960-8524.

[44] B. Wu, S. Yi, and A. G. Fane, "Effect of substrate composition (c/n/p ratio) on microbial community and membrane fouling tendency of biomass in membrane bioreactors," *Separation Science and Technology*, vol. 47, no. 3, pp. 440–445, 2012.

[45] G. Sochacki, A. Abdulali, and F. Iida, "Mastication-enhanced taste-based classification of multi-ingredient dishes for robotic cooking," *Frontiers in Robotics and AI*, vol. 9, 2022, ISSN: 2296-9144.

[46] D. T. K. Hue, T. Shiba, Y. Maeda, and N. Takenaka, "A 3-step chemiluminescence method for chemical oxygen demand measurement with dichromate oxidizing reagent," eng, *Analytical methods*, vol. 9, no. 39, pp. 5797–585, 2017, ISSN: 1759-9660.

[47] C. Niu, Y. Zhang, Y. Zhou, K. Shi, X. Liu, and B. Qin, "The potential applications of real-time monitoring of water quality in a large shallow lake (lake taihu, china) using a chromophoric dissolved organic matter fluorescence sensor," eng, *Sensors (Basel, Switzerland)*, vol. 14, no. 7, pp. 11 580–11 594, 2014, ISSN: 1424-8220.

[48] E. Biemans-Oldehinkel, N. A. Mahmood, and B. Poolman, "A sensor for intracellular ionic strength," *Proceedings of the National Academy of Sciences*, vol. 103, no. 28, pp. 10 624–10 629, 2006.

[49] I. Kelly, O. Holland, and C. Melhuish, "Slugbot: A robotic predator in the natural world," in *Proceedings of the Fifth International Symposium on Artificial Life and Robotics for Human Welfare and Artificial Liferobotics*, Citeseer, 2000, pp. 470–475.

From Disembodiment to Embodiment in Artificial Intelligence and Psychology – Parallels in Thinking

Sheila L Macrine[1] and Jennifer MB Fugate[2]

[1]University of Massachusetts-Dartmouth, USA
[2]Kansas City University, USA

smacrine@umassd.edu

Abstract. This paper briefly traces how both Computer Science and Psychology grew- in tandem- to share similar notions of Embodied Cognition. It concludes an analysis of how the two fields are uniting in their work on Embodied Artificial Intelligence, and future challenges for the field.

1. Introduction

In 1936, Alan Turing first posited that if humans use available information and reason to solve problems and make decisions, then why can't machines do the same thing? Turing went on to discuss how to build intelligent machines and how to test computer intelligence (Turing, 1950). Five years later, the term "Artificial Intelligence" (AI) was coined at Dartmouth College (McCarthy et al., 1955, 2006). These ideas, written over 72 years ago, are still considered a watershed event in computer science and continue to influence current AI research (Anyo, 2017; Leake, 2006). In this paper, we briefly trace how both Computer Science and Psychology grew- in tandem- to share similar notions of Embodied Cognition. We end with how the two fields are uniting in their work on Embodied Artificial Intelligence, and future challenges for the field.

2. Artificial Intelligence and Embodied Cognition

The Turing computer became the central model for cognitive science during the 1950s and 1960s (Sprevak, 2017), and was reinforced by Putman's Computational Theory of Mind (CMT) (1961). Not long after, Jerry Fodor (1975, 1983) advocated for mental processes consisting of computational operations that were symbolic representations within the brain. Ultimately, his ideas were dominated largely by psychological nativism, and his legacy became synonymous with the modularity of the brain and domain specificity (Fodor, 1975, 1983, 1990). During this time, developments in linguistics (see Chomsky, 1965), in cognitive science and in artificial intelligence joined forces to usher in the cognitive revolution (Smelser & Baltes, 2001).

In 1980s, John Hopfield and David Rumelhart promoted *deep learning* techniques that allowed computers to learn using experience/input (Hopfield, 1984; Rockwell, 2017). Deep learning and the idea of parallel distributed processing (Rumelhart & McClelland, 1987; Rumelhart & Ortonty, 1977; Rumelhart et al., 1986) became the basis for questioning traditional models of cognition (disembodied) in favor of models of embodied cognition. Similarly, deep learning solved AI's central problem of representation (Hopfield, 1984) and combated the failings of representational methods (Duffy & Joue, 2000). Specifically, this meant that learning could emerge from multilayered distributed processing models (through rapid accelerations and decelerations) and exhibit stage-like effects, suggesting that emergence could occur from gradual processes (McClelland, 1989). Finally, the development of dynamical systems approaches (Smith & Thelen, 2003; Thelen & Smith, 1996) advanced the field further.

Andy Clark and David Chalmers (1998) put forth the position that human cognition is not brain-bound, but rather that cognitive processes are based on the active role of the environment. Further, Clark (2008), elaborating on J.J. Gibson's work on affordances (Gibson, 1979, see below), said there is no need for mental representations of the world if a body's actions allow it to navigate the world directly. He added an organism's thinking doesn't happen only in its head but that "certain forms of [human] cognizing include inextricable tangles of feedback, feed- forward and feed-around loops: loops that promiscuously crisscross the boundaries of brain, body, and world" (Clark, 2008, p. xxviii). Presciently, in 2001, Clark wrote that the new millennium should radically reconfigure our image of human rationality within the Cognitive Sciences, drawing together embodied, distributed, and enacted cognition in the real-world with robotics and artificial intelligence.

Similarly, Brooks (1991, 1999), also influenced by J.J. Gibson's work, challenged the view that intelligence was the result of bigger brains and larger central processors to encode elaborate representations. Rather, he argued that millions of years of evolution within the environment was responsible for the complexity of human responding. He went on to champion the position that robots could be designed in a bottom-up manner, with sensors and movement through their environment to navigate the world (see Morgan, 2018).

3. Psychology and Embodied Cognition

Much like AI, Psychology in the 1960s, ushered in the 'cognitive revolution,' challenging behaviourist theory and endeavouring to discover how the brain functions and learns. The mind was now thought of as running on the hardware of the brain (Dennett, 2014). Central to these ideas was that of "representation", in which internal symbols "stood in" for the environment outside the body of the organism. Moreover, such mental representations within the brain were usually thought of as abstractions of the original information. This computer analogy, an updated version of the age-old Cartesian model, continued to promote the disembodied view of brain-bound cognition (see Fugate et al., 2019; Macrine & Fugate, 2020, 2021, 2022; Thompson & Cosmelli, 2011). The body was seen as a "passive" observer of the brain, and necessary only in the execution of motor actions (Macrine & Fugate, 2020).

Beginning in the late 1970s, however, this view began to change, but not without input from some famous predecessors. John Dewey (from works between 1925-1953), echoing William James (1890), suggested that higher-order cognitive functions were adaptations generated by interactions with the world. J.J. Gibson's "ecological theory" (1979) married both phenomenological (i.e., the *subjective experience*) and naturalistic perspectives. Gibson (1979) argued that perception was direct, and the environment meaningful. He advocated against representation and argued rather that organisms make sense of their world through direct interaction with it. For Gibson, affordances allowed an organism to interact in specific "smart" ways with its world: An organism and its adaptive environment have a "goodness of fit", and bodies, as well as their perceptual and motor functions, did not evolve independently of one another. Importantly, this co-adaptation ensured little need for an internal executive processing system or representation of information, and the cognitive burden was offloaded to the environment (see also Anderson, 2003; Clark, 1999, Nöe, 2009; Wilson, 2002). Both James and Dewey rejected the "rational psychology" drawn from Cartesian Dualism.

The ideas of action-based learning also mirrored those of earlier developmental and educational psychologists (e.g., Kolb, 1984; Montessori, 1969; Piaget, 1952, 1968; Rogoff, 1990; Vygotsky, 1978). Early work on action-on-thinking can also be seen in sociocultural psychology (e.g., Vygotsky, 1978), as well as in activity theory (e.g., Gal'perin, 1992; Leontiev, 1978), Apprenticeship in Thinking (e.g., Rogoff,

1990), and by a variety of perspectives of learning, activity, and knowledge appropriation (e.g., Brown et al., 1989; Lave & Wenger, 1991; Robbins & Aydede, 2009; Rogoff, 1990; Wilson & Foglia, 2017).

In 1991, Varela, Thompson, and Rosch introduced the world to embodied action and formalized what many considered the original view of Embodied Cognition (EC). According to their theory: 1) Cognition is dependent on experiences that come from having a particular body design (and differences in design among animals); 2) Experiences of the body are embedded in a more encompassing biological, psychological, and cultural context. Specifically, Varela et al. (1991) write: "The perceiver, with its body and sensory and motor systems, will determine what is available to that organism, such that as an organism moves through the world, its motion will produce opportunities for new perceptions and new perception will reveal opportunities for new actions!" (p. 173).

Today, there are many versions of EC. Most reject the computational metaphors of the brain in favour of ones that stress the intertwined nature of sensory and perceptual information with action (Brooks, 1990; Grosso et al., 1995; Pfeifer, 1999). EC argues that an organism's cognition is grounded in both perception and action, such that perception is for action, and deeply dependent upon features of the organism's physical body (e.g., Barsalou, 1999, 2008; see also Clark, 2008; Golonka & Wilson, 2013; Lakoff & Johnson, 1999; Pfeifer & Bongard, 2007; Shapiro, 2011; Willems Francken, 2012). In addition, most theories suggest that bodily sensations interact with physical environments, and - in the case of humans- with their social and cultural contexts (Barsalou, 2010; Núñez & Lakoff, 2005; Wilson, 2002).

While different theories of EC differ in the extent and nature of such representations, central to most of them is the idea of amodal abstraction, in which the latter refers specifically to the representation outside of the sensory-motor areas. To help explain how such activation is used conceptually, Barsalou (1999) developed the Perceptual Symbols Systems (PSS). Accordingly, thinking about action will evoke the same visual stimuli, motor movement, and tactile sensations, etc., that occur during the action itself (Barsalou, 1999, 2003, 2008). The experience can be later recreated (through simulation) without action (i.e., when just thinking).

During the last few decades, theories of EC have created a paradigm shift in cognitive psychology from a reactive model of the brain to a generative and predictive processing model (McNerney, 2011). Today, the mind is no longer conceived of as a set of logical/abstract functions, but as a biological system rooted in bodily experience and interconnected with bodily action with other individuals and the environment (Adenzato & Garbarini, 2012). From this perspective, action and representation are no longer interpreted in terms of the classic physical–mental state dichotomy but are closely interconnected (Marmeleira & Duarte Santos, 2019). Embodied cognition frameworks include embodied (body-based) thinking; embedded (situated) thinking, extended (in the environment) thinking, and enacted (dynamic interaction) thinking (Gallagher's 4Es, as cited in Rowlands, 2010; Newen et al., 2018). Expanding on these ideas, radical enactivism suggests that the body is not just a means but also "an end of having a cognitive system" (di Paolo & De Jaegher, 2017; see also Hutto & Abrahamson, 2022), Thus, cognition is less a matter of mental reasoning and abstract problem solving, than a matter of adaptive self-regulation (di Paolo & De Jaegher, 2017) and enacting in an environment (Hutto & Abrahamson, 2022). Accordingly, the exact way an organism's body performs certain interactions with the environment leads to perceptual recognition (Kumar, 2008).

4. Parallels Between Psychology and AI in Embodied Cognition

The recent developments in cognitive science and neuroscience have contributed to "predictive processing" theories of mind and cognitive function (Clark, 2013; Friston, 2011; Hohwy, 2013). Such approaches are similar in that they emphasize top-down, error-minimizing predictions as the key locus of information processing, as opposed to feed-forward feature recognition, common in more classical computational theories (Allen & Friston, 2018). This embodied turn in AI indicates that the body shapes thinking, and thinking is bound and realized through our embodiment (Pfeifer & Bongard, 2007). Therefore, when thinking is divorced from the body, critical elements are lost. For example, intelligent behaviors can be rapidly learned by agents whose morphologies are adapted to the affordances in the environment (Brooks, 1991; Bongard, 2014; Gupta et al., 2021; Pfeifer & Scheier, 2001). At a more proximal level, this involves consideration of what sensors an agent can use, how its actuators operate, and how its perceptions and the dynamics of its movements are affected by its physical body (Iida & Nuraman, 2016). Pfeifer and colleagues (2007) wrote that the material properties of an organism's musculoskeletal system, the sensor morphology, and the interaction with the environment, all provide an "alternative avenue for tackling the challenges faced by robotics. The tasks performed by the controller in the classical approach are now partially taken over by morphology and materials in a process of self-organization" (p.1098). Thus, true AI will only be achieved when robots can connect sensory and motor skills through a body (see Brooks,1991; Pfeifer, 2001, 2006).

The direction of AI has changed from a purely computational discipline into a highly transdisciplinary system based on the paradigm shift of EC. "With the fundamental paradigm shift from a computational to an embodied perspective, the nature of research topics, the theoretical and technical issues, and the disciplines involved have changed dramatically, or in other words, the "landscape" has changed completely" (Pfeifer & Iida 2004, p.1) This embodied focus has led to the creation of the field of Embodied Artificial Intelligence (EAI). EAI rejects computationalism and its limitations, instead focusing on intelligent systems established through the interaction of an agent with the environment through dynamical predictive processing of sensory-motor activity (Bermudez, 2021; Brooks, 1990; Pfeifer, 1999). The idea here is that intelligence *emerges*, rather than being learned exclusively from datasets of text and images (Duan et al., 2022; Gupta et al., 2021). Supporting this thesis, Hinton (2013) argued that EC continues to have an increasingly influential perspective in the user-experience design fields and stands to fundamentally change the way we think about and design human-computer interaction.

5. Summary and Future Directions: EAI and Evolutionary Robotics

More recently, EAI researchers are looking at human behavior and embodiment, to try to give both hard and virtual agents life-like capabilities (Floreano, et al., 2014, p.1). Pfeifer and Iida (2004) argue that one of the major challenges in EAI is an ecological balance- for example, in both hard and virtual agents there should be a match in the complexity of the sensory, motor, and neural control systems. In the case of hard robotics, they are built of hard materials and electrical motors and lack a sophisticated sensory-motor system and still require an enormous amount of computation (Pfeifer & Iida, 2004).

Our discussion would not be complete without a quick mention of the field of Evolutionary robotics (ER). Within the last ten years, ER as a field provides the tools for simulating embodied agents in a realistic environment to study evolutionary constraints on behaviors as they evolved, namely with respect to the interactions of an organism's morphology with the environment over time. Thus, ER addresses successful adaptations as pre-programmed mutations which lead to the end goal – discarding and replacing failed designs with mutations and combinations for better designs. The idea is to generate an accurate "design landscape" through the application of high-throughput techniques and "tightly coupled digital-experimental systems" (Howard et al., 2022, p.1). Whereas early evolutionary robotics were mainly virtual, today with 3D printing, robotic "bodies" can handle more advanced body-based processing. Several issues still exist within the field, however, including the problem of evolving in "real-time" to a rapid, ever-changing environment to drive learning (Howard et al., 2022). Alongside this push there needs to be a focus on task-based interactive questions, in which an embodied agent performs specific tasks in the real world to unlock new insights (Duan et al., 2022). For instance, an agent might have to boil an egg to learn how long it takes to do so. From this, the action learned would explicate the knowledge to develop generalized task planning, which is crucial to developing general all-purpose AI simulations in the real world (Duan et al., 2022). Understanding the limits of what is possible for an ER is a critical question that remains to be addressed. For example, understanding exactly how morphology and sensory modality impact the ability to learn (i.e., the tradeoffs between power, mass, computing) are still not fully understood, and how these create inductive biases in learning, especially because many robots are built for one environment and fail when presented to a new environment (Roy et al., 2021).

According to Kaelbing (2021), there has been a complex and fruitful historical interplay between human cognition and artificial intelligence fields. Although EAI and ER are in great demand, the idea that AI can learn the way humans interact with their environments from an egocentric perception has not been a contemporary view in this field until quite recently (Duan et al., 2022). That said, Bird and colleagues (2020) argue that neither EAI or ER are meant to replace humans or human decision-making, but rather to improve quality of life, improve education, policy, and practice as well as moral and ethical considerations for the greater good of society. We believe that the guiding principles of EC and EAI can continue to share and enhance the future for understanding intelligent humans and robots.

Table 1

Human Embodied Cognition (EC)	***Embodied Artificial Intelligence (EAI)***
→Rejects computer model	*→Rejects computational model*
→Rejects disembodiment (Cartesian Dualism)	*→Rejects traditional disembodied AI*
→Gibsonian affordances (Gibson, 1979)	*→ iGibson-generative functions*
→Affordances (Borghi, 2012; Mazzuca, et al., 2021)	*→ Affordances in AI , (Chen et al., 2020)*
→Behavioral Adaptations	*→ Adaptations, Modeling, and Simulation*
→ Categorization (learning models) (Rosch et al., 1976)	*→ Categorization - image classification (Krizhevsky, et. al. 2017)*
→Vision perception, visuo-motor activation & object recognition in Embodied Cognition (Friston, 2018, 2022; Matheson et al., 2015)	*→ Embodied visual recognition (Zhu et al., 2021; Fei-Fei & Krishna, 2022).* *→ ImageNet - visual object recognition (Fei-Fei, 2021)*
→Mind-body-environment; multimodal systems (e.g., gestures, gaze, language) (e.g., Goodwin, 2003; 2010)	*→Simulated agents and robots learning through their environment (Muratore et al., 2022)*
→Merleau-Ponty (1978) *→Embodied Cognition (Shapiro, 2011)* *→Embodied Learning through Adaptations, Affordances, and Assimilations*	*→ Embodied Artificial Learning through Modeling and Simulation* *→ Demonstrates human-like adaptation of affordances (e.g., Laio et al., 2022)*
→ Embodied Language, sensorimotor simulations of natural language processing (e.g., Glenberg & Kaschak, 2002). *→ Embodied Semantics (Aziz-Zadeh & Ivry, 2009)*	*→ Rejects fixed morphology in favor of Natural Language Processing (NLP) – in situ morphological adaptation (Nygaard et al., 2021)* *→ Data labeling: Informative labels provide context to facilitate future learning (e.g., Miller, 2021)*
→Neuroscience-inspired approaches to neural brain functioning (predictive processing *→ Embodied Cognition and Mirror Neurons* *→Sensory and motor systems (e.g., Clark, 2008);* *→Neural/psychological primitives (Barrett, 2017)* *→Offloading cognition (Clark & Chalmers, 1998; Wahn, et al., 2023)* *→Embodied Emotion theories (Cañamero, 2021; Fugate & Wilson-Mendenhall , 2022)*	*→EAI dynamical neural networks(e.g., Bartolozzi et al., 2022), artificial neural networks -ANNs (Chauhan et al., 2018)* *→Morphological computation, sensory–motor coordination, and developmental embodied cognition (Cangelosi, Bongard, Fischer, & Nolfi, 2015).* *→Morphological computation (Müller & Hoffman, 2017).* *→Emotion AI; Emotion Robotics;*
→ Evolutionary-based cognitive adaptations (e.g., Lakatos & Janka, 2008; L. Barrett, 2018) *→ Evolutionary Human Intelligence* *→Interoceptive inference (Owens et al, 2018)* *→Evolutionary Robotics (Bongard, 2013)*	*→Evolution of Embodied AI Intelligence: Deep Evolutionally Reinforced Learning (DERL)* *→Unimals, evolutionary virtual robots, self-learning, in-silico simulation; evolving diverse agent morphologies (Gupta et al., 2021a,b)*
→Embodied Human Intelligence (Hughes & Abdulali, 2022; Pfiefer & Bongard, 2006)	*→ Embodied Artificial Intelligence (Pfiefer & Bongard, 2006)*
→Active Inference (Friston, 2012, Friston et al, 2017) *→The Free Energy Principle (Parr, et al., 2022)* *→Bayesian model -predictive coding (Blei, et. al, 2017)*	*→ Active inference algorithms* *→The Free Energy Principle- bio-inspired robotics* *→ Bayesian model predictive coding (BMPC)- (Lanillos et. al, 2021)*

→Embodied Design for Learning (Abrahamson, Worsley, Pardos & Ou, 2021) *Embodied Cognition and Enactivism (Clark & Chalmers, 1998)*	*→Embodied Artificial Intelligence for Learning (Gupta, et al., 2021)* *→Building and Working in Environments for Embodied AI(Sims, 2022)* *→Embodied Design of robots with neuromorphic technologies (Bartolozzi, et al., 2022)*

References

Abrahamson, D., Worsley, M., Pardos, Z., & Ou, L. (2021). Learning analytics of embodied design: Enhancing synergy. International Journal of Child-Computer Interaction, 100409. https://doi.org/10.1016/j.ijcci.2021.100409

Adenzato M., & Garbarini, F. (2012). Embodied Cognition. In N.M. Seel (Eds.), *Encyclopaedia of the sciences of learning.* Springer. https://doi.org/10.1007/978-1-4419-1428- 6_900

Allen, M., & Friston, K.J. (2018). From cognitivism to autopoiesis: Towards a computational framework for the embodied mind. *Synthese*, *195*(6), 2459-2482.

Anderson, M.L. (2003). Embodied cognition: A field guide. *Artificial intelligence*, *149*(1), 91- 130.

Anayo, R. (2017, August 28). The History of Artificial Intelligence. *Science in the News, Harvard Graduate School of the Arts and Sciences Special Edition on Artificial Intelligence.* https://sitn.hms.harvard.edu/flash/2017/history-artificial-intelligence/

Aziz-Zadeh, L., & Ivry, R. B. (2009). The human mirror neuron system and embodied representations. *Advances in experimental medicine and biology*, *629*, 355–376. https://doi.org/10.1007/978-0-387-77064-2_18

Barrett, L. F.(2018). The Evolution of Cognition: a 4E Perspective. In Newen, Albert, Leon De Bruin, and Shaun Gallagher (eds), *The Oxford Handbook of 4E Cognition*, Oxford Library of Psychology. https://doi.org/10.1093/oxfordhb/9780198735410.001.0001

Barrett, L.F. (2017). The theory of constructed emotion: an active inference account of interoception and categorization, *Social Cognitive and Affective Neuroscience*, Volume 12, Issue 1, January 2017, Pages 1–23, https://doi.org/10.1093/scan/nsw154

Barsalou, L.W. (1999). Perceptual symbol systems. *Behavioural and Brain Sciences*, *22*(4), 577-660.

Barsalou, L.W. (2008). Grounded cognition. *Annual Review of Psychology*, *59*(1), 617-645.

Barsalou, L.W. (2010). Grounded cognition: Past, present, and future. *Topics in Cognitive Science*, *2*(4), 716-724.

Bartolozzi, C., Indiveri, G., & Donati, E. (2022). Embodied neuromorphic intelligence. *Nature Communications*, *13*(1), 1024.

Bermudez, L. (2021, February 9). Overview of Embodied Artificial Intelligence. Retrieved August 1, 2022. https://medium.com/machinevision/overview-of-embodied-artificial- intelligence-b7f19d18022.

Bird, E., Fox-Skelly, J., Jenner, N., Larbey, R., Weitkamp, E., & Winfield, A. (2020). The ethics of artificial intelligence: Issues and initiatives. European Parliament: Directorate-General for Parliamentary Research Services. *https://www.europarl.europa.eu/RegData/etudes/STUD/2020/634452/EPRS_STU(2020)634452_ EN.pdf*

Blei, D.M., Kucukelbir, A. , & McAuliffe, J.D. (2017) Variational Inference: A Review for Statisticians, Journal of the American Statistical Association, 112:518, 859-877, DOI: 10.1080/01621459.2017.1285773

Bongard, J. C. (2013). Evolutionary Robotics. Commun. ACM. 56, 74–85.

Bongard, J. (2014). Why morphology matters. *The Horizons of Evolutionary Robotics*, *6*, 125- 152.

Brooks, R.A. (1990). Elephants don't play chess. *Systems (1-2), 3-15.* https://doi.org/10.1016/S0921-8890(05)80025-9

Brooks, R.A. (1991a). New approaches to robotics. *Science,* 253, (5025), 1227–1232.

Brown, J.S.., Collins, I. and Duguid, P. (1989). Situated Cognition and the Culture of Learning. *Educational Researcher,* 18(1), (pp. 32-42).

Cañamero, L. (2021). "Embodied Robot Models for Interdisciplinary Emotion Research," in *IEEE Transactions on Affective Computing*, vol. 12, no. 2, pp. 340-351, 1 April-June 2021, doi: 10.1109/TAFFC.2019.2908162.

Cangelosi A, Morse A, Di Nuovo A, Rucinski M, Stramandinoli F, Marocco D, De La Cruz V, Fischer K. Embodied language and number learning in developmental robots Conceptual and Interactive Embodiment: Foundations of Embodied Cognition. 2: 275-295. DOI: 10.4324/9781315751962

Chauhan, N., Yadav, N., & Arya, N. (2018). Applications of artificial neural network in textiles. *International Journal current microbiol applied sciences*, *7*, 3134-3143.

Chen, Y. L. N., Li, P., & Grant, S. (2020) Embodied cognition and language learning in virtual environments. *Educational Technology Research and Development*. https://doi.org/10.1007/s11423-015-9401-x

Chomsky, N. (1965). *Aspects of the theory of syntax*. Cambridge, MA: MIT Press.

Clark, A. (1999). An embodied cognitive science? *Trends in Cognitive Sciences*, *3*(9), 345-351.

Clark, A. (2001). Reason, robots, and the extended mind. *Mind and Language 16*(2),121-145.

Clark, A. (2008). *Supersizing the mind: Embodiment, action, and cognitive extension*, Oxford University Press.

Clark A. (2013). Whatever next? Predictive brains, situated agents, and the future of cognitive science. *The Behavioral and Brain Sciences*, *36*(3), 81–204. https://doi.org/10.1017/S0140525X12000477

Clark, A., & Chalmers, D. (1998). The extended mind. *Analysis, 58*(1), 7–19. http://www.jstor.org/stable/3328150

Chauhan, R., Ghanshala, K. K., and Joshi, R. (2018). "Convolutional neural network (CNN) for image detection and recognition," in *2018 First International Conference on Secure Cyber Computing and Communication (ICSCCC)* (IEEE), 278–282.

Dennett, D. (2014). The software/wetware distinction Comment on "Toward a computational framework for cognitive biology: Unifying approaches from cognitive neuroscience and comparative cognition" by W. Tecumseh Fitch. *Phys Life Rev*, *11*, 367-368.

Dewey, J. (1985). *The later works, 1925-1953* (Vol. 3). SIU Press.

Di Paolo, E., & De Jaegher, H. (2017). Neither individualistic nor interactionist. In: C. Durt (Ed.) *Embodiment, enaction, and culture: Investigating the constitution of the shared world* (pp. 87-105). MIT Press.

Duan, J., Yu, S., Tan, H.L., Zhu, H., & Tan, C. (2022). A survey of embodied AI: From simulators to research tasks. *IEEE Transactions on Emerging Topics in Computational Intelligence*, 6, 230-244.

Duffy, B. R., & Joue, G. (2000, December). Intelligent robots: The question of embodiment. In *Proc. of the Brain-Machine Workshop.*

Fei-Fei, Li., & Krishna, R. (2022). Searching for Computer Vision North Stars. *Daedalus*, *151*(2), 85–99. https://doi.org/10.1162/DAED_a_01902

Floreano, D., Ijspeert, A. J., & Schaal, S. (2014). Robotics and neuroscience. Current biology: CB, 24(18), R910–R920. https://doi.org/10.1016/j.cub.2014.07.058

Fodor, J.A. (1975). The Language of Thought. Harvard University Press.
Fodor, J. (1980). Methodological solipsism considered as a research strategy in cognitive science. *Behavioural and Brain Sciences, 3*, 63-73.

Fodor, J.A. (1983). *The modularity of mind.* MIT Press. Fodor, J.A. (1987). *Psychosemantics.* Bradford Books.

Fodor, J.A. (1990). *A theory of content and other essays*. Cambridge, MA: MIT Press.

Friston, K.J. (2011). Functional and effective connectivity: A review. *Brain Connectivity*, *1*(1), 13–36. https://doi.org/10.1089/brain.2011.0008

Fugate, J.M.B., Macrine, S.L., & Cipriano, C. (2019). The role of embodied cognition for transforming learning. *International Journal of School & Educational Psychology, 7*(4), 274 – 288.

Fugate, J. M. B., & Wilson-Mendenhall, C. D. (2022). Embodied emotion, emotional granularity, and mindfulness: Improved learning in the classroom. In S. Macrine & J. M. B. Fugate (Eds.), *Movement matters: How embodied cognition informs teaching and learning.* Cambridge, MA: MIT Press.

Gal'perin, P.I. (1992). The problem of activity in Soviet psychology. *Journal of Russian & East European psychology*, *30*(4), 37-59.

Gibson, J.J. (1979). *The Ecological Approach to Visual Perception.* Boston: Houghton Mifflin.

Gibson, J.J. (2014). The theory of affordances (1979). In: *The people, place, and space reader* (pp. 90-94). Routledge.

Glenberg, A. M., & Kaschak, M. P. (2002). Grounding language in action. *Psychonomic Bulletin & Review*, *9*(3), 558-565.

Goodwin, C. (2003). The semiotic body in its environment. In J. Coupland & R. Gwyn (Eds.), *Discourses of the body* (pp. 19–42). Palgrave/Macmillan.

Goodwin, C. (2010). Multimodality in human interaction. Calidoscópio 8(2):85-98. DOI: 10.4013/cld.2010.82.01

Grosso, E., Manzotti, R., Tiso, R., & Sandini, G. (1995, November). A space-variant approach to oculomotor control. In: *Proceedings of International Symposium on Computer Vision-ISCV* (pp. 509-514). IEEE.

Gupta, A., Anpalagan, A., Guan, L., & Khwaja, A. S. (2021a). Deep learning for object detection and scene perception in self-driving cars: Survey, challenges, and open issues. *Array, 10, Article* 100057.

Gupta, A., Savarese, S., Ganguli, S., & Fei-Fei, L. (2021b). Embodied intelligence via learning and evolution. *Nature Communications*, *12*(1),1-12. https://doi.org/10.1038/s41467-021- 25874-z

Hinton, A. (2014). Understanding context: Environment, language, and information architecture. O'Reilly Media, Incorporated. ISBN: 9781449326562

Hopfield, J.J. (1984). Neurons with graded response have collective computational properties like those of two-state neurons. *Proceedings of the national academy of sciences of the United States of America, 81,* 3088-3092. https://doi.org/10.1073/pnas.81.10.3088

Howard, D., Collins, J., & Robinson, N. (2022). Taking shape: A perspective on the future of embodied cognition and a new generation of evolutionary robotics. *IOP Conference Series: Materials Science and Engineering*, *1261*(1), 012018. https://doi.org/10.1088/1757- 899X/1261/1/012018

Hohwy, J. (2013). *The predictive mind.* Oxford Academic Press. https://doi.org/10.1093/acprof:oso/9780199682737.001.0001

Hughes, J., Abdulali, A., Hashem, R., and Iida, F. (2022). Embodied Artificial Intelligence: Enabling the Next Intelligence Revolution. *IOP Conf. Ser.: Mater. Sci. Eng.* 1261 012001

Hutto, D., & Abrahamson, D. (2022). Embodied, enactive education: Conservative versus radical approaches. In: S.L. Macrine & J.M.B. Fugate (Eds.) *Movement Matters*: *How embodied cognition informs teaching and learning.* MIT Press.

IEEE Computer Society (2022). Statement on conference on computer vision and pattern recognition. https://embodied-ai.org/

Iida, F., & Nurzaman, S.G. (2016). Adaptation of sensor morphology: an integrative view of perception from biologically inspired robotics perspective. *Interface focus, 6:* 20160016. http://dx.doi.org/10.1098/rsfs.2016.0016

James, W. (1983). *Essays in psychology* (Vol. 13). Harvard University Press.

Kaelbling, L.P. (2021). Natural science and engineering science. *The Learning & Intelligent Systems Group*.https://lis.csail.mit.edu/the-engineering-science-of-embodied- intelligence

Kolb, D.A. (1984). *Experiential learning: Experience as the source of learning and development.* Englewood Cliffs, NJ Prentice Hall.

Krizhevsky, A., Sutskever, I., & Hinton, G. E. (2017). Imagenet classification with deep convolutional neural networks. *Communications of the ACM*, *60*(6), 84-90.

Kumar, A. (2018). The mind and the body towards true artificial intelligence. Becoming human. *Artificial Intelligence Magazine.* https://becominghuman.ai/embodied-cognition-the-road-towards-true-artificial-intelligence- d0306c514670

Liao, Y.C., Todi, K., Acharya, A., Keurulainen, A., Howes, A.,& Oulasvirta, A. (2022). Rediscovering Affordance: A Reinforcement Learning Perspective. In Proceedings of the 2022 CHI Conference on Human Factors in Computing Systems (CHI '22). Association for Computing Machinery, New York, NY, USA, Article 362, 1–15. https://doi.org/10.1145/3491102.3501992

Lakoff, G., Johnson, M., & Sowa, J.F. (1999). Review of philosophy in the flesh: The embodied mind and its challenge to Western thought. *Computational linguistics*, *25*(4), 631-634.

Lakatos, L., & Janka, Z. (2008). Az emberi agy és intelligencia evolúciója [Evolution of human brain and intelligence]. *Ideggyogyaszati szemle*, *61*(7-8), 220–229.
Lanillos, P. et al (2021). Active inference in robotics and artificial agents: Survey and challenges. arXiv preprint arXiv:2112.01871.

Lave, J., & Wenger, E. (1991). *Situated learning: Legitimate peripheral participation.* Cambridge University Press.

Leake, D.B. (2006). Editorial Introduction. *AI Magazine*, *27*(4) 3. https://doi.org/10.1609/aimag.v27i4.1903

Leontiev, A. N. (1978). *Activity, consciousness, and personality.* Prentice-Hall Englewood Cliffs

Macrine, S.L., & Fugate, J.M.B. (2020). Embodied Cognition. In: *Oxford Research Encyclopaedia of Education*. Oxford University Press. https://doi.org/10.1093/acrefore/9780190264093.013.885

Macrine, S.L., & Fugate, J.M.B. (2021). Translating embodied cognition for embodied learning in the classroom. *Frontiers in Education, 6.* https://doi.org/10.3389/feduc.2021.712626

Macrine, S.L. and Fugate, J.M. (Eds.). (2022). *Movement matters: How embodied cognition informs teaching and learning.* MIT Press.

Marmeleira, J., & Duarte Santos, G. (2019). Do not neglect the body and action: The emergence of embodiment approaches to understanding human development. *Perceptual and Motor Skills, 126*(3), 410–445. https://doi.org/10.1177/0031512519834389

Matheson, H., White, N., & McMullen, P. (2015). Accessing embodied object representations from vision: A review. *Psychological Bulletin, 141*(3), 511–524. https://doi.org/10.1037/bul0000001

McCarthy, J., Minsky, M.L., Rochester, N., & Shannon, C.E. (1955; 2006). A proposal for the Dartmouth summer research project on Artificial Intelligence, August 31, 1955. *AI Magazine*, *27*(4), 12. https://doi.org/10.1609/aimag.v27i4.1904

McClelland, J.L. (1989). Parallel distributed processing: Implications for cognition and development. In: R. G. M. Morris (Ed.), *Parallel distributed processing: Implications for psychology and neurobiology* (pp. 8– 45). Clarendon Press.

McNerney, S. (2011). A brief guide to embodied cognition: Why you are not your brain – *Scientific American Blog Network.* https://blogs.scientificamerican.com/guest-blog/a-brief- guide-to-embodied-cognition-why-you-are-not-your-brain/

Millier, B. (2021). Data Labeling: What It Is and Why It Matters
Improve your machine learning models by creating high-quality data labels. Exchange: Scale Virtual Events Blog.

Montessori, M.M. (1969). The Four Planes of Development. *AMI Communications.* (2/3), 4–10.

Morgan, D.L., (2018). Skinner, Gibson, and embodied robots: Challenging the orthodoxy of the impoverished stimulus. *Journal of theoretical and philosophical psychology*, *38*(3), Advance online publication. http://dx.doi.org/10.1037/teo0000083

Müller, V. C., & Hoffmann, M. (2017). What Is Morphological Computation? On How the Body Contributes to Cognition and Control. *Artificial life*, *23*(1), 1–24. https://doi.org/10.1162/ARTL_a_00219

Muratore, F., Ramos, F., Turk, G., Yu, W., Gienger, M., & Peters, J. (2022). Robot Learning from Randomized Simulations: A Review. In *Frontiers in Robotics and AI* (Vol. 9). Frontiers Media S.A. https://doi.org/10.3389/frobt.2022.799893

Newen, A., De Bruin, L., & Gallagher, S. (Eds.). (2018). *The Oxford handbook of 4E cognition.* Oxford University Press.

Noë, A., (2009). Extending our view of mind. *Trends in Cognitive Sciences*, *13*(6), 237-238.

Núñez, R., & Lakoff, G. (2005). The cognitive foundations of mathematics: The role of conceptual metaphor. In: J. I. D. Campbell (Ed.), *Handbook of mathematical cognition* (pp. 109– 124). Psychology Press.

Nygaard, T.F., Martin, C.P., Torresen, J. *et al.* (2021). Real-world embodied AI through a morphologically adaptive quadruped robot. *Nature Machine Intelligence* **3**, 410–419. https://doi.org/10.1038/s42256-021-00320-3

Owens, A. P., Allen, M., Ondobaka, S., & Friston, K. J. (2018). Interoceptive inference: From computational neuroscience to clinic. *Neuroscience and biobehavioral reviews*, *90*, 174–183. https://doi.org/10.1016/j.neubiorev.2018.04.017

Parr, T., Pezzulo, G., & Friston, K (2022) Active Inference: The Free Energy Principle in Mind, Brain, and Behavior. Cambridge, MA: MIT Press. DOI: 10.7551/mitpress/12441.001.0001

Pfeifer, R. (1999). September. Dynamics, morphology, and materials in the emergence of cognition. In *Annual conference on Artificial Intelligence* (pp. 27-44). Springer.

Pfeifer, R. (2001). Embodied artificial intelligence 10 years back, 10 years forward. In *Informatics* (pp. 294-310). Springer.

Pfeifer, R. & Bongard, J.(2006). *How the body shapes the way we think: a new view of intelligence*. Cambridge, MA: MIT Press.

Pfeifer, R., Bongard, J., & Grand, S. (2007). *How the body shapes the way we think: A new view of intelligence*. MIT Press.

Pfeifer, R., Lungarella, M., & Iida, F. (2007). Self-organization, embodiment, and biologically inspired robotics. *Science*, *318*(5853), 1088-1093.

Pfeifer, R., & Iida, F. (2004). Embodied Artificial Intelligence: Trends and challenges. In: F. Iida, R. Pfeifer, L. Steels, & Y. Kuniyoshi (Eds.), *Embodied Artificial Intelligence* (pp. 1–26). Springer. https://doi.org/10.1007/978-3-540-27833-7_1

Pfeifer, R. & Scheier, C. (2001). *Understanding intelligence*. MIT Press.

Piaget, J. (1952). *The origins of intelligence in children*. Translated by M. Cook. W.W. Norton.

Piaget, J. (1968). *Six psychological studies*. Translated by A. Tenzer. Putnam.

Putnam, H. (1961). Brain and behaviour, originally read as part of the program of the American Association for the Advancement of Science. *Section L (History and Philosophy of Science)*.

Robbins, P., & Aydede, M. (Eds.). (2008). *The Cambridge handbook of situated cognition*. Cambridge University Press.

Robbins, P., & Aydede, M. (2009). A short primer on situated cognition. In: M. Aydede & P. Robbins (Eds.), *The Cambridge Handbook of Situated Cognition* (pp. 3-10). Cambridge University Press.

Rogoff, B. (1990). *Apprenticeship in thinking: Cognitive development in social context.* Oxford University Press.

Rosch, E., Mervis, C. B., Gray, W. D., Johnson, D. M., & Boyes-Braem, P. (1976). Basic objects in natural categories. *Cognitive Psychology*, *8*(3), 382-439.

Rumelhart, D.E., Hinton, G., & McClelland, J.L. (1986). A general framework for parallel distributed processing. In: J.L. McClelland, & D.E. Rumelhart (Eds.), *Parallel distributed processing: explorations in the microstructures of cognition* (Vol. 2) (pp. 45-77). Bradford Books: MIT Press.

Rumelhart, D.E., &McClelland, J.L. (1987). Learning internal representations by error propagation. In: J.L. McClelland, & D.E. Rumelhart (Eds.), *Parallel distributed processing: explorations in the microstructures of cognition* (Vol. 2) (pp. 318-362). Bradford Books: MIT Press.

Rumelhart, D.E., & Ortony, A. (1977). The representation of knowledge in memory. In: R.C. Anderson, R.J. Spiro, & W.E. Montague (Eds.), *Schooling and the acquisition of knowledge* (pp. 99-135). Erlbaum.

Shapiro, L.A., (2011). Embodied cognition: Lessons from linguistic determinism. *Philosophical Topics*, *39*(1), 121-140.

Sims, M (2022). Self-Concern Across Scales: A Biologically Inspired Direction for Embodied Artificial Intelligence. Front. Neurorobotics., 25 April 2022
Volume 16 - 2022 | https://doi.org/10.3389/fnbot.2022.857614

Smelser, N.J., & Baltes, P.B. (2001). *International encyclopaedia of the social & behavioural sciences* (1st ed.). Elsevier.

Smith, L.B., & Thelen, E. (2003). Development as a dynamic system. *Trends in Cognitive Sciences*, *7*(8), 343–348. https://doi.org/10.1016/s1364-6613(03)00156-6

Sprevak, M. (2017). Turing's model of the mind. In: *The Turing guide*. Oxford University Press.

Thelen, E., & Smith, L.B. (1994). *A dynamic systems approach to the development of cognition and action*. MIT Press, Bradford Book.

Thompson, E., & Cosmelli, D. (2011). Brain in a vat or body in a world? Brainbound versus enactive views of experience. *Philosophical Topics*, 163-180.

Turing, A.M. (1950). Computing Machinery and Intelligence. *Mind, 49,* 433-460.

Varela, F.J., Thompson, E., & Rosch, E. (1991). *The embodied mind: Cognitive science and human experience*. The MIT Press.

Vygotsky, L.S. (1978). *Mind in society: The development of higher psychological processes.* W.W. Norton & Co.

Willems, R.M., & Francken, J.C. (2012). Embodied cognition: Taking the next step. *Frontiers in Psychology*, *3*, Article 582. https://doi.org/10.3389/fpsyg.2012.00582

Wilson, M., (2002). Six views of embodied cognition. *Psychonomic Bulletin & Review*, *9*(4), 625-636.

Wilson, R. & Foglia, L. (2012), "Embodied Cognition," *Stanford Encyclopaedia of Philosophy* (Summer 2021 Edition), Edward N. Zalta (ed.), https://plato.stanford.edu/archives/sum2021/entries/embodied-cognition/

Wilson, A. D., & Golonka, S. (2013). Embodied cognition is not what you think it is. *Frontiers in Psychology*, *4*, 58.

Zhu, F., Zhu, Y., Lee, V. C., Liang, X., & Chang, X. (2021). *Deep Learning for Embodied Vision Navigation: A Survey*. http://arxiv.org/abs/2108.04097

IOP Conf. Series: Materials Science and Engineering 1292 (2023) 012019 doi:10.1088/1757-899X/1292/1/012019

Intelligence Offloading and the Neurosimulation of Developmental Agents

B Alicea[1,2], A Pang[1], and J Parent[1]

[1] Orthogonal Research and Education Laboratory, Champaign-Urbana, IL

[2] OpenWorm Foundation, Boston, MA.

E-mail: bradly.alicea@outlook.com

Abstract. Cognitive offloading occurs when environmental affordances expand cognitive capacity while facilitating spatial and social behaviors. Capacity-related constraints are also important, particularly as embodied agents come online during development. Vast differences in brain size and offloading capacity exist across the tree of life. We take from multiple perspectives to understand the proportional contributions of internal models (brain) and externalized processing (offloading) in developing embodied computational agents. As developing nervous systems scale with body size and/or functional importance, offloading is also driven by neural capacity. Cognitive capacity is ultimately determined by various innate and environmental constraints. We propose a similar model for computationally developing cognitive agents. A regulatory model of cognition is proposed as a means to build cognitive systems that interface with biologically-inspired substrates. Multiple tradeoffs result from energetic, innate, and informational constraints, and determine the proportion of internal to external information processing capacity. As growth of a biologically-inspired substrate accelerates or decelerates over developmental time, it changes the acquisitional capacity of the agent. Our agent's capacity limitations determine externalization potential, which is characterized by three parameters and two mathematical functions. The neurosimulation approach to intelligence offloading can be applied to a broad range of agent-based models and Artificial Intelligences.

1. Introduction

The interactions between the nervous system and environment are an important component to simulating intelligent behavior. While environmental realism (in the form of enrichment) is certainly important [1,2], some aspects of cognition are embedded in the environment itself. Let us consider the case of phototaxis in single-celled organisms. Light is acquired from the external environment, and serves as information for the internal nervous system. All of the computation is performed in the cell: morphological asymmetries and properties of the cell body are of primary importance [3]. In the algae *Chlamydomonas reinhardtii*, the cell is steered towards or away from a light source by its flagellum based on the accumulation of light signals [4]. In algae, this integration occurs via biochemical processes. Among marine annelids such as polychaetes [5] and the genus *Platynersis* [6], biochemical control of simple behaviors is replaced by small connectomes of 68 and 71 neurons, respectively to enable more complex behaviors. In both of these examples, complexity of the internal model increases with information processing needs, but also clearly requires interactions with body size and the environment. Likewise, intelligent behavior in insects is not only determined by changes in the size of the mushroom bodies [7], but also environmental cues and communication with conspecifics. Reliance on a purely connectionist approach results in an incomplete account of even the simplest forms of cognitive processing.

These examples have obvious lessons for models of intelligent behavior mimicked in computational agents. Yet rather than simply scaling up the size of a nervous system for more complex behaviors, we propose a different principle: computational offloading as a consequence of agent development. Developmental approaches offer a novel approach to agent learning and intelligence more generally. For example, it has been shown that human babies and toddlers extract information from the world much differently than our current methods for training machine learning models [8]. Human infants intensively explore their environment, but also intently explore specific objects. This provides a flexible form of experience that can be alternatively wide and deep. Early experience with the environment serves as a scaffold for what will come later on in the individual's lifetime. This has been demonstrated computationally through autotelic agents [9]. Autotelic agents are defined by intrinsically-motivated acquisition, and use these internal mechanisms to master skills and tasks in a flexible, open-ended manner.

To better understand how this might lead to general principles, we must first understand the nature of cognitive offloading, particularly with respect to the regulation of agent behaviors. We assume that cognition consists of both an internal model and environment which are regulated together. Building upon this assumption, we then consider the concept of tradeoffs and associated factors in determining the share of internal processing to offloading that characterizes increases in cognitive capacity. A developmental approach with a model of innateness is an excellent way to consider both information processing along with the role of a substrate during acquisition. When viewing cognition from a regulatory perspective, we must consider what drives innate capacity for intelligence, and how this system might balance internal capacity for cognitive processing with offloading and related forms of cognitive externalization. To understand this, we introduce a quantitative theoretical model based on heterochrony. Heterochrony refers to changes in the proportion of agent body size to brain (or internal model) size. This connects the themes of regulatory cognition with embodied agents, and why embodiment plays a critical role in determining offloading capacity in computational agents.

1.1 *Defining Cognitive Offloading*

Risko and Gilbert [10] provide an operational definition of cognitive offloading in humans. Their definition of cognitive offloading is the "use of *physical action* to *alter the information processing requirements* of a task so as to *reduce cognitive demand*". Unpacking this definition provides us with a basis for applying this definition to the world of computational agents. Physical action is represented by "development and externalization". Alter(ation) is represented by "development and externalization". Reduc(tion of)e Cognitive Demand is represented by tradeoffs in response to developmental constraints.

Expanding this definition to non-human animals, Andy Clark [11] provides an example of cognitive offloading using affordances in the environment. Bluefin dolphins utilize vortex rings (hydrodynamic features of the water column) to augment their muscle power during swimming. While bluefin dolphins produce more than ample muscle power to propel themselves through the water column [12], vortex rings are produced by the organisms to provide an external computational mechanism for hunting and refining swimming behavior. Critically, the vortex ring provides a means for environmental feedback, and even allows for complex signals to be stored as affordances.

Affordances are environmental features that enable computation or ease of behavior. To fully understand how these affordances are useful for the development of artificially intelligent agents, we must consider the developmental and evolutionary origins of offloading.

1.2 *Cognitive Offloading in Development*

Now we will review two studies that demonstrate the role of development in the emergence of cognitive offloading ability. The first involves a shift towards offloading over the course of development in humans [13]. As human babies interact with their environments, they learn how to utilize the affordances available in the world, thus adding cognitive capacity as they mature. A complementary study [14] suggests that reactive behavior coupled with offloading acts to reinforce complex cognition in agents. In both of these cases we can see a clear role for developmental contingency [15], which involves context-specific cycles of interaction dependent on the outcome of past events. Therefore, we can see that change and plasticity are hallmarks of developmental approaches to understanding the emergence of offloading. This implies that a stable biological baseline is also important to include in a computational model of offloading. A computational model of innate processes [16-19] provides the ability to generate more complex abilities as the developmental process unfolds.

2. Models and Definitions

2.1 *Offloading as an Intelligent Embodied System*

A cybernetic-inspired diagram of our embodied intelligent system is shown in Figure 1. Contrary to models of cognition as sequential information processing [20-21], this model highlights the regulatory aspects of naturalistic cognition [22, 23]. The agent phenotype (an example shown in its embryonic form) is influenced by a model of innateness and direct feedback from a model of the environment. In computational terms, the model of innateness serves as an initial condition for the agent, while the environment is a representation constructed gradually by the agent. Innateness is defined in terms of both body and behavior: the shape, growth condition, and sensor/effector units of the body, along with behavioral primitives such as phototaxis, parameters for stimulus sensitivity, or environmental pre-training. Interactions over developmental time contribute to an internal model of the world that becomes iteratively refined. The model of environment consists of selected features outside of the agent's own development. An agent's operational ensemble of environmental features are in part defined through indirect feedback from affordances that conform to innate capabilities such as grasping, rolling, or sensing light. Affordances also occur in association with their fellow agents and serve to inform meta-representations [24, 25], shaping how an agent interacts with its environment.

The sophistication of the agent's internal processing and model of innateness, in addition to the quality of available affordances and environmental model determine how much information is retained in the internal model and how much is distributed to the environment. This will be defined in more concrete biological terms when we discuss developmental tradeoffs in the next section. The sociality of the agent is a secondary factor which we will also revisit in the next section in the form of proximal development, or the advantages of social interactions from the environment.

2.2 *Two timescales of naturalistic cognition*

Development and cognition alike unfold over multiple timescales. There are two timescales of particular concern: morphogenetic and dynamical. The morphogenetic timescale involves physical changes and transformations with respect to the biologically-inspired substrate. Changes in the size, shape, and connectivity of the internal model or agent's body fall into this category. By contrast, dynamical time involves the processing of cognitive information processing, which generally happens in concert with an internal model and environmental structure. Changes in information acquisition and offloading capacity fall into this category.

Temporal changes related to the biologically-inspired components of a developing agent occur in *morphogenetic time*. Morphogenetic time also includes the emergence of the capacitance properties and architectures involved with information processing via the body's sensors. Morphogenetic time presents a unique challenge to models of cognition. Namely, the rate and extent of growth of the agent's body can determine the capacity of the internal model. With this constraint in hand, intelligent behavior might draw from the richness of the environment. Depending on the presence of patterns in the environment, or the ability to utilize affordances, a particular agent will have an externalization capacity that helps them overcome internal model capacity limitations.

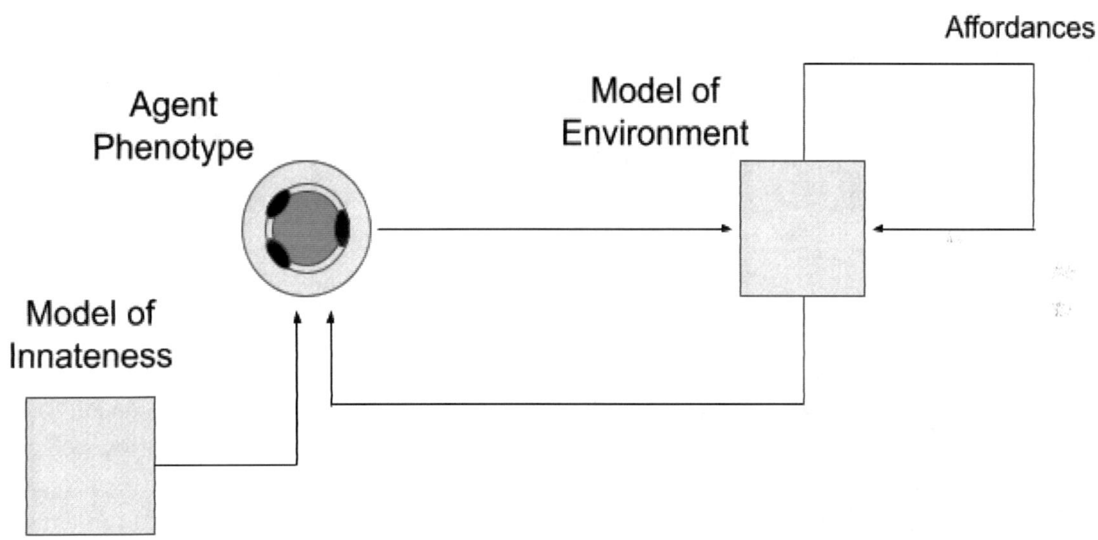

Figure 1. Model of regulation between the agent phenotype and the environment in which the agent is embedded. Agent phenotype is shown in its developmental state. Affordances are highlighted as an incremental feedback. Arrows refer to the direction of interaction: the model of innateness provides instructions to growth and development of the agent phenotype, while the agent phenotype interacts (bidirectionally) with models of the environment. Our models of the environment also interact with affordances, providing a feedback loop that shapes the model composition.

Dynamical time is the processing of information in the internal model, in concert with interactions between the agent and features of the environment. This includes feedback between the internal model and the external environment. The rate of acquisition is constrained by the unfolding of morphogenetic time: the rate of morphogenetic change determines both the ability of an agent's body to acquire information, as well as the capacity of their internal model. The rate of morphogenetic change is determined by the heterochronic rates of growth and form [26, 27] determined as one of an agent's innate factors.

2.3 *Developmental tradeoffs*

For a theoretical understanding of offloading and its relationship to the internal model, we will consider a model of tradeoffs. A developmental tradeoff involves prioritizing one factor at the expense of others given a finite amount of developmental potential. Amongst primate species, metabolic constraints due to food source and nutrition provide a selection mechanism for tradeoffs between body size and number of neurons in the brain [28]. In this case, brain growth may be limited. Yet if extending cognition leads to a better diet, it may not negatively impact cognition [29]. There also exist functional tradeoffs which prioritize certain cognitive functions over others [30]. For example, different cognitive functions such as attention and memory take priority based on the needs of a specific cognitive context [15]. Tradeoffs can also include behavioral strategies such as speed and accuracy [31], or higher level phenomena such as robustness, resilience, and performance [32]. In the case of speed-accuracy tradeoffs, the balance of contributors involves multiple scales of organization (e.g. physiology, cognitive representation, and social interactions). While these examples may confound us with great complexity in terms of feedback and balances, a typical agent-based implementation might involve only a simple tradeoff that favors either a greater or lesser amount of offloading. Some of these factors include growth and energetics, information, affordances and direct and perception.

Environmental tradeoffs in brain size is critical to our perspective on offloading and development. Evidence from mammals suggests that there is a relationship between brain size, intelligence, and lifespan [33]. While increased cognitive demands select for larger brain sizes, factors such as degree of social interactivity can mitigate increases in brain size [34]. These findings suggest that while the length of development and enriched environments are important for developing intelligent behavior, the capacity for intelligent behavior can be offloaded in strategically interesting ways. Critically, the process of epigenetics [15] governs the inherent variation of the many factors that can contribute to the tradeoffs proposed here. While we will not consider a formal model of epigenetics here, such a model can help control how this process unfolds in agent development.

One advantage of embodied agents with specified innate components is the ability to control the sensors, affectors, and flow of information. Different configurations of these agent properties provide a means to control and understand information acquisition both dynamically and in terms of contingency. Dynamic information acquisition examines the cumulative effects of exposure to the environment. The accumulation of environmental observations is also based on developmental contingency: repeated observations using the available inputs over time determines how the environment is perceived and (more importantly) utilized. Contingent information processing involves configurations that narrow the possibilities over time.

Taken together, direct perception and affordances can include the body (e.g. peripheral motor system), which can function like a statistical prior and information pre-filter [35]. This also helps to integrate the environment with the biologically-inspired components of cognitive regulation, and serves as a conduit of offloading. Stigmergy [36, 37] involves cues for collective behaviors, and as such serve as affordances that are particularly relevant to coordinating social and other collective behaviors. In social insects, features in the environment serve as a common landmark or reference point to coordinate collective behavior. Stigmergic mechanisms enable distributed communication across fellow agents, and provide a means for emergent self-organization of more complex behaviors [38, 39]. Heylighen [37] defines stigmergy as a trace left by an action that stimulates subsequent action, providing an indirect but powerful feedback mechanism for collective behavior. We can also view stigmergy as providing the information needed by individual agents to understand how an object in the environment can be used [40]. It is this environmental structure that fosters offloading [41], although this is a necessary but not sufficient condition for promoting offloading behavior.

As they map out their environment, developing agents engage with the social milieu. This results in direct feedback to individual agents from both the environment and other agents. This feedback arises through stigmergic constraints, but stigmergic interactions themselves arise from a shared environment that is easier to structure with meaningful information. More generally, proximal development in the form of emerging social interactions can drive shared cognitive resources, which in turn can increase the potential for further offloading, and even larger internal models (for an example from Primate brains, see [42]).

2.4 *Examples of tradeoff based Regulatory Systems*

The combination of factors that go into a developmental phenotype and their effect on the resulting features of a developmental phenotype is called a tradeoff based regulatory system. An example of such a system for a developmental agentis shown in Figure 2. In the computational agent context, tradeoffs occur at two levels: physiological and cognitive. In our regulatory system example, there is no duality between these two domains. This is supported by the observation of temporal tradeoffs in information processing during psychophysical tasks [43]. Yet it is not as simple as an equivalence: one feature of this type of regulatory system is that the developmental mechanisms observed in the organismal physiology can be tied to cognitive development. A similar relationship is relevant to computational agents. To simulate the growth and development of a biological system, a scaling model for growth (energetics) and acquisition (information) will be introduced. The relationship between physiology and information processing is particularly interesting for understanding the potential internal conditions that lead to offloading.

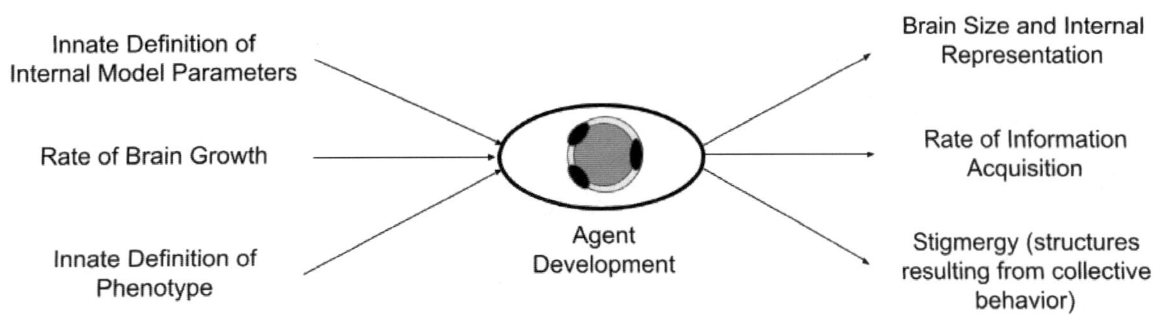

Figure 2. An example of a tradeoff-based regulatory system. Inputs to the developmental agent (both from feedforward innate and feedback environment) include the innate definition of internal model parameters, rate of brain growth, and the innate definition of the phenotype. Outputs from the agent's development include a mature brain size and internal representation, rate of information acquisition, and stigmergy (cues for collective behaviors).

An example of how these tradeoffs play out in biological systems can be found in how neural tissue expansion is enabled by increased energetic input, which in turn leads to increased neural processing. This can be demonstrated in the context of Primate fovea [44]. The outcomes of different causal factors are partially determined by behavioral strategies such as the explore-exploit tradeoff [45]. But it comes down to this: as a developing agent explores their world, neuronal tissue must have the capacity to compute and represent an organism's cognitive needs. If not, the organism must find ways (such as affordances) to encode information in the environment.

Biological systems also enforce constraints through the interactions of these tradeoffs. Earlier, we discussed the role of development constraints as defined by developmental systems theory. But there is another definition of developmental constraints that is more specific to growth in the biological substrate. This model of constraints predicts that brain (neural tissue) size grows proportionally with the body, and this proportionality is expected to be consistent throughout development [46]. Depending on environmental demands, neural tissue growth increases via an increased growth rate relative to the dimensions of the body. As the growth rate of neural tissue increases, it becomes limited by energetic demands, which in turn dampens the increasing growth rate. This type of nonlinear positive feedback [47] results from dynamically changing biological/energetic tradeoffs over time. As an aside, this is also consistent with what is known about offloading: using an affordance (such as the internet for reference information) encourages further offloading behavior, and strengthens the reliance on offloading over time via positive feedback [48].

3. Heterochronic Model of Offloading

Now we can map the evidence from neural tissue growth during development to computational agents. We maintain that much of what is said about organisms and neural tissue is also true of computational agents and their internal models (or internal representations). The two main issues involve building a substrate for cognitive capacity and enabling regulation of information acquisition and processing through a regulatory system that includes the environment. Thus, characterizing the potential for substrate growth is important for observing and perhaps even controlling how offloaded information processing proceeds.

Neural architectures are formed and maintained through an interaction between physiological conditions and information processing. The capacity of a neural architecture is related to its size and ability to adapt. Both of these factors are the product of development, and are governed by a scaling relationship known as heterochrony. Put simply, heterochrony involves changes in the timing and duration of growth in development. When growth slows down relative to a baseline, the developmental process yields a less mature and smaller mature phenotype. When growth speeds up, the opposite is true, yielding a more mature and larger phenotype. Since this relationship can be used to determine the potential for internal processing, it also serves as an indicator of whether external processing is needed or will even become predominant.

Our heterochronic model for determining the offloading potential of a given agent is shown in Figure 3. Figure 3A shows that acceleration and deceleration are characterized as changes in the value of parameter k (growth rate). This graph also demonstrates changes in k over developmental time (t). Removing k from its temporal context, Figure 3B shows the relationship between k and offloading potential (o). Generally, this is the rate of internal acquisition, with a negative relationship between k and o (low values of o predict high values of k and vice versa). Finally, Figure 3C shows o as a component of developmental time (t) and plotted against k. Characterized by the function $f(t)$ alongside the relationship shown in Figure 3B, this relationship is predicted to decrease with respect to k until an inflection point is reached for moderate values of o and t. Beyond this point, both o and k have a positive relationship. This captures the notion that offloading can be synergistic with large brains, which becomes particularly pronounced at later stages of development. Why this inflection point occurs in Figure 3C is not clear, but we might expect this scenario to occur due to behavioral plasticity later in development.

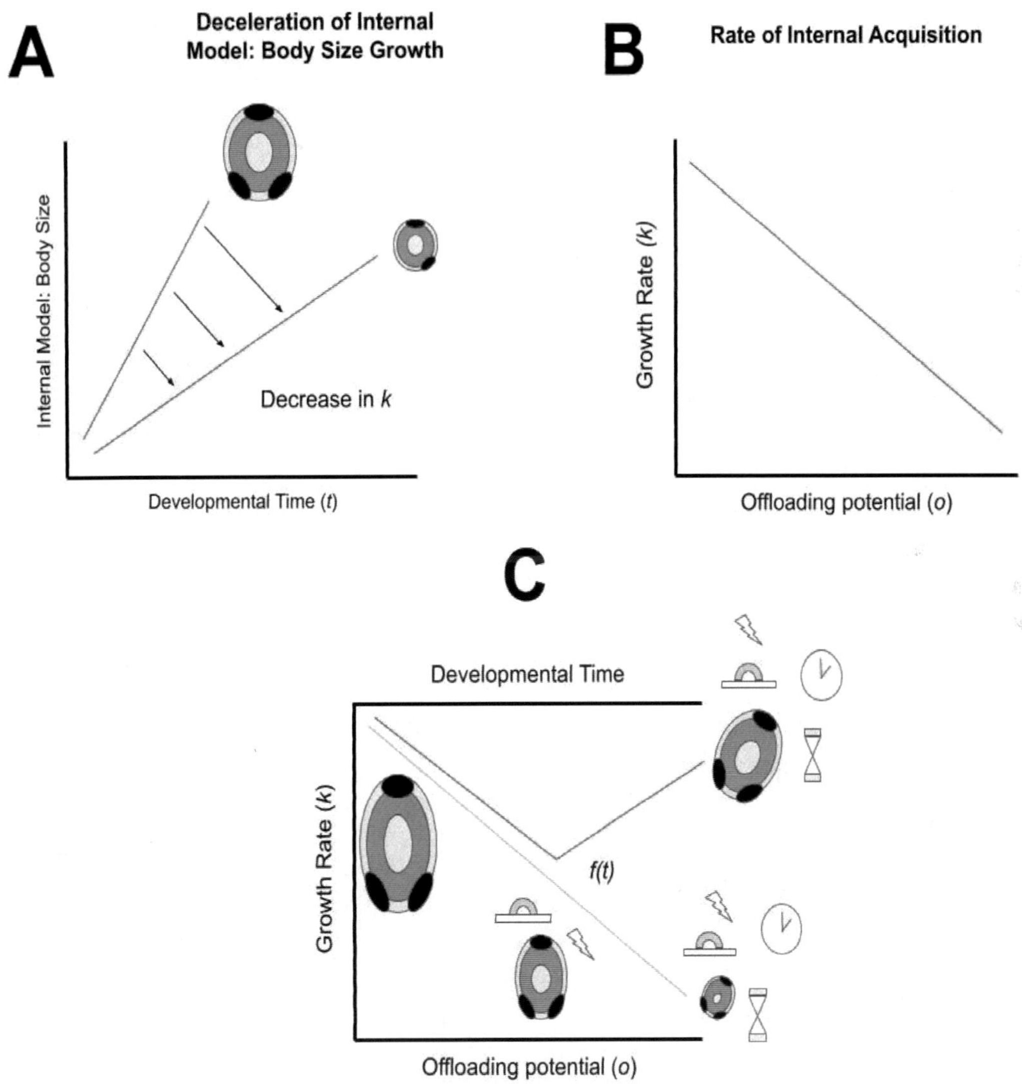

Figure 3. An example of a heterochronic model incorporating growth and acquisition. A: an example of deceleration of internal model:body size growth due to environmental constraints. B: baseline offloading potential in terms of the rate of acquisition given heterochronic trends. C: co-development of internal and external acquisition given the potential for behavioral plasticity. Gray: baseline from Figure 3B. Red: an increase in agent size with the same offloading potential. The icons adjacent to the

agents in Figure 3C represent affordances such as tools and cultural artifacts that help shape the gray and red functions.

There is yet another function that allows us to characterize the developmental potential called continuous Gibsonian Information (GI). GI [49] is characterized mathematically by the distribution of sensory structure encountered over time by an agent. The function *gi(t)* approximates the spatiotemporal structure of the sensory environment in terms of affordances, which can be mapped to parameter *o* in a positive manner: when the value of *gi(t)* increases, the value of *o* also increases, and leads to a greater share of offloading as part of an agent's overall cognitive capacity.

3.1 *Heterochrony, Acquisition, and Neural Diversity*

As information acquisition is the opposite of offloading, the basic version of our model predicts that the acquisition of information and heterochrony have a positive relationship. When the neural architecture gets larger (with a corresponding increase in sensory resolution), the acquisition of information from the environment also increases. Furthermore, the configurations of agent properties can be controlled by developmental growth trends. In agents, the growth of neural processing architectures, the development of sophisticated sensory organs, or size and shape properties of the body can all be determined by the length of time and rate of growth and reconfiguration.

We can also conceptualize development as a discrete set of acquisitions that occur in a specific order. Sequence heterochrony is the relative shifting of these events in time [50]. In a biological context, this is called sequence heterochrony [51, 52], and characterizes development as a series of interrelated steps. For agent-based development, sequence heterochrony can be thought of as developmental contingency in stages. This modifies our continuous growth model in Figure 3 only a bit, as development out of sequence can be disruptive or even dysfunctional. Indeed, sequence heterochrony can be used to develop a viable model of innateness, and allows us to test the compositional sequence of substrate development.

Given the heterochronic rules of development for our agent, development of a specific agent or population of agents follows a specific developmental trajectory. Developmental trajectories can be characterized by epigenetic landscapes [53, 54], where a series of contingencies and the path of least resistance produces the mature phenotype. An epigenetic landscape of computational agents provides a set of configurations that allow us to develop specialized phenotypes specialized for various cognitive tasks and social interactions.

4. Further Considerations

Rather than asking why internal representations are established, a theory of offloading reliant upon proportional growth and biologically-inspired embodiment recasts the notion of internal models as scalable information processing models that extend into the environment depending upon the constraints of the computational task in question. The scalable component is rooted in agent embodiment and rooted in an model of innateness that provides baseline conditions for the implementation of a wide range of architectures. In the computational context, using a biologically-inspired model allows for offloading capacity to become both predictable and controllable.

Returning to Figure 3, we demonstrate how an innate model of growth and transformation leads to the composition of an agent phenotype that serves to both embody cognition and require a certain degree of cognitive offloading. The potential for offloading, and the continuum mentioned previously, results from the tuning of three parameters that summarize agent-environment interactions. Changes in growth rate (k), offloading potential (o) and its relationship to k over time provide a means to introduce or inhibit offloading under specific conditions represented by locations on the epigenetic landscape.

To consider the distinctions and connections between representationalism and offloading, let us consider interface theory on the one hand and radical embodied cognition on the other. Interface theory suggests that the world is constructed in the mind through natural selection, while the external environment is not at all as we make sense of it [55]. This is quite different from radical embodied cognition, which argues that our interactions with the external world (and not mental representation) is the primary driver of our cognitive world [56]. We can use these approaches as two extremes in a continuum ranging from zero to absolute reliance on offloading.

5. Conclusions and Future Directions

We have introduced a model of offloading for computational agents. This is done from a developmental perspective, which allows us to both compose agents in new ways and view the externalization of cognition from an emergent standpoint. In the course of introducing a model of developmental offloading, we introduce a number of concepts. The first is a regulatory model of cognition, and how offloading and affordances play an integral role in this perspective. Second, we stress the importance of developmental tradeoffs in determining how much of cognition gets offloaded during an agent's cognition. Our model stresses the potential for multiple causal factors. Finally, we introduce a heterochronic model for the embodied and cognitive growth and development of cognitive agents. Utilizing the theory of heterochrony in the context of our regulatory approach to cognition allows us to define the nature of our tradeoffs and exactly how our agent is embodied.

Viewing cognition as a regulatory phenomenon allows us to characterize our agent's development as an inherently naturalistic set of interactions, and thus evaluate how offloading in the form of augmenting or compensating cognitive capabilities generated by the internal model (or brain). While developmental resources can be specified in terms of energetic or geometric constraints, they generally act to limit the size and capacity of the agent's internal model. Figure 4 shows the developmental context of these three parameters and two functions: growth rate, or k; time, or t, offloading potential, or o; the codevelopment of internal and external acquisition represented by function $f(t)$; and the distribution of affordances and other sensory features during direct perception represented by the function $gi(t)$. This provides a parameterized counterpart to our boxes and arrows model of a regulatory cognition. In fact, one advantage of our heterochronic perspective is the use of three parameters and two continuous functions that describe the relationship between the development of an agent's body, the proportion of developmental resources, and the development of information acquisition by the agent's internal model.

One aspect of our environmental characterization that deserves further investigation is the distinction between affordances and the concept of stigmergy. An environment contains affordances, which enable a number of opportunities for offloaded cognition, including stigmergy. Since stigmergic behavior represents feedback between offloaded cognitive markers and the internal model, it is interesting from both a regulatory and a direct perception standpoint. Beyond the connections to direct perception, a Markov blanket representation [57] of Figure 1 might be useful in identifying which external features are used in tandem with internal processing. Furthermore, the role of externalized cognition can play a role in encouraging the formation of social groups [34]. Energetic constraints on brain size among Hominids have been mitigated by externalized forms of cultural practices such as cooking, which in turn enables greater social interactivity [58]. This might be useful to social robotics applications where a heterochronic model might be optimized for agent groups of certain sizes.

5.1 *Future Directions*

To expand the usefulness of this work, we should consider the implementation of our approach using specific computational paradigms. Such paradigms include Partially Observable Markov Decision Processes (POMDPs - [59]) and Reinforcement Learning (RL - [60]). Both POMDPs and RL models are a natural fit to our developmental approach to offloading. Specifically, they are easy means to model our regulatory conception of cognition, which requires both an accounting of partially obscured environmental states (or world states in the parlance of POMDP) as well as feedback between the environment (or policy in the parlance of RL) and an internal model. Genetic Algorithms (GAs - [61]) also provide an acceptable means to implement regulatory cognition, and are actually the preferred means to model the innate features of the agent. Furthermore, POMDP models are compatible with multiagent RL (MARL) models [62], which allows for models representing the development of offloading potential to be extended to populations of interacting agents.

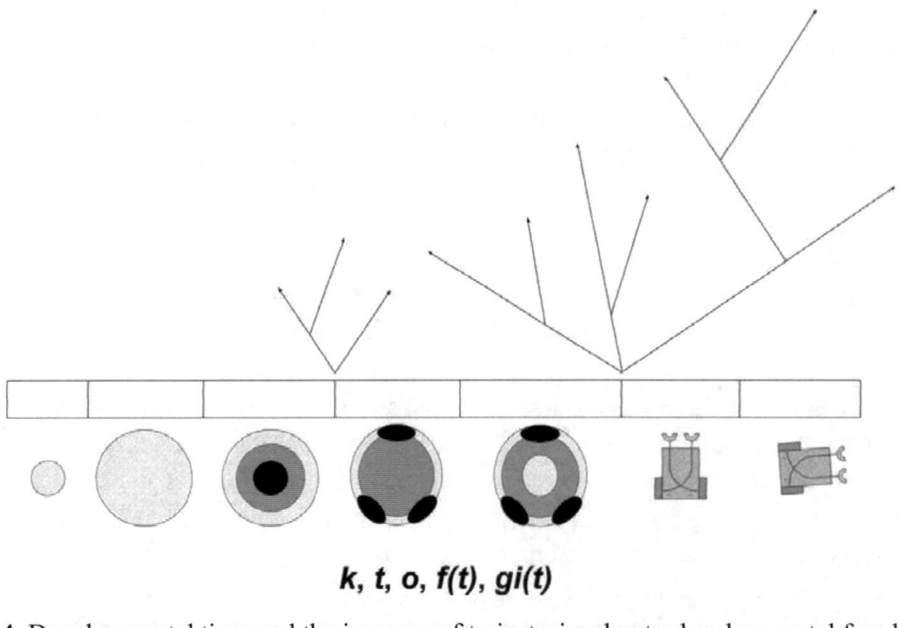

Figure 4. Developmental time and the increase of trajectories due to developmental freedom. The series of letters at the bottom represent the three parameters and two functions that represent our model of offloading potential and heterochronic growth. Boxes represent periods within the developmental process, denoted underneath by caricatures of developing agent bodies. The right-hand boundary of select boxes features a branching point. Branching structures above the adjacent boxes represent developmental trajectories and their associated contingencies that result. Developmental freedom occurs both later in development and when the number of possible developmental trajectories proliferate.

In a typical biological setting, constraints restrain more mature phenotypes from exploring the possibilities of novel form [63]. This is a guiding principle of our biologically-inspired model of growth and development. Indeed, our model of innateness proposes that the agent's body be constrained in exactly this way. Yet we must return to the difference between this type of constraint and what is posited by developmental systems theory [15, 64]; namely, that contingencies are conceptualized as a continual set of interacting cycles between the body, brain, and environment. But this is actually consistent with our regulatory approach to cognition by examining the effect of developmental rules on behavioral plasticity that become manifest later in the development process. Returning to Figure 4, we can see that as the possibility space shrinks in terms of what the body, sensors, and effectors look like (or as parameter k driven by t reaches maturity), the amount of developmental freedom, or number of possible trajectories enabled in part by parameters $gi(t)$ and o, and characterized by function $f(t)$, increases. Stated in terms of our proposed theory: developmental freedom is the increase in the number of information processing alternatives despite the winnowing of physiological configurations [19]. Developmental freedom allows for critical periods of neural refinement [65, 66], in addition to the incorporation of affordances and other externalized objects into a common cognitive system [67].

References

[1] Togelius J 2016 AI Researchers, Video Games Are Your Friends! *arXiv* 1612.01608

[2] Beyret B Hernández-Orallo J Cheke L Halina M Shanahan M and Crosby M 2019 The Animal-AI Environment: training and testing animal-like artificial cognition *arXiv* 1909.07483

[3] Dieckmann C and Mittelmeier T 2016 Phototaxis: life in focus *eLife* doi:10.7554/eLife.14169

[4] Arrieta J Barreira A Chioccioli M Polin M and Tuval I 2017 Phototaxis beyond turning: persistent accumulation and response acclimation of the microalga *Chlamydomonas reinhardtii Sci Rep* **7** 3447

[5] Windoffer R and Westheide W 1988 The nervous system of the male *Dinophilus gyrociliatus* (Polychaeta, Dinophilidae): II. Electron microscopical reconstruction of nervous anatomy and effector cells *J Computa Neurol* **272** 475–488.

[6] Randel N Asadulina A Bezares-Calderon LA Veraszto C Williams EA Conzelmann M Shahidi R and Jekely G 2014 Neuronal connectome of a sensory-motor circuit for visual navigation *eLife* **3** e02730

[7] Peng F and Chittka L 2017 A simple computational model of the bee mushroom body can explain seemingly complex forms of olfactory learning and memory *Curr Biol* 27 224-230

[8] Smith, L.B. and Stone, L.K. (2017) A Developmental Approach to Machine Learning? *Front Psychol*, 8, 2124.

[9] Colas C Karch T Sigaud O and Oudeyer, P-Y 2022 Autotelic Agents with Intrinsically Motivated Goal-Conditioned Reinforcement Learning: A Short Survey *J Artif Intell Res* **74** 1159-1199

[10] Risko EF and Gilbert SJ 2016 Cognitive Offloading *Trends Cogn Sci* **20** 676-688

IOP Conf. Series: Materials Science and Engineering 1292 (2023) 012019 doi:10.1088/1757-899X/1292/1/012019

[11] Clark A 1996 Being There: Putting Brain, Body, and World Together Again (Cambridge, MA: MIT Press)

[12] Fish FE Legac P Williams TM and Wei T 2014 Measurement of hydrodynamic force generation by swimming dolphins using bubble DPIV *J Exp Biol* **217(2)** 252–260

[13] Armitage KL Bulley A and Redshaw J 2020 Developmental origins of cognitive offloading *Proc Royal Soc B* **287** 20192927

[14] Carvalho JT and Nolfi S 2016 Cognitive offloading does not prevent but rather promotes cognitive development *PLoS One* **11(8)** e0160679

[15] Oyama, S. (1985). The Ontogeny of Information: developmental systems and evolution (Durham, NC: Duke University Press)

[16] Zador A 2019 A critique of pure learning and what artificial neural networks can learn from animal brains. *Nat Commun* **10(1)** 1-7

[17] Loulakov A Shuvaev S and Zador T 2021 Encoding innate ability through a genomic bottleneck *bioRxiv* doi:100.1101/2021.03.16.435261

[18] Barabási DL Beynon T Katona A Perez-Nieves N 2023 Complex computation from developmental priors *Nat Comm 14(1)* 2226

[19] Alicea B Chakrabarty R Gopi A Lim A and Parent J 2021 Embodied continual learning across developmental time via developmental Braitenberg vehicles *arXiv* 2103.05753

[20] Simon HA 1979 Information processing models of cognition *Ann Rev Psych* **30** 363-396

[21] Rabinovich MI and Varona P 2018 Discrete sequential information coding: heteroclinic cognitive dynamics *Front Computa Neurosci* **12** 73

[22] Di Paolo EA 2010 Robotics Inspired in the Organism *Intellectica* **53(1)** 129-162

[23] Kennedy A 2022 The what, how, and why of naturalistic behavior *Curr Opin Neurobiol* **74** 102549

[24] Shea N 2012 Reward prediction error signals are meta-representational *Nous* **48(2)** 314–341

[25] McDonough KL Costantini M Hudson M Ward E and Bach P 2020 Affordance matching predictively shapes the perceptual representation of others' ongoing actions *J Exp Psych: Hum Percept Perf* **46(8)** 847–859

[26] Thompson DW 1942 On Growth and Form (Cambridge, UK: Cambridge University Press)

[27] Briscoe J and Kicheva A 2017 The physics of development: 100 years after D'Arcy Thompson's "On Growth and Form" *Mech Dev* **145** 26-31

[28] Fonseca-Azevedo K and Herculano-Houzel S 2012 Metabolic constraint imposes a tradeoff between body size and number of brain neurons in human evolution *PNAS* **109(45)** 18571-18576

[29] Huang HJ Kram R and Ahmed AA 2012 Reduction of metabolic cost during motor learning of arm reaching dynamics *J Neurosci* **32** 2182-2190

[30] Del Giudice M and Crespi BJ 2018 Basic functional tradeoffs in cognition: An integrative framework *Cognition* **179** 56-70

[31] Heitz RP 2014 The speed-accuracy tradeoff: history, physiology, methodology, and behavior. *Front Neurosci* **8** 150

[32] Inouye BD Brosi BJ Le Sage EH and Lerdau MT 2021 Tradeoffs among resilience, robustness, stability, and performance and how we might study them *Integ Comp Biol* **61(6)** 2180–2189

[33] Gonzalez-Lagos C Sol D and Reader SM 2010 Large-brained mammals live longer *J Evol Biol* **2(3)** 1064–1074

[34] Reséndiz-Benhumea GM Sangati E Sangati F Keshmiri S and Froese T 2021 Shrunken Social Brains? A minimal model of the role of social interaction in neural complexity *Front Neurorobot* **15** 634085

[35] Ma Z Turrigiano GG Wessel R and Hengen KB 2019 Cortical circuit dynamics are homeostatically tuned to criticality *in vivo Neuron* **104(4)** P655-P664

[36] Camazine S Deneubourg JL Franks NR Sneyd J Theraula G and Bonabeau E 2003 Self-organization in biological systems (Princeton, NJ: Princeton University Press)

[37] Heylighen F 2016 Stigmergy as a universal coordination mechanism I: Definition and components. *Cogn Syst Res* **38** 4-13

[38] Dorigo M Bonabeau E and Theraulaz G 2000 Ant algorithms and stigmergy *Future Gener Comput Syst* **16(8)** 851-871

[39] Khuong A Gautrais J Perna A Sbai C Combe M Kuntz P Jost C and Theraulaz G 2015 Stigmergic construction and topochemical information shape ant nest architecture. *PNAS* **113(5)** 1303-1308

[40] D'Angelo, A., Pagello, E. (2016). From Stigmergy to Affordance: The Mechanical Basis of Robot Motion Control. In: Menegatti, E., Michael, N., Berns, K., Yamaguchi, H. (eds) Intelligent Autonomous Systems 13. Advances in Intelligent Systems and Computing, vol 302. Springer, Berlin.

[41] Berry EDJ Allen RJ Mon-Williams M and Waterman AH 2019 Cognitive Offloading: structuring the environment to improve children's working memory task performance. *Cogn Sci* **43(8)** e12770

[42] Dunbar RIM 2009 The social brain hypothesis and its implications for social evolution *Ann Hum Biol* **36(5)** 562-572

[43] Barack DL and Gold JI 2019 Temporal tradeoffs in psychophysics *Curr Opin Neurobiol* **37** 121-125

[44] Hsu A Smith RG Buchsbaum G and Sterling P 2000 Cost of cone coupling to trichromacy in primate fovea J Opt Soc Am **17(3)** 635-640

[45] O'Farrell S Sanchirico JN Spiegel O Depalle M Haynie AC Murawski SA Perruso L and Strelcheck A 2019 Disturbance modifies payoff in the explore-exploit tradeoff *Nat Commun* **10** 3363

[46] Finlay BL and Darlington RB 1995 Linked regularities in the development and evolution of Mammalian brains *Science* **268(5217)** 1578-1584

[47] Streidter GF 2005 Principles of Brain Evolution (Sunderland, MA: Sinauer)

[48] Storm BC Stone SM and Benjamin AS 2017 Using the Internet to access information inflates future use of the Internet to access other information *Memory* **25(6)** 717-723

[49] Alicea B Cialfi D Lim A and Parent J 2022 Gibsonian Information: a new approach to quantitative information *Stud Comput Intell* **1032** 2

[50] Goswami A 2007 Cranial modularity and sequence heterochrony in mammals *Evol Dev* **9(3)** 290-298

[51] Smith KK 2002 Sequence heterochrony and the evolution of development *J Morphol* **252** 82–97

[52] Bininda-Emonds ORP Jeffrey JE and Richardson MK 2003 Is sequence heterochrony an important evolutionary mechanism in Mammals? *J Mamm Evol* **10** 335–361

[53] Newell KM Liu Y-T and Mayer-Kress G 2003 A dynamical systems interpretation of epigenetic landscapes for infant motor development *Infant Behav Dev* **26(4)** 449-472

[54] Griffiths PE 2017 Genetic, epigenetic and exogenetic information in development and evolution *Interface Focus* **7(5)**

[55] Hoffman DD Singh M and Prakash C 2015 The Interface Theory of Perception *Psych Bull Rev* **22(6)** 1480-1506

[56] Chemero AP 2009 Radical Embodied Cognitive Science (Cambridge, MA: MIT Press)

[57] Kirchhoff M Parr T Palacios E Friston K and Kiverstein J 2018 The Markov blankets of life: autonomy, active inference and the free energy principle *J Royal Soc Interface* **15(138)** 20170792

[58] Herculano-Houzel S 2016 The Human Advantage: How Our Brains Became Remarkable. (Cambridge, MA: MIT Press).

[59] Kurniawati H 2022 Partially Observable Markov Decision Processes and Robotics *Annu Rev Control Robot Auton* **5** 253-277

[60] Nussenbaum K and Hartley CA 2019 Reinforcement learning across development: What insights can we draw from a decade of research? *Dev Cogn Neurosci* **40** 100733

[61] Yang Y-P and Tian X 2019 Combining users' cognition noise with interactive genetic algorithms and trapezoidal fuzzy numbers for product color design *Comput Intell Neurosci* **2019**

1019749

[62] Hernandez-Leal P Kartal B and Taylor ME 2019 A survey and critique of multiagent deep reinforcement learning. *Auton Agents Multi-Agent Syst* **33** 750–797

[63] Newman SA Forages G and Muller GB 2006 Before programs: the physical origination of multicellular forms *Intl JDev Biol* **50** 289-299

[64] Griffiths PE and Tabery J 2013 Developmental Systems Theory: what does it explain, and how does it explain it? *Adv Child Dev Behav* **44** 65-94

[65] Watson RE DeSesso JM Hurtt ME and Cappon GD 2006 Postnatal growth and morphological development of the brain: a species comparison *Birth Defects Res B: Dev Reprod Toxicol* **77(5)** 471-484

[66] Petralia RS Mattson MP and Yao PJ 2014 Communication Breakdown: the impact of aging on synapse structure *Aging Res Rev* **14** 31-42

[67] Maravita A and Iriki A 2003 Tools for the body (schema) *Trends Cogn Sci* **8(20)** 79-86

IOP Conf. Series: Materials Science and Engineering 1292 (2023) 012020 doi:10.1088/1757-899X/1292/1/012020

Understanding Lived Experience: Bridging Artificial Intelligence and Natural Language Processing with Humanities and Social Sciences

Roxana Girju

Departments of Linguistics and Computer Science, University of Illinois at Urbana-Champaign, Urbana IL 61802

E-mail: girju@illinois.edu

Abstract. True human-centered Artificial Intelligence (AI) is impossible without addressing the inherent and diverse aspects of humanness. Deep learning models have achieved remarkable success in some tasks in vision and language processing, and few can deny it. However, as it moves forward, the field cannot continue to pretend it can do it all by itself, especially when we advertise it as 'human-centered AI'. It has come the time to open up the stage for methodological pluralism in the interest of critical and democratic science, and for the benefit of society. In this paper, I want to draw particular attention to the aspect of *lived (subjective) experience*, one research area highly misunderstood and hugely neglected in AI, and especially in Natural Language Processing (NLP). Our intensions, selfhood, autonomy, emotions, feelings, sensory knowledge, cultural history are integral components of our intelligence. Thus, the future AI and NLP models will need to more closely align with the embodied component of human intelligence. As we push the limit of creativity and innovation in AI, we need to develop a new way of looking at human experience, with a better scientific understanding of intelligence and its own practices, at the intersection of many disciplinary fields.

1. Introduction

In the last decade, Artificial Intelligence (AI) has seen a resurgence in almost all its core areas, including vision, natural language processing (NLP), speech recognition, image and video generation, multi-agent systems, and robotics. This has impacted a wide number of applications in a variety of domains like games, medical diagnosis, autonomous driving, language translation, and interactive personal assistance. People, businesses, and devices alike are contributing to the unprecedented data explosion while computing power has strongly and steadily advanced. Natural language processing, the oldest area of AI, has also seen major advances in quantiative reasoning, question answering, machine translation, text classification, speech recognition, and chatbots, across academic and industry sectors. A notable achievement is represented by large language models like ELMo, GPT, mT5, and BERT [1, 2] which can learn grammatical relations, predict word meaning, and capture basic world knowledge from statistically relevant patterns in naturally occurring text. Large quantities of data are used to tune billions of parameters such of language models[1]. By predicting the next word in a sequence, these models can generate new

[1] GPT-3, for instance, uses over one trillion words.

text with impressive mastery of language, being able to reach close to human levels for various genres like news articles, essays, fiction and computer code.

Such NLP technologies, however, do not come without challenges. The NLP community has struggled to find and use quality data especially for resource-poor languages and for specialized applications. Moreover, these state-of-the-art large language models do not exhibit a deep understanding of the texts they process which limits their applicability. Most research and tools are still mostly focused on narrow benchmark tasks and situated scenarios where "their incomplete mastery still provides value" [3]. In addition, these models' exponentially growing size and processing power impose unaffordable computational costs for real-world applications [4]. Take, for instance, Google Duplex, a recent over-the-phone natural conversation technology designed for narrow, specific but real world tasks, like restaurant reservations, which was advertised as being able to make the conversation experience as natural as possible, in human-like voice. However, putting aside the inherent ethical issues[2], most of such bots today are successful "so long as the customer stays on the 'happy path,' the ideal scenario where the customer says all the right things" [5]. Yet, our natural way of communicating is versatile and embodied, multimodal and multisensory [6, 7], requiring more seamless coupling between the NLP technology and the human-computer interface design for the tool to find its market purpose and better address customer needs.

The limitations of current language technologies like such conversational interfaces are even more apparent in critical applications like healthcare. Although AI has brought much needed solutions to various biomedical applications like drug discovery and basic health operations, it significantly lacks in dedicated resources in diagnosis and practices of care within and outside clinics and hospitals. The need to provide optimized, individualized, and person-centered care is growing. Thus, addressing these competing needs and complex problems requires novel and creative approaches to health care. Specifically, dedicated AI tools could aid in the development of innovative, effective, and person-centered solutions to health challenges, supporting the realization of a future for health care that is personalized and participatory in nature [8, 9, 10].

Another major focus has been on the inevitable ethical concerns about the different and inherent types of human and machine bias, as well as about the kind of moral decisions that should guide machine behaviour [11]. As expected, these issues have spurred wide debates in the humanities and social sciences circles, with more universities and private companies being forced to consider more seriously such dimensions. However, the dialogue across AI and NLP, and the humanities and social sciences (as well as other fields) has weakened over time. This is an interesting paradox: the more AI is becoming part of our daily life, the more isolated it seems to get. Recent studies which examined data from 1950 to 2018 [12] have shed new light on this tendency of AI to face social and cultural questions on its own, disregarding the importance of and situated expertise in the humanities and social sciences. In fact, the disconnect is mutual as scholars of the social sciences, physical sciences, and humanities appear to be falling behind the quick development of AI. Moreover, these statistics show a shift in AI research domination today, with the more recent concentration of AI research in private industry due to easy access to expensive infrastructure and compute power resources. To these we also add conceptual and financial obstacles that drive these communities even further apart [13].

As a researcher, teacher, and mentor, I have been fortunate to work and collaborate across a variety of disciplines that allowed me to look at AI problems from a wide range of perspectives. I, too, join the growing number of researchers in different fields to argue for the crucial importance of collaborative practices that have the potential to foster greater synergy among diverse communities. Such an approach will have a huge impact on tech innovation, leading to the much needed human-centric vision of future AI technologies. Even though there have been some

[2] For instance, the system does not identify itself as a bot, nor does it make any disclosure of how the the recorded data will be used.

IOP Conf. Series: Materials Science and Engineering 1292 (2023) 012020 doi:10.1088/1757-899X/1292/1/012020

initiatives that highlighted perspectives from various academic fields as well as across public-private sectors to further our understanding of the key emerging issues [14, 13, 15], they are still relatively rare.

As such, in this paper, I want to draw particular attention to the aspect of *lived (subjective) experience*, one research area highly misunderstood and hugely neglected in AI, and especially in NLP. One major reason is the incorrect representation and data assumptions we make in AI and NLP about natural intelligence, which is tightly linked to our bodies. Our intensions, selfhood, autonomy, emotions, feelings, sensory knowledge, cultural history are integral components of our intelligence [16]. Thus, the future of AI models will need to more closely align with the embodied component of human intelligence. As we push the limit of creativity and innovation in AI, we need to develop a new way of looking at human experience, with a better scientific understanding of intelligence and its own practices. In spite of recent advancements in the emergent field of embodied intelligence, we are still looking for answers when it comes to issues like how to represent and make sense of human instinct, social awareness and interaction, and human-centric machine collaboration [17]. Research along such dimensions will point to directions for creating more robust, trustworthy, and perhaps actually intelligent AI systems which align more closely with the biological aspect of human intelligence.

The correction I am suggesting in this paper is, indeed, intended to foster innovation in the technology sector by diversifying the rather narrowly established practices in AI and NLP. Given the highly interdisciplinary nature of these fields, it is rather odd to see the trends, paradigms, and key insights of other cognate fields (like psychology, philosophy, sensory anthropology, history) which have not been taken on board by AI researchers. As such, we should not disregard the important disciplinary turns and methodologies in other relevant input communities of practice and areas of research [18], especially when it comes to meaning and felt experience. It is thus important for NLP, in particular, to call for *relevance* – for language technologies to be more open to interdisciplinary collaboration, more accessible, and in effect, socially consumable.

I argue that such a phenomenological approach to AI and NLP has exciting implications for how to design and build semantic systems that get us closer to felt experiences in context — meaning, what people know and feel, and how deeply they are engaged sensorially. It also allows us to investigate the areas of intersection between AI and NLP, on one side, and human communication, perceptual needs and cultural values, on the other.

The paper is structured as follows. I start with a brief history of AI and scientific inquiry as a disembodied, objectivist view of the world, followed by a discussion on the disconnect between this approach to science and human experience (Section 3). In Section 4, I focus in particular on the phenomenon of lived experience, and argue for the crucial role of AI and NLP in its systematic, empirical investigation (Section 5). In Section 6, I outline some challenges, suggest some ways in which AI and NLP and other disciplinary fields can work together toward building such a framework of understanding, and offer some future directions.

2. Science as a Disembodied, Objectivist View of the World

The Turing test [19] is a famous method of inquiry in AI for determining whether or not a computer system is intelligent – meaning, capable of thinking like a human being. It is attributed to the mathematician Alan Turing whose ideas led to the early versions of modern computing and the foundations of the field. He was, arguably, the first to analyze the role of embodied intelligence, not only locating the complexity of AI in the sophistication of the hardware and software, but also taking into consideration that language is embodied, as part of the larger context of the human-computer interaction.

In 1950, Turing proposed 'The Imitation Game' - an engineering solution to the question "Can Machines Think?" [20]. Instead of providing a definition of intelligence, Turing referred to two key intellectual human abilities: reasoning and generating illusions. As such, he suggested to imitate the human intellect in five areas: various games; language learning; language translation; cryptography; and mathematics. Turing defended the thesis that '[t]he extent to which we regard something as behaving in an intelligent manner is determined as much by our own state of mind and training as by the properties of the object under consideration'.

Turing's remarks, however, have been largely simplified and overlooked by subsequent AI research. Since the statistical turn of the nineties, AI has shifted more and more toward what is known as 'weak AI', using practical approaches that infer statistical correlations from ample datasets, moving away from the need for a more comprehensive understanding of human experience and behavior [21, 22, 23]. It has been a series of shortcuts, trade-offs, and simplified assumptions that led to the narrow-scope success of current AI systems. While the research community is divided into roughly two camps that have followed different paths towards building intelligent systems (symbolic and non-symbolic AI), it is clear that the current NLP practices tend to focus on scaling and benchmarking [24], while the state-of-the-art models do not have a good grasp on issues like common sense, knowledge acquisition, or how we use language to engage with both our external and internal world.

The cognitive revolution of the 1950's was an intellectual movement that focused on an interdisciplinary study of the mind and its processes. New voices at the time (like Herbert Simon, Noam Chomsky, Marvin Minsky, John McCarthy), put forth ideas that would later serve as the main principles of modern cognitive science, linguistics, and the nascent field of artificial intelligence [25]. The main intuition behind it was that human intelligence resembles computation in its essential characteristics. To date, the scientific method is built around testable hypotheses and it relies on facts established through repeated measurements and agreed upon universally by what is considered an independent observer. Here, "observer" refers to a physical system capable of acting as a measuring device (i.e., the third-person view of the world).

Under such a scientific understanding of reality, the observer of the experience is "a disembodied eye, looking objectively at the play of phenomena" [25]. The fundamental assumption of this view is that we inhabit a pre-given world with particular properties (e.g., length, color, etc.) which we, as objective observers, represent internally as a set of symbols and manipulation operations on these symbols. Under such a theory, intelligent behavior presupposes the ability of the agent to act by representing relevant features of the pre-given world. The agent's behavior is considered successful, to the extent to which her representation of the world (or situation) is accurate. In other words, the evaluation of such systems relies on the representation and manipulation of symbols as accurately capturing some pre-defined features of the real world, and the information processing leads to a successful solution of the given problem.

3. The Disconnect between Science and Experience

This tacit, unquestioned commitment to an objectivist, disembodied view of how reality manifests, how we come to know the world, and what we are, has dominated the AI world pretty much from its inception.

Two of the major critiques of such a world view contend that symbol processing does not seem to be the right approach for representing experience and that operations on symbols can be specified using only their physical form, not their meaning – i.e., "[t]he symbolic level is not reducible to the physical level" [25]. However, it is clear that the capability of the symbols to represent objects/ideas does not depend on how they are referenced in the world (e.g., the word 'tree' - has nothing to to with the way we identify and refer to it in the physical world). Thus, in such computational systems, words are regarded as ungrounded symbols, given the disconnect between them and the world in which they are used and experienced [26]. Yet, we do not treat words as arbitrary symbols. Instead, we rely on the affordances of the word, in an embodied way — i.e., how people evaluate meaning and for what purpose in terms of their own body, actions, and goals. In this respect, language understanding models in NLP and AI are no less than disembodied computational accounts of meaning.

One domain where the ungrounded representation issue has become evident is the scientific study of experience, since "all distinctions we make in relation to the world around us are manifested primarily in our experience" [27]. Yet, concepts like Self and Other, and the gap between them have been a major subject of inquiry in philosophy and cognitive science [28] throughout the years – as one cannot experience the world through other's body, their subjective context, and their point of view. As such, to compensate for this gap, the focus has shifted toward the complex construct of empathy, as an attempt to understand the other [29].

This disconnect has lead to what is known as the "Explanatory Gap" [30] between perceiving the human from without (i.e., through causal explanations, using third-person methods of behavioral and neuroscience approaches) and perceiving the human from within (i.e., how it feels, captured by the first-person nature of lived experience). Given that lived experience is subjective and directly accessible only from first-person position of the person who lives it, its scientific study has presented challenges that cannot be adequately (and fully) addressed with the objectivist view of scientific knowledge, understanding and validity of experience. One reason is that this view requires that the act of observing, describing, reflecting on, and remembering the (past) lived experience[3] to actually 'represent' it in the present. Thus, according to Western science, since we cannot access our past experience with precision, studying experience cannot lead to scientific results that can be validated intersubjectively.

In early 1990's, non-cognitivist theories like enactivism have started to disrupt cognitive science, giving birth to novel ideas, including embodied cognition, the Bayesian brain, active inference and the free energy principle — having as main focus a generative perception wherein observers actively shape the worlds they perceive. The field has since gradually opened up to the study of lived experience after previously missing from its research agenda for many years. According to famous philosophers like Nagel [31] and Chalmers [32], while most studied cognitive science phenomena (like perception, memory, thinking, emotions) manifest in our lived experience, the field has failed so far to account for the "what it is like" – that is, the experiential or phenomenally conscious character of lived experience. Both Nagel and Chalmers address the subjectivity of conscious experience, in particular the so-called 'phenomenal consciousness' which focuses on properties like perceptions, sensations, feelings, emotions, thoughts, and wants which are different from any cognitive, intentional or functional properties of access consciousness [33].

All this time, Western science, including artificial intelligence, has favored the third-person perspective approach, thus neglecting the study of first person experience labeling it as unreliable and prone to bias. This direction has lead to a paucity of research on lived embodied experience and knowledge practices that allow us to get in contact with it. This tension between science and experience has deepened especially in artificial intelligence (and NLP, in particular) given their highly interdisciplinary nature – standing at the crossroads between many cognate fields that are indispensable in defining what is human and how we experience our inner and outer world.

4. The Science of Lived Experience

The investigation of experience from a first-person perspective can no doubt complement and enrich existing research from a third-person perspective. However, making the first-person investigation a research program is a relatively recent development. In the past few decades, different research fields have undergone changes, leading a new view of the scientific role of the observer and her embodied experience in the generation of knowledge. One such notable proposal is the enactive approach to cognition, which reached popularity with the famous Varela,

[3] Past experience is described in retrospect.

Thompson, and Rosch's study of the Embodied Mind [25]. Varela's claim was that only the development of a Science of Consciousness can handle the primacy and irreducible nature of lived experience necessary to any understanding of the mind. Thus, he called for a dialogue between mainstream third-person approaches and a disciplined, systematic and empirical exploration of experience from a first-person perspective, with its own specialized methods, procedures, and validation criteria. In fact, in his view of lived experience, first-person (i.e., phenomenological: lived through the body) and third-person (i.e., behavioral and physiological) methods, each represent one side that can and should inform and constrain investigations of the other side in terms of research questions, design, protocols of data acquisition, analysis, and validation of results.

According to enactivism, observer and world co-evolve through reciprocal interaction. Here, perception and action are not disconnected, but tightly linked as our perceptions guide our actions and our actions determine what we perceive. According to Varela, a disciplined first-person account of experience cannot be acquired with theoretical, observational practices of psychological investigation through introspection, like simple psychological questionnaires and other limited techniques still very popular in cognitive science today. The reason is that they are inadequate to describe the 'what it is like' of experience as it is lived in first person, focusing instead on reproducing people's beliefs about their experience and the world along a number of pre-determined categories of inquiry.

I, too, believe that we need to open up the study of lived experience at the intersection of many research fields, promoting the development of special methods and procedures of first-person perspective inquiry.

5. The Role of AI and NLP in the Systematic, Empirical Investigation of Lived Experience

Human communication is, in fact, multimodal and multisensory, integrating multiple sensory and effector modalities such as speech, gestures, body posture and movement, and even senses like touch, smell, and taste. Our natural ways of thinking, interacting with and experiencing the world, and the knowledge generated this way are all embodied. This idea is somewhat connected to Whorf's 1956 linguistic determinism which asserts that the particular language one speaks influences the way one thinks about reality [34]. These views influenced research in linguistics, but also psychology, ethnology, anthropology, sociology, philosophy, as well as the natural sciences. Of course, the concept of embodied experience is more complex and sophisticated than Whorf's hypothesis suggests, but the main idea is that cognition goes beyond the simple manipulation of formal symbols, including sensing and interacting with the environment and one's own body.

As we move toward immersive experiences in augmented reality, virtual reality, mixed reality and want to build technologies that extend human capabilities, it becomes clear that we need to ground abstractions to familiar experiences so that they are meaningful. Concrete concepts are mainly grounded in sensory-motor information, whereas abstract word meanings are underpinned predominantly. For instance, if we take a word like 'pencil' and look it up in a dictionary, we get definitions like "a thin cylindrical pointed writing implement; a rod of marking substance encased in wood" [35]. Yet, by grounding the concept, it might also bring back a blend of sensorimotor experiences of seeing, touching, holding, smelling such pencils. In this way, the concept PENCIL will be grounded for us in ways to encapsulate these lived experiences. So, thinking about and remembering concrete and even abstract concepts includes, rather automatically, the way we use them and experience them sensorially [26].

5.1. Appropriate Methodologies for the Collection of both first- and third-person data
In Chalmers' views, the purpose of a science of consciousness is to connect first-person data to third-person data [36] (investigating behaviour, brain processes and environmental interactions accessible through known scientific methods). First person data are about conscious experience like various sensory experience (e.g. vision, auditory, tactile, smell and taste), bodily experience (e.g. pain), mental imagery (e.g. recalled visual images), emotional experience (e.g. happiness, fear), and concurrent thought (e.g. the experience of deciding) [37]. Chalmers has argued that reductive strategies to explain conscious experiences are doomed to fail [32, 38]. Even if science could provide a complete functional explanation, we are still left with the question of why is this functioning associated with the particular subjective experience. First-person data are not data about objective functioning, as "to live in accordance with scientific theory [..] undermines our genuine subjectivity" [39].

Scientists are thus facing the problem of finding good methodologies for collecting the data (both first-and third-person), expressing them in suitable language, and finding connecting principles. Thus, scientists have to develop methods in both domains.

The last few decades have seen a fascinating development of methods in psychology and especially in neuroscience (e.g. brain imaging, single cell studies), as well as impressive improvement in computer science and engineering with data-driven statistical models. Thus, while the science behind the third-person domain is well understood, unfortunately, this is not the case when it comes to lived experience, whose domain is missing well-developed methods for gathering and analyzing first-person data. Traditions that have largely disappeared from the scientific study of experience (i.e., introspection and philosophical phenomenology) are now making a comeback, offering new first-person approaches to the study of consciousness [40]. In fact, many new first person methods today seem to merge ideas from introspectionism and phenomenology. Introspection, which was introduced by Wilhelm Wundt in the 19th century (see [41]), is a psychological method where "one attends carefully to one's own sensation and reports them as objectively as possible" [42] – thus, describing the felt sensation and not the stimulus that provoked it. On the other hand, for early phenomenologists like Brentano, the mental acts (i.e., judging, sensing, imagining, or hearing) reflect a sense of direction and purpose [43]. Under such a framework, the phenomenal structure of experience involves intentionality, in addition to sensory qualities of the object of purpose. Thus, one cannot study phenomena like thoughts and judgement except by taking into account one's inner phenomenal experience [44].

6. Some Challenges of the Study of Experience
To describe an experience, one needs first to 'get in touch' or reconnect with it (or with some aspects of it), observe it (in retrospect), self-reflect, and/or communicate it to others. All these aspects bring up the issue of validity, which under an objectivist scientific approach, reduces to determining the extent to which we can compare them against the 'right' experience (i.e., the 'ground truth') 'as it was'. In this section, I will take a look at some important questions encountered in the scientific study of experience, identifying challenges centered around three main topics: (a) memory, (b) description and expression, and (c) intersubjectivity of lived experience [45].

Despite the major progress in AI in almost all its standard areas powered by large-scale data and high compute power resources, it still needs a significant amount of work in order to imitate, extend, simulate and augment many aspects of human embodied intelligence. Specifically, any investigation of lived experience requires a framework of both theoretical orientations, practices of knowledge generation, and empirical validation that account for this experiential dimension. In the subsections corresponding to each challenge, I will suggest some ways in which AI and NLP can help toward building such a framework of understanding.

6.1. Memory: Reconnecting and Recounting Experience

"This naked moment may well be apprehended with greater acuity in retrospect, but how can we know if what we view in hindsight will ever have truly been?" [46]. This quote brings up one important fact about experience - that it is always lived in the retrospect. This aspect poses a significant challenge to any first-person methodology that seeks to capture all its dimensions. Even in the case of recent (i.e., just-elapsed) experience, there is always some delay, forcing us to observe, describe, and reflect upon it as it has already happened. Thus, any study of experience requires an understanding of the phenomenon of memory. For this, we need a set of conceptual and epistemological orientations accompanied by appropriate methodological approaches to lived experience. Such resources have the potential to lead to a better understanding of the process of relating to, connecting with, and recalling past experience. In addition, such an understanding is directly coupled with first-person data collection, the language we use in recalling the past, and the methods of validity of first-person research.

One major scientific, observer-independent challenge of the memory aspect of experience is that it cannot be accessed with precision – meaning, we do not have a 'ground truth' of the exact experience event 'as it happened' (i.e., due to inherent "temporal" and "interpretative" distortions [47] involved in the process). In fact, memory researchers have expressed concerns about human confabulation, self-deception, and tendency toward "false memories" [48, 49]. In response, some researchers have started developing specific methodological tools to mitigate such inherent memory challenges. For instance, tools like DES (Descriptive Experience Sampling) [50] and MMP (Micro-phenomenology) [51] are aimed at minimizing memory distortions (e.g., the temporal gap between the past experience and its examination in the present) and decreasing the possibility of false memories [52], as well as determining the extent to which the person remembering and reliving the past experience is "in contact" with it in the present [47]. Of course, retrospective evaluations can lead to memory bias, so such research should be done with caution. For instance, some studies [53] have found discrepancies between individuals' recall of affective experiences and their momentary report during that time. Additionally, different people show different abilities to describe their experiences, and this may be related to different abilities that lead to specific experiences (i.e., reacting to different memory triggers).

Contrary to the objectivist approach, more recent (enactive) accounts seem to converge on the ideas that the only valid starting point for examining the process and epistemology of relating to, remembering, and even 'reliving' the past experience is that only from present experiencing can we relate to past experience [47]. This is only possible from what makes sense to us of that particular experience as we investigate it in the present moment. However, this recognition has so far not been fully validated theoretically and empirically. Although psychology and cognitive science have long recognized the dynamic and constructive nature of memory [54], mainstream cognitive science accounts of memory in the scientific community still largely regard it as storage space, and remembering as a retrieval mechanism that should "correctly" recover the content stored in the past as it was.

It is clear that a more complete theoretical and empirical understanding of memory is desperately needed in the field of first-person studies if we want to better understand (past) experience. Specifically, I argue that, the few promising contemporary enactive theoretical accounts in cognitive science (see, for instance, [55, 56, 57]), must be validated by rigorous empirical, data-driven investigations in artificial intelligence and natural language processing. With few exceptions coming from computer vision and human-computer interaction (like [58, 59]), AI (and especially NLP) have not even scratched the surface when it comes to first-person research and lived experience. The AI community should and can bring its own contribution to the science of experience through new practices, tools, computational platforms, and methods of validity which address the specific peculiarities of experience.

6.2. Expressing and Describing Experience through Language

Once the past experience is retrieved and potentially re-lived to a significant extent, the next step is to express and communicate it to others. This way, any scientific first-person research can undergo intersubjective validation. Although expressing or communicating an experience can be done in a number of modalities, most of the time we communicate our experiences via language. In fact, some researchers believe that first-person research require "a linguistic description of the content intended by the reflecting act" [60].

Under the mainstream scientific approach, verbal descriptions of evoked experiences are measured against accurate descriptions of how they happened in the past (i.e., an ideal 'ground truth'). Even assuming people are able to reconnect to and recall the past experience as accurately as possible, their ability to describe and share it can vary considerably. This is not only due to social, cultural, and educational differences in vocabulary use, but also to our capacity for self-reflection, to mention a few factors. In fact, a large portion of our experience remains unnoticed, hidden, and pre-reflective [61] due to our tendency to focus on the 'what' of our experience, instead of how it unfolds and feels to us. Thus, like an iceberg immersed in water, most of it is not readily and even easily accessible to awareness, let alone to verbal description and communication. This not only complicates the process of comparing the inherent linguistic variations across people and experiences, but also the validation process of describing experience.

Nevertheless, this does not mean that one cannot have any access to past experience from the present moment, even partially. We simply need a set of resources and tools that help us become aware of such hidden lived aspects of the lived experience, and then describe it in detail and more accurately. There have been some attempts in this direction in cognitive science [51, 50], but these methods have been tested in a rather solipsistic framework and have not yet been empirically validated on a large scale.

I argue that advances in NLP, deep learning, human-computer interaction, and other areas of AI should be seriously considered not only to test such methods, but to improve them with large-scale data-driven approaches.

6.3. Validating the Scientific Inquiry into Lived Experience

Research on experience, as any kind of research, requires intersubjective validation. Therefore, although one can investigate their own experience without the requirement to reflect on and communicate the outcome of that inquiry, in scientific research, there can be no inquiry without description or communication. Given the subjective and rather personal dimension of experience, it cannot be easily corroborated, verified, and even less replicated by mainstream scientific practices. Nevertheless, many believe [62, 47] it is possible to communicate experience and make it accessible to others through shareable descriptions and expressions of experience. These translate into specialized resources and tools that can assist along all stages of the introspection from reconnecting to past experience, focusing attention to various aspects of it, describe it (through language or other means), acquire first-person data about it, all the way to building dedicated tools and resources to interpret and validate it.

Given the special qualitative character of experience which is always subjective and from first-person perspective, Nagel and Chalmers have argued that strategies for explaining experiences must be different from mainstream objectivist approaches. Previous attempts in psychology to explain experience scientifically (e.g., introspectionism, philosophical phenomenology) reached a dead end given their results could not be intersubjectively verifiable, bringing doubts about such methods.

However, the state-of-the-art in cognitive science research on this aspect has not passed the brainstorming phase. I, again, argue here that, in our quest for a methodological and intersubjective framework of lived experience, we need a cross-method approach coming from a wider set of cognate fields. Artificial intelligence, and natural language processing in particular, can and should offer a suite of resources and tools for the methodological and especially empirical exploration and validation of subjective experience. Specifically, we need tools and resources to widen our sensory abilities and to develop our own sensory capacity for fine-grained descriptions of our inner emotional processes.

7. Conclusion

True human-centered artificial intelligence is impossible without addressing the inherent and highly diverse aspects of humanness. Deep learning models have achieved remarkable success in some tasks in vision and language processing, and few can deny it. However, as it moves forward, the field cannot continue to pretend it can do it all by itself, especially when we advertise it as 'human-centered AI'. It has come the time to open up the stage for methodological pluralism in the interest of critical and democratic science, and for the benefit of society. Specifically, AI and NLP need a serious reassessment of existing theories, methodologies, and foundational principles, along with the so much taken-for-granted scientific assumptions. The community also needs to facilitate the exploration of new methodologies. This way language processing can take its rightful place toward a diversity of approaches in NLP/AI.

In this paper I have called for a broader dialogue between AI and NLP, on one side, and humanities and social sciences on the other side when it comes to the future of true human-centered technologies. Specifically, I addressed the versatile topic of lived experience which needs a highly-interdisciplinary approach for a more accurate, robust, and ultimately, more meaningful understanding of the practices of being human.

Such diversity and methodological pluralism will lead to the creativity and innovation so much needed to restore the harmony of the AI ecosystem.

References

[1] Peters M E, Neumann M, Iyyer M, Gardner M, Clark C, Lee K and Zettlemoyer L 2018 *CoRR* **abs/1802.05365** (*Preprint* 1802.05365) URL `http://arxiv.org/abs/1802.05365`

[2] Devlin J, Chang M, Lee K and Toutanova K 2018 *CoRR* **abs/1810.04805** (*Preprint* 1810.04805) URL `http://arxiv.org/abs/1810.04805`

[3] Littman M L, Ajunwa I, Berger G, Boutilier C, Currie M, Doshi-Velez F, Hadfield G, Horowitz M C, Isbell C, Kitano H *et al.* 2021 *Stanford University. Retrieved: September* **16** 2021

[4] Bender E M, Gebru T, McMillan-Major A and Shmitchell S 2021 *Proceedings of the 2021 ACM Conference on Fairness, Accountability, and Transparency* FAccT '21 (New York, NY, USA: Association for Computing Machinery) p 610–623 ISBN 9781450383097 URL `https://doi.org/10.1145/3442188.3445922`

[5] Horn S 2018 *Parsing the turing test* (VentureBeat) URL `https://venturebeat.com/ai/google-duplex-is-a-giant-leap-for-ai-powered-bots/`

[6] Girju R 2021 *Rethinking the Senses: A Workshop on Multisensory Embodied Experiences and Disability Interactions, the ACM CHI Conference on Human Factors in Computing Systems* (arXiv preprint arXiv:2110.15054)

[7] Girju R and Girju M 2022 *Proceedings of the Second Workshop on Bridging Human–Computer Interaction and Natural Language Processing* (Seattle, Washington: Association for Computational Linguistics) pp 21–27 URL `https://aclanthology.org/2022.hcinlp-1.3`

[8] Matheson G O, Pacione C, Shultz R K and Klügl M 2015 *American journal of preventive medicine* **48** 472–479

[9] Chan K 2018 *World Medical & Health Policy* **10** 111–119

[10] Göttgens I, Oertelt-Prigione S *et al.* 2021 *JMIR mHealth and uHealth* **9** e28102

[11] Awad E, Dsouza S, Kim R, Schulz J, Henrich J, Shariff A, Bonnefon J F and Rahwan I 2018 *Nature* **563** 59–64

[12] Frank M R, Wang D, Cebrian M and Rahwan I 2019 *Nature Machine Intelligence* **1** 79–85

[13] McGillivray B, Alex B, Ames S, Armstrong G, Beavan D, Ciula A, Colavizza G, Cummings J, De Roure D, Farquhar A *et al.* 2020 *The challenges and prospects of the intersection of humanities and data science: A white paper from The Alan Turing Institute* (The Alan Turing Institute)

[14] National Academies of Sciences, Engineering, and Medicine and others 2018 *The integration of the humanities and arts with sciences, engineering, and medicine in higher education: Branches from the same tree* (National Academies Press)
[15] The National Humanities Center 2021 *The National Humanities Center* (The National Humanities Center) URL https://nationalhumanitiescenter.org/wp-content/uploads/2021/10/NHC-Artificial-Intelligence-Teachin
[16] Genovese S 2020 *ORDO* **71** 444–449
[17] Roy N, Posner I, Barfoot T, Beaudoin P, Bengio Y, Bohg J, Brock O, Depatie I, Fox D, Koditschek D *et al.* 2021 *arXiv preprint arXiv:2110.15245*
[18] Boddice R and Smith M 2020 *Emotion, Sense, Experience* Elements in Histories of Emotions and the Senses (Cambridge University Press)
[19] Turing A M 1956 *The world of mathematics* **4** 2099–2123
[20] Turing A M 2009 *Parsing the turing test* (Springer) pp 23–65
[21] Dreyfus H L 1992 *What computers still can't do: A critique of artificial reason* (MIT press)
[22] Steels L 2007 *Fifty years of AI: From symbols to embodiment-and back* (Springer) pp 18–28
[23] Cristianini N 2019 Shortcuts to artificial intelligence
[24] Michael J, Holtzman A, Parrish A, Mueller A, Wang A, Chen A, Madaan D, Nangia N, Pang R Y, Phang J *et al.* 2022 *arXiv preprint arXiv:2208.12852*
[25] Varela F J, Thompson E and Rosch E 2017 *The embodied mind : cognitive science and human experience* revised edition ed (Cambridge, Massachusetts ; London England: The MIT Press) ISBN 978-0-262-52936-5
[26] Nathan M J 2021 *Foundations of embodied learning: A paradigm for education* (Routledge)
[27] Maturana H R and Varela F J 1987 *The tree of knowledge: The biological roots of human understanding.* (New Science Library/Shambhala Publications)
[28] Zahavi D 2014 *Self and other: Exploring subjectivity, empathy, and shame* (Oxford University Press, USA)
[29] Levinas E 1979 *Le temps et l'autre [Time and the Other]* (Paris, France: PUF)
[30] Levine J 1983 *Pacific philosophical quarterly* **64** 354–361
[31] Nagel T 1974 *The philosophical review* **83** 435–450
[32] Chalmers D J 1995 *Journal of consciousness studies* **2** 200–219
[33] Block N 1995 *Behavioral and brain sciences* **18** 227–247
[34] Whorf B L 1956 *Language, thought, and reality: selected writings (Edited by John B. Carroll)* (Technology Press of MIT)
[35] Fellbaum C 2010 *Theory and applications of ontology: computer applications* (Springer) pp 231–243
[36] Chalmers D J 1999 *First-person methods in the science of consciousness* (Consciousness Bulletin)
[37] Chalmers D J 2004 *There's something about Mary: Essays on phenomenal consciousness and Frank Jackson's knowledge argument* 269
[38] Chalmers D J 1996 *The conscious mind: In search of a fundamental theory* (Oxford Paperbacks)
[39] Froese T 2022 *Irruption Theory: A Realist Framework for the Efficacy of Conscious Agency* (PsyArXiv)
[40] Shear J and Varela F J 1999 *The view from within: First-person approaches to the study of consciousness* (Imprint Academic)
[41] Hatfield G 2005 *Introspective evidence in psychology* (John Hopkins University Press)
[42] Gardner H 1987 *The mind's new science: A history of the cognitive revolution* (Basic books)
[43] Fréchette G 2020 *Franz Brentano and Austrian Philosophy* (Springer) pp 101–115
[44] Markič O 2012 *Interdisciplinary Description of Complex Systems: INDECS* **10** 213–222
[45] Valenzuela-Moguillansky C and Demšar E 2021 *Adaptive Behavior* 10597123221084739
[46] Shanker A 2019 *Inward and toward: Volume III* (lulu.com) p 718
[47] Petitmengin C and Bitbol M 2009 *Journal of Consciousness studies* **16** 363–404
[48] Johansson P, Hall L, Sikstrom S and Olsson A 2005 *Science* **310** 116–119
[49] Johnson M K and Raye C L 1998 *Trends in cognitive sciences* **2** 137–145
[50] Hurlburt R T and Akhter S A 2006 *Phenomenology and the Cognitive Sciences* **5** 271–301
[51] Petitmengin C, Remillieux A and Valenzuela-Moguillansky C 2019 *Phenomenology and the Cognitive Sciences* **18** 691–730
[52] Petitmengin C, Remillieux A, Cahour B and Carter-Thomas S 2013 *Consciousness and cognition* **22** 654–669
[53] Zygar-Hoffmann C and Schönbrodt F D 2020 *Collabra: Psychology* **6**
[54] Wagoner B 2021 *Journal: The Palgrave Encyclopedia of the Possible* 1–6
[55] Brouillet D 2020 *Frontiers in psychology* **11** 114
[56] Caravà M 2021 *Phenomenology and the Cognitive Sciences* **20** 703–722
[57] Hutto D D and Peeters A 2018 *New directions in the philosophy of memory* (Routledge) pp 97–118
[58] Taylor S 2018 *Proceedings of the 2018 Annual Symposium on Computer-Human Interaction in Play Companion Extended Abstracts* CHI PLAY '18 Extended Abstracts (New York, NY, USA: Association for Computing Machinery) p 85–89 ISBN 9781450359689 URL https://doi.org/10.1145/3270316.3270598

[59] Győrbíró N, Larkin H and Cohen M 2010 *Proceedings of the 7th International Conference on Advances in Computer Entertainment Technology* ACE '10 (New York, NY, USA: Association for Computing Machinery) p 124–125 ISBN 9781605588636 URL https://doi.org/10.1145/1971630.1971673
[60] Depraz N E, Varela F J and Vermersch P E 2003 *On becoming aware: A pragmatics of experiencing.* (John Benjamins Publishing Company)
[61] Petitmengin C 2021 *Journal of Phenomenological Psychology* **52** 36–77
[62] Hurlburt R T 2011 *Investigating pristine inner experience: Moments of truth* (Cambridge University Press)

Open to Evolve Embodied Intelligence

W B Langdon

Department of Computer Science, University College London, Gower Street, WC1E 6BT, UK

E-mail: `w.langdon@cs.ucl.ac.uk`

Abstract. For the goal of automatically evolving Embodied Intelligence (EI), we investigate an open software architecture inspired by the high surface area to volume ratio of animal lungs, which aims to avoid information theoretic limits on long term evolution experiments (LTEE) encountered with monolithic genetic programming trees. Instead individuals are teams composed of 1023 trees whose inputs and outputs are linked by a low entropy loss branching data (air) pathway. Most trees are shallow and software engineering's failed disruption propagation (FDP) is observed in the small fraction of deep trees. After initial search, most improvements are at intermediate depths and performance is still rising even after 100 000 generations.

Despite the use of double precision for the bifurcating data interconnect, some information loss is seen, particularly in early generations. The static optimisation benchmark, appears to encourage early convergence, which locks the population into possibly sub-optimal phenotypes. Later thousands of small improvements, sometimes in large bloated ensemble members, appear to compensate for early overfitting.

Using tournament fitness selection and subtree crossover, we target pure nested side-effect free floating point functions, which are known to have low FDP, and high fidelity data paths, in the hope of generating code which is not too robust so as to prevent on going improvement. However, we again find genetic changes deep within trees are silent. For single precision, we find a maximum evolvability sweet spot with trees of depth 10 to 100. Accordingly, we suggest to evolve very large very complex programs needed for Embodied Intelligence, an open structure with a high surface area permitting most mutation sites to be within 10–100 levels of the organism's environment, and many better placed test oracles to monitor the impact of mutated code, will be needed.

1. Open Complex System

Efforts to evolve artificial intelligence are running into information theoretic limits on program depth. Instead can we evolve software systems which are like cell interiors (e.g. Figure 1). These could have processing concentrated in thin computing membranes in a permeating data interconnect environment. The high surface area membranes might be composed of very many small adjacent programs each of limited depth placed side-by-side. The membranes forming an open structure with many gaps between them. The gaps themselves supporting high bandwidth communication with no, or little, processing ability and consequently little information loss. Figure 2 shows 1300 small programs arranged in an open structure.

Rich Lenski, in his long-term evolution experiment (LTEE) [3, 4] has demonstrated that Nature can continue to innovate in a static environment for more than 75 000 generations. We have shown genetic programming [5] can do similarly for at least 100 000 generations. However, when evolving deep structures, progress slows dramatically and therefore we feel monolithic

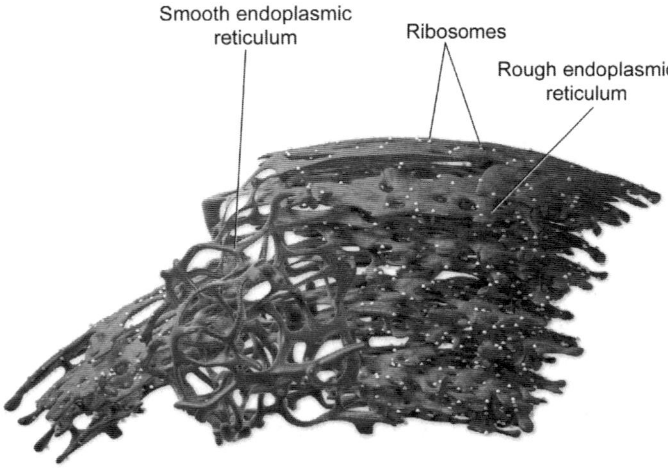

Figure 1. Small fraction of one of the active membranes within eukaryotic cells [1].

Figure 2. Lung like open complex evolving system composed of 1300 individual genetic programming functions (average height 9.22). These compute element are placed side-by-side to form an open structure. The gaps promote short-cut side effects between functions' input and outputs and the environment [2].

deep structures will not be sufficient to automatically evolve complex systems. Instead an open structure like Figure 2 may be needed.

Section 3 describes initial experiments using an architecture inspired by our lungs. Section 4 considers the discovery of the maximum evolvability sweet spot and how it may change over time or in other experiments. In Section 5 we consider ways to improve these experiments, and in Section 6 we conclude that although we do see continued fitness improvement, there is still a long way to go to demonstrate automatic Embodied Intelligence. But first, the next section describes recent work based on information theory, which lead to the view that failed disruption propagation is inevitable in both human designed and automatically evolved digital software, and therefore the need to control the depth of nesting in Embodied Intelligence.

2. Background: Evolution and Failed Disruption Propagation in Nested Software

Taking advantage of modern high performance parallel computers, we have been investigating the long term evolution of genetic programming (GP). Firstly, using Poli's submachine code GP [6, 7], to evolve large binary Boolean trees [8] and more recently exploiting SIMD Intel AVX and multi-core parallelism to evolve floating point GP [9, 10]. Running for up to a million generations without size limits has generated, at two billion nodes, the biggest programs yet evolved and forced the development [11, 12, 13] of, at the equivalent of more than a trillion GP operations per second, the fastest GP system [14, 15, 16]. It has also prompted detailed analysis of programs [17], including from an information theoretic [18] perspective [19, 20, 21]. (Of course information theory has long been used with evolutionary computing, e.g. [22].)

One immediately applicable result has been the realisation that in deep GP trees most changes have no impact on fitness and once this has been proved, for a given child, its fitness evaluation can be cut short and fitness simply copied from the parent. This can lead to enormous speed ups [23].

Without insight [24] into the construction of the test suite (i.e. in black box testing), and where fitness testing is limited by failed disruption propagation (FDP) [25], simply increasing the number of tests suffers diminishing returns and their joint effectiveness at *best* increases by a slow logarithmic $O(\log n)$ factor when tests are independent [21]. Instead FDP suggests, if possible, fitness tests should be augmented by checking values as close to the genetic change as possible. Alternatively to alleviate FDP, we could move the genetic changes closer to the existing test oracle [26] (at the tree's root). However this risks generating huge amounts of "dead" non-evolving deep code, which once created becomes fossilised and is never subsequently modified.

We have also considered traditional (human written) imperative programs and shown these to are much more robust than is often assumed [27, 28]. Indeed we suggest that information theory provides a unified view of the difficulty of testing software [29, 30, 25], particularly positioning test oracles [31, 32, 20].

The question of why fitness is so often exactly inherited [33, 34, 35], despite brutal genetic change, is answered by the realisation that without side effects, the disruption caused by mutation or crossover must be propagated up the tree through a long chain of irreversible operations to the root node. Being irreversible, each function in the chain can loose entropy. In many cases deeply nested functions progressively loose information about the perturbation as the disruption fails to propagate to the program's output giving rise to a deep neutral network. Thus the mutations and crossovers become invisible to fitness testing and their utility cannot be measured. Without fitness selective pressure, evolution degenerates into an undirected random walk [8, 10].

In bloated structures information loss leads, from an evolutionary point of view, to extremely high resilience, robustness and so stasis. From the engineering point of view this is problematic, as then almost all genetic changes have no impact and evolutionary progress slows to a dawdle.

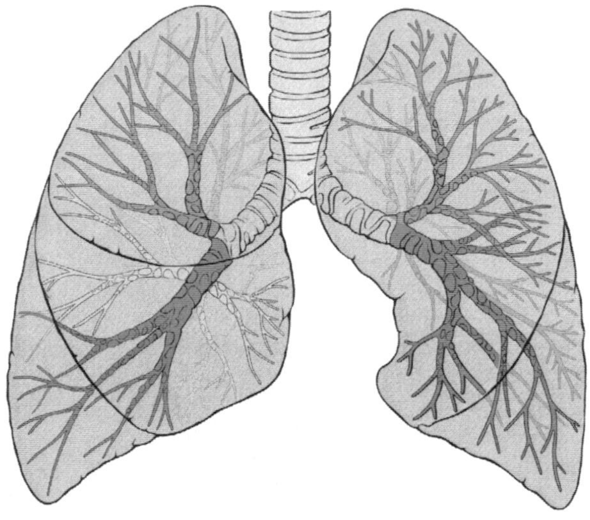

Figure 3. Human lungs [38].

Since all digital computing is irreversible [36], it inherently losses information and so without short cuts, must lead to failed disruption propagation (FDP). We suggest in order to evolve large complex systems it must be possible to measure the impact of genetic changes, therefore we must control FDP and suggest in the next section that to evolve large systems, they be composed of many programs of limited nesting depth and structured to allow rapid communication of both inputs and outputs to the (fitness determining) environment.

3. Experiments with a High Surface Area Lung Like Architecture

We seek a software architecture with a large interface between thin evolvable code and I/O data. We take inspiration from lungs (Figure 3), which hold a large surface area between the animal's body, particularly its blood vessels, and the air, in order to permit exchange of gases such as carbon dioxide and oxygen out of and into the animal. Although the lungs of different mammals vary enormously in size, the airways show a common bifurcating pattern with the diameter of the airways decreasing as $2^{-\frac{n}{3}}$ for the first 17 branches. Branching continues to n=24 but with a slower power law [37].

To limit computational costs, in our initial experiment we limit our "lungs" to a bifurcation depth of n=10, giving 511 branch points (bronchioles) and 512 alveoli ends (total 1023, Figure 4). We anchor a traditional genetic programming (GP) tree [5] at each (giving 1023 evolvable trees per individual). We think of each tree as having the task of regulating the passage of data through the data channel beneath it. The output of each tree is discharged into the channel. In addition to the usual GP primitives (see Table 1), trees at branch points can monitor data upstream of them, data in the channel to their left and to their right. All the trees can read how deep they are in the "lung" using the special leaf D, which has a value 1 to 10. The 512 end point trees (D=10) in each individual are evaluated first. These are followed by the 256 trees closer to the "lung" "output" (D=9). Thus level 9 trees can access left, right and current values, which have been calculated from the outputs of the neighbouring level 10 trees. These are followed by the 128 level 8 trees, and so on until the level 1 tree.

To minimise entropy loss, double precision is used throughout the data channel, e.g. for calculating "current". However as usual, normal single precision floating point precision is used inside the trees and they generate single precision outputs. Double precision is also used when calculating the mean of the 1023 outputs during fitness calculation.

IOP Conf. Series: Materials Science and Engineering 1292 (2023) 012021 doi:10.1088/1757-899X/1292/1/012021

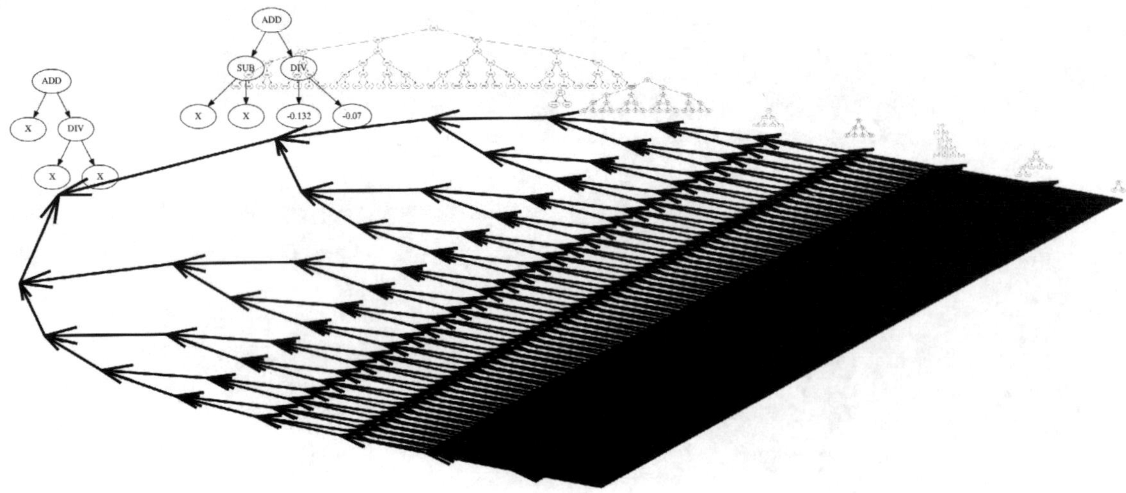

Figure 4. Lung like data structure. Data flows from 512 alveoli ends (right) collecting outputs of evolving trees and providing them with addition inputs. Ten of 1023 trees shown.

Table 1. Genetic programming lung experiment with Sextic polynomial symbolic regression

Tree leafs:	X D and 246 constants between -0.995 and 0.997
	(bronchioles, D=1 to D=9) trees have additional leafs: left right current
Tree internal nodes:	MUL ADD DIV SUB
Fitness test cases:	48 fixed inputs -0.97789 to 0.979541 (randomly selected from -1.0 to +1.0)
	Target (Sextic polynomial) $y = xx(x-1)(x-1)(x+1)(x+1)$
Selection:	Tournament size 7 with fitness $= \frac{1}{48}\sum_{i=1}^{48} \lvert GP(x_i) - y_i \rvert$
Population:	500 teams each of 1023 trees. Population is fully mixed (panmictic), non-elitist, with separate, non-overlapping, generations.
GP parameters:	Initial trees are each created with ramped half and half [39], depth between 2 and 6. 100% unbiased subtree crossover. 100 000 generations. No size or depth limits.

DIV is protected division `(y!=0)? x/y : 1.0f`

For each test case, the GP's output is the mean of all 1023 trees. Fitness is the sum over the 48 test cases of absolute difference between the mean output and the target value. Occasionally an individual tree gives a very bad non-finite value. This invalidates the mean of all the trees, but such individuals have very poor fitness and are seldom selected to be parents and are excluded from the population without any special processing.

3.1. Initial Results

Figure 5 shows, as hoped, the new architecture is able to solve the problem in a reasonable time[1] and continues to show improvements even after many thousands of generations[2].

The runs show some aspects of GP convergence [10]. Although Figure 6 shows there is variation between runs, they all show large increase in size (bloat). Across the ten runs, 86% of trees in the best individual in the population do not change throughout the last 90% of the run. The distribution of these "converged" trees has a long tail but most are small and contain only

[1] On average the ten runs each took less than two days on a single Intel i7-4790 3.60 GHz core.

[2] On average 608 children that were better than their mums were created in the last 1000 generations.

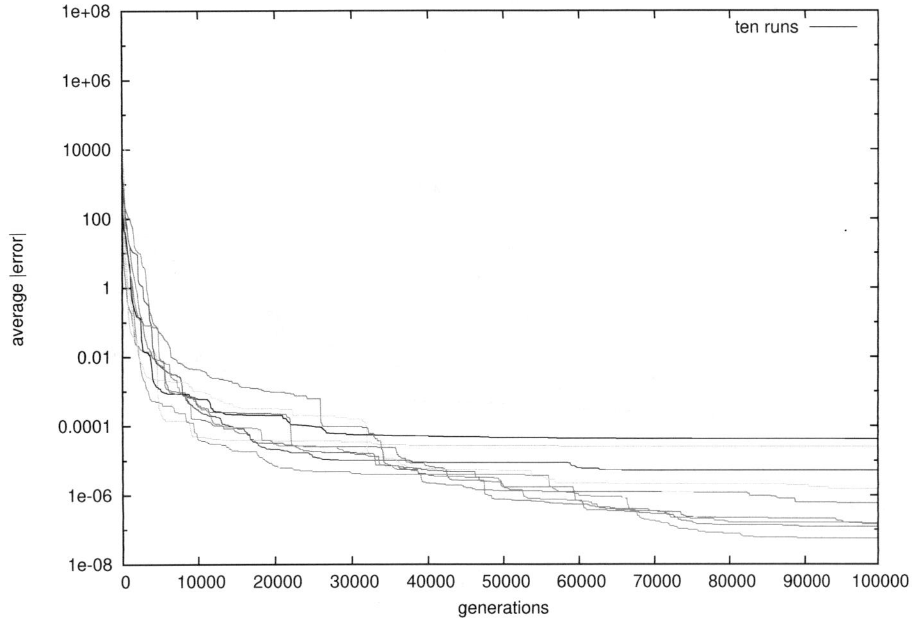

Figure 5. Evolution of mean absolute error in ten runs of Sextic polynomial [39] with population of 500, each team composed of 1023 trees. One crossover per team per generation, i.e. 1022 trees copied without modification from the 1st parent (mum).

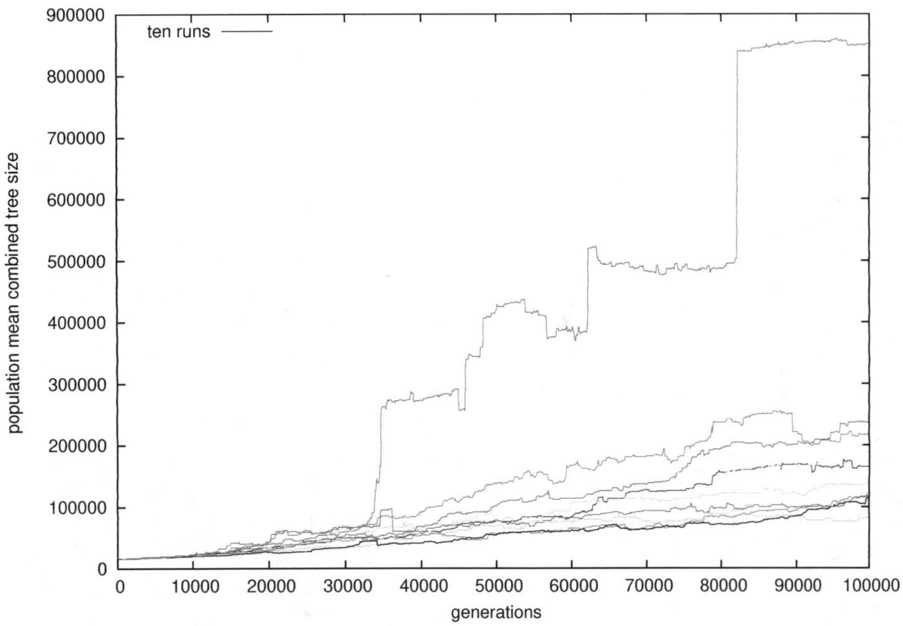

Figure 6. Evolution of total ensemble [40] size in ten runs. (Same colours as runs in Figure 5.)

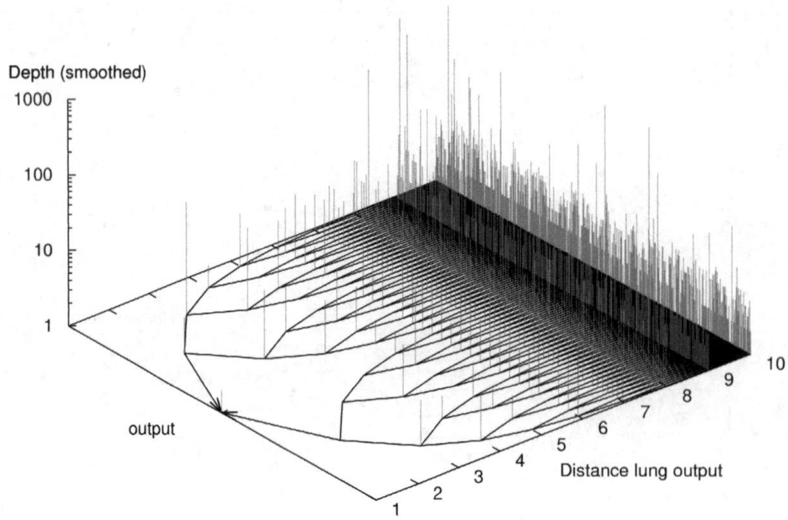

Figure 7. Depth of best in population GP trees after 100 000 generations in a typical run.

1, 3 or 5 nodes (i.e. up to two functions and three leafs). Indeed, at the end of the runs, about half the best trees contain five or fewer nodes. In contrast, of those tree which do change their height in the last 90% of the run, 41% (5% of all trees) contain on average more than 10 levels by the end of their run. It is the largest trees amongst this 5% that give the great increase in size seen in Figure 6.

The location of the large trees is different in each run, and appears to be essentially random. Figure 7 shows for the best individual in the first run the depth and location of its constituent trees.

We expect our large evolved "lung" trees to behave as other large binary GP trees, so that on average the distance from crossover points to the trees' output is about half the depth of the tree [41, 42, 10]. For deep trees this means any disruption caused by crossover may fail to propagate to the root node and so the child's fitness will be equal to its parent's. Figure 8 reports failed disruption propagation (FDP) in the deep trees. The solid curve in Figure 8 is the empirical fit for FDP which suggests at least half of crossovers deeper than 63 levels will make no difference. (With other mixtures of functions, e.g. containing logic variables or conditional statements, we expect the depth needed to be much less [8, 20].)

In the case of shallow trees, crossovers must be near the root node, however Figure 9 shows that at the end of the runs, although almost all shallow crossovers do disrupt fitness only a tiny fraction improve fitness. Indeed by the end of the 10 runs the 2300 trees composed of a single leaf were sufficiently converged, that in the last 1000 generations all 1 122 850 crossovers gave back the same child. As expected with trees containing at least one function (i.e. size 3) most crossovers are disruptive (until trees are deep enough to suffer FDP). These small converged trees are the most resistant to improvement with only about 0.1% of children being better. However intermediate converged trees are more evolvable as deeper crossovers become possible and the fraction of improvements increases to about 0.5%. The deepest improvement is 335 levels from the root node (Figure 9). But only three of the 6 845 crossovers deeper than 200 improve fitness.

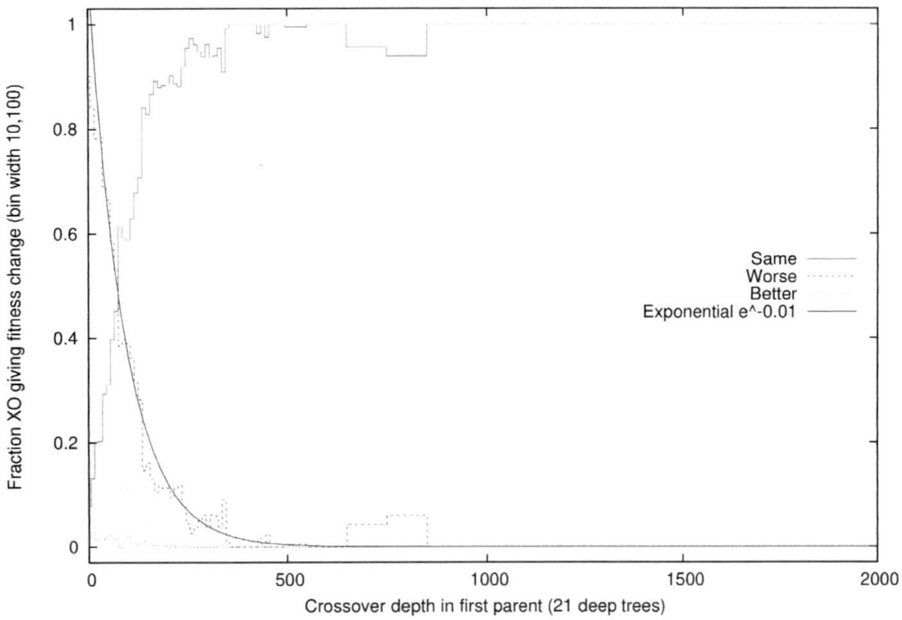

Figure 8. Solid red line shows the fraction of crossovers in 21 trees deeper than 300 levels that do not change fitness. Dashed blue make it worse. Dotted purple improve fitness. Grouped by distance from crossover site to the root node. Above 500, the bin size is increased from 10 to 100. The smooth curve (black) is the best RMS fit of an exponential decay to the fraction of crossovers which do change fitness. (Data gathered from last 1000 generations in all ten runs.)

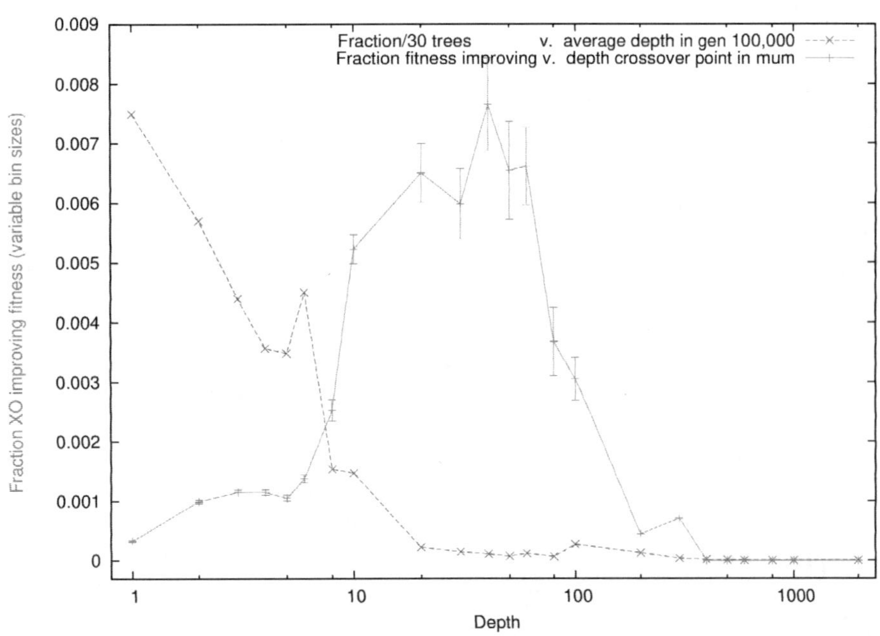

Figure 9. Solid purple line gives, for the ten runs in the last 1000 generations, the fraction of crossovers whose disruption not only propagates to the root node but also improves fitness (see also Figure 8). For comparison the dashed blue line gives the fraction of trees by their average depth at the end of the ten runs (rescaled by 1/30 to plot on same axis). Note log horizontal scale.

IOP Conf. Series: Materials Science and Engineering 1292 (2023) 012021 doi:10.1088/1757-899X/1292/1/012021

4. Evolution of Maximum Evolvability

In the last 1000 generations there is more chance of crossovers at least 7 levels deep being successful (see Figure 9). It appears that this is because this late in the run the population is now good and so only has scope for small fitness improvements. It seems that deeper changes may find small phenotype changes which are sometimes beneficial, whereas code changes close to the root node may make larger phenotypic changes, which could overshoot the target output. This trend seems to continue to about depth 20. Still deeper, failed disruption propagation (FDP) starts preventing many crossovers changing the phenotype at all.

We anticipate over considerably extended evolution, perhaps many millions of generations, continued improvement of the ensemble, will mean the fitness step size will continue to decrease, moving the "left cliff" of the evolvability sweet spot in Figure 9 to the right. However the "right FDP" edge will not move. Leading eventually to a narrowing of the maximum evolvability range. Possibly this will be accompanied by more team members moving into that narrowing range?

With a different mix of ingredients in our trees, especially if they include conditionals or Boolean operators, there may be considerably more entropy loss, so causing much more FDP. This will reduce the maximum depth limit on the location of maximum evolvability. (I.e. move the "FDP cliff" to the left.) To avoid pinching out our evolvability bubble, we now need to consider again its left hand edge. We need to ensure that the fitness function is not too discrete. Instead, the fitness calculation must be sufficiently graded, so that as we make progress and fitness improves, it is still possible for genetic changes to find small fitness improvements.

5. Discussion

In theoretical work on small digital circuits Wright and Laue [43] argue that digital evolution supports an "arrow of complexity" whereby maximum Kolmogorov complexity increases with time. Note that by using Kolmogorov complexity they measure not increase in program size (i.e. bloat) but changes in functionality. Here we see huge overfitting and so are seeing Kolmogorov complexity increase. It is doubtful that this is useful, but our goal is not to solve the benchmark problem, but to use it to explore ways and difficulties of extended evolution [44, 14].

Another interesting recent experiment by Kelly, Smith, Heywood and Banzhaf [45], considers extended genetic programming evolution (up to 89 days) using their "Tangled Program Graphs" representation. They evolve agents to play aspects of the video game ViZDoom.

In terms of our experiments, it seems we need more open ended benchmarks. We have stuck with just one genetic change per fitness evaluation [46]. However, with increasing numbers of team members this begins to look infeasible. Certainly in human genetic we see genomes being split into multiple diploid chromosomes with multiple crossover points and mutation sites and a more equal contribution of identifiable genes from the two parents. In terms of computational experiments it may be feasible to use multiple fitness measurements, possibly include cheap ways [47] to abort complete fitness evaluation early [48], to complement increased number of crossovers or mutations per individual per generation.

Although we should be wary of building too many preconception into any system which hopes to show extended evolution, having studied the smothering blanket of failed disruption propagation (FDP) in pure floating point code and also in integer [20] and logic [8] expressions, perhaps it is time to adopted more directed choices of genetic changes. Alternatively rather than cooking in our choices, perhaps such as system could learn from its own past and be self adaptive.

6. Conclusions

In all runs we see a few trees growing to enormous depth, which become resistant to phenotypic change due to failed disruption propagation. It may be that future evolving "thin membrane" systems will have to include mechanisms to prevent such growths or include mechanisms to exorcise them. We also see shallow trees quickly discovered but then become less evolvable as the team as a whole improves. Nevertheless we have that demonstrated a new many tree architecture, inspired by the branching air pathways in lungs, is capable of supporting extended evolution. As with many evolutionary systems (and indeed life on earth) it shows early lock-in followed by many small evolutionary improvements. For our pure floating point expressions, typically about 70 trees in the ensembles evolve into the sweet spot of maximum evolvability and have a depth between 10 and 100 nested levels.

Acknowledgement

This work was initially inspired by conversations at Dagstuhl Seminar 18052 on Genetic Improvement of Software [49]. Supported by the Meta Oops project.

References

[1] Blaus B 2014 *WikiJournal of Medicine* **1** ISSN 2002-4436 URL `http://dx.doi.org/10.15347/wjm/2014.010`
[2] Langdon W B 2022 *SIGEVOlution newsletter of the ACM Special Interest Group on Genetic and Evolutionary Computation* **15** ISSN 1931-8499 URL `http://dx.doi.org/10.1145/3532942.3532945`
[3] Lenski R E *et al.* 2015 *Proceedings of the Royal Society B* **282** ISSN 0962-8452 URL `http://dx.doi.org/10.1098/rspb.2015.2292`
[4] Good B H *et al.* 2017 *Nature* **551** 45–50 URL `http://dx.doi.org/10.1038/nature24287`
[5] Poli R *et al.* 2008 *A field guide to genetic programming* (Published via `http://lulu.com` and freely available at `http://www.gp-field-guide.org.uk`) (With contributions by J. R. Koza) URL `http://www.gp-field-guide.org.uk`
[6] Poli R and Langdon W B 1999 *Advances in Genetic Programming 3* ed Spector L *et al.* (MIT Press) chap 13, pp 301–323 ISBN 0-262-19423-6 URL `http://www.cs.ucl.ac.uk/staff/W.Langdon/aigp3/ch13.pdf`
[7] Poli R and Page J 2000 *Genetic Programming and Evolvable Machines* **1** 37–56 ISSN 1389-2576 URL `http://dx.doi.org/10.1023/A:1010068314282`
[8] Langdon W B 2017 *GECCO '17 Companion* ed Wagner M *et al.* (Berlin: ACM) pp 235–236 URL `http://dx.doi.org/10.1145/3067695.3075965`
[9] Langdon W B 2019 *GECCO '19 Companion* ed Doerr C *et al.* (Prague, Czech Republic: ACM) pp 63–64 URL `http://dx.doi.org/10.1145/3319619.3326770`
[10] Langdon W B 2022 *Genetic Programming and Evolvable Machines* **23** 71–104 ISSN 1389-2576 URL `http://dx.doi.org/10.1007/s10710-021-09405-9`
[11] Langdon W B 2020 Fast generation of big random binary trees Tech. Rep. RN/20/01 Computer Science, University College, London Gower Street, London, UK URL `https://arxiv.org/abs/2001.04505`
[12] Langdon W B 2020 *SIGEVOLution newsletter of the ACM Special Interest Group on Genetic and Evolutionary Computation* **13** 2–4 ISSN 1931-8499 URL `http://dx.doi.org/10.1145/3430913.3430914`
[13] Langdon W B 2021 *Genetic Programming Theory and Practice XVIII* ed Banzhaf W *et al.* (East Lansing, MI, USA: Springer) pp 143–164 URL `http://dx.doi.org/10.1007/978-981-16-8113-4_8`
[14] Langdon W B and Banzhaf W 2022 *Artificial Life* **28** 173–204 ISSN 1064-5462 invited submission to Artificial Life Journal special issue of the ALIFE'19 conference URL `http://dx.doi.org/10.1162/artl_a_00360`
[15] Langdon W B *ACM Transactions on Evolutionary Learning and Optimization* ISSN 2688-299X just accepted URL `http://dx.doi.org/10.1145/3539738`
[16] Langdon W B 2022 A trillion genetic programming instructions per second ArXiv URL `https://arxiv.org/abs/2205.03251`
[17] Langdon W B 2022 *Complex Systems* **31** ISSN 0891-2513 forthcoming
[18] Renyi A 1987 *A Diary on Information Theory* Probability and Statistics, Applied Probability and Statistics Section (Chichester: John Wiley and Sons) ISBN 0 471 9971 8
[19] Langdon W B *et al.* 2021 *5th Workshop on Landscape-Aware Heuristic Search* GECCO '21 Companion ed Veerapen N *et al.* (Internet: ACM) pp 1683–1691 URL `http://dx.doi.org/10.1145/3449726.3463147`
[20] Langdon W B 2022 *GECCO '22 Companion* ed Trautmann H *et al.* (Boston, USA: ACM) URL `http://dx.doi.org/10.1145/3520304.3528878`
[21] Langdon W B *et al.* 2022 *GECCO '22* ed Rahat A *et al.* (Boston, USA: ACM) pp 964–972 URL `http://dx.doi.org/10.1145/3512290.3528738`
[22] Deschaine L M *et al.* 2011 *Journal of Mathematical Machines and Systems* 50–61 URL `http://www.immsp.kiev.ua/publications/eng/2011_2/`
[23] Langdon W B 2021 *EuroGP 2021: Proceedings of the 24th European Conference on Genetic Programming*

(*LNCS* vol 12691) ed Ting Hu *et al.* (Virtual Event: Springer Verlag) pp 229–246 URL http://dx.doi.org/10.1007/978-3-030-72812-0_15

[24] Petke J *et al.* 2015 *IEEE Transactions on Software Engineering* **41** 901–924 URL http://dx.doi.org/10.1109/TSE.2015.2421279

[25] Petke J *et al.* 2021 *ESEC/FSE 2021, Ideas, Visions and Reflections* ed Avgeriou P and Zhang D (Athens, Greece: ACM) pp 1475–1478 URL http://dx.doi.org/10.1145/3468264.3473133

[26] Bruce B R *et al.* 2019 *IEEE Transactions on Software Engineering* **45** 1150–1169 ISSN 0098-5589 URL http://dx.doi.org/10.1109/TSE.2018.2827066

[27] Langdon W B and Petke J 2015 *Complex Systems Digital Campus E-conference, CS-DC'15* Proceedings in Complexity ed Parrend P *et al.* (Springer) pp 203–211 invited talk URL http://dx.doi.org/10.1007/978-3-319-45901-1_24

[28] Langdon W B 2016 *Software is not fragile* vol 5 (Dagstuhl, Germany: Schloss Dagstuhl–Leibniz-Zentrum fuer Informatik) chap 4.7, pp 100–101 URL http://dx.doi.org/10.4230/DagRep.5.10.89

[29] Langdon W B *et al.* 2021 Information loss leads to robustness IEEE Software Blog URL http://blog.ieeesoftware.org/2021/09/information-loss-leads-to-robustness-w.html

[30] Clark D *et al.* 2020 Software robustness: A survey, a theory, and some prospects Presented at Facebook Testing and Verification Symposium 2020 URL https://fbresearchevents.bevylabs.com/events/details/facebook-tav-symposium-division-facebook-testing-and-verification-symposium-presents-dress-rehearsal-facebook-tav-symposium-2020/

[31] Jahangirova G *et al.* 2016 *Proceedings of the 25th International Symposium on Software Testing and Analysis (ISSTA'16)* (Saarbruecken, Germany: ACM) pp 247–258 URL http://dx.doi.org/10.1145/2931037.2931062

[32] Terragni V *et al.* 2020 *ESEC/FSE 2020* ed Cohen M B and Zimmermann T (Sacramento, California, USA) pp 1178–1189 URL http://dx.doi.org/10.1145/3368089.3409758

[33] Petke J *et al.* 2019 *7th edition of GI @ GECCO 2019* ed Alexander B *et al.* (Prague, Czech Republic: ACM) pp 1715–1721 URL http://dx.doi.org/10.1145/3319619.3326870

[34] Petke J and Blot A 2020 *The First International Workshop on Automated Program Repair (APR@ICSE 2020)* ed Shin Hwei Tan *et al.* (internet: ACM) pp 13–14 URL http://dx.doi.org/10.1145/3387940.3392180

[35] Pimenta C G *et al.* 2020 *EvoCOP 2020* (*LNCS* vol 12102) ed Paquete L and Zarges C (Seville, Spain: Springer) pp 114–130 URL http://dx.doi.org/10.1007/978-3-030-43680-3_8

[36] Langdon W B and Poli R 2008 *Natural Computing* **7**(1) 21–43 invited contribution to special issue on Unconventional computing URL http://dx.doi.org/10.1007/s11047-007-9044-x

[37] Iber D 2021 *Cellular Networks in Development* (*Current Topics in Developmental Biology* vol 143) ed Affolter M (Academic Press) chap 6, pp 205–237 URL http://dx.doi.org/10.1016/bs.ctdb.2021.02.002

[38] Lynch P J and Jaffe C C 2006 Lungs diagram with internal details Yale University School of Medicine, Center for Advanced Instructional Media URL https://en.wikipedia.org/wiki/Lung

[39] Koza J R 1992 *Genetic Programming: On the Programming of Computers by Means of Natural Selection* (MIT Press) ISBN 0-262-11170-5 URL http://mitpress.mit.edu/books/genetic-programming

[40] Langdon W B *et al.* 2003 *Applications of Evolutionary Computing, EvoWorkshops2003* (*LNCS* vol 2611) ed Raidl G R *et al.* (Essex, UK: Springer) pp 87–98 URL http://dx.doi.org/10.1007/3-540-36605-9_9

[41] Sedgewick R and Flajolet P 1996 *An Introduction to the Analysis of Algorithms* (Addison-Wesley) ISBN 0-201-40009-X

[42] Langdon W B 2000 *GECCO-2000* ed Whitley L D *et al.* (Las Vegas, Nevada, USA: Morgan Kaufmann) pp 451–458 ISBN 1-55860-708-0 URL http://gpbib.cs.ucl.ac.uk/gecco2000/GA069.pdf

[43] Wright A H and Laue C L 2021 *GECCO '21* ed Chicano F *et al.* (internet: ACM) pp 840–848 URL http://dx.doi.org/10.1145/3449639.3459393

[44] Altenberg L 1994 *Advances in Genetic Programming* ed Kinnear, Jr K E (MIT Press) chap 3, pp 47–74 URL http://dynamics.org/~altenber/PAPERS/EEGP/

[45] Kelly S *et al.* 2021 *ACM Transactions on Evolutionary Learning and Optimization* **1** ISSN 2688-299X URL http://dx.doi.org/10.1145/3468857

[46] Langdon W B 1998 *Genetic Programming and Data Structures* (Kluwer) ISBN 0-7923-8135-1 URL http://dx.doi.org/10.1007/978-1-4615-5731-9

[47] Maxwell III S R 1994 *IEEE World Congress on Computational Intelligence* vol 1 (Orlando, Florida, USA) pp 413–417a ISBN 0-7803-1899-4 URL http://dx.doi.org/10.1109/ICEC.1994.349915

[48] Tackett W A 1994 *Foundations of Genetic Algorithms 3* ed Whitley L D and Vose M D (Estes Park, Colorado, USA: Morgan Kaufmann) pp 271–297 ISBN 1-55860-356-5 published 1995 URL http://dx.doi.org/10.1016/B978-1-55860-356-1.50017-0

[49] Petke J *et al.* 2018 *Dagstuhl Reports* **8** 158–182 ISSN 2192-5283 URL http://dx.doi.org/10.4230/DagRep.8.1.158

IOP Conf. Series: Materials Science and Engineering 1292 (2023) 012022 doi:10.1088/1757-899X/1292/1/012022

Two notions of medium and their implications for intelligence

Tetsushi Nonaka

Graduate School of Human Development and Environment, Kobe University, Kobe 6578501, Japan

E-mail: tetsushi@people.kobe-u.ac.jp

Abstract. In biology, the term *medium* is defined as a substance that a biological system exists or grows in or that it travels through. In physics, the term *medium* is defined as a substance that propagates and transmits the energy from one location to another. These two notions of medium in biology and physics are distinct, yet in reality, their referents frequently coincide in the same material systems in the environment. The air, a medium for terrestrial animals in the sense defined in biology, is a medium that transmits light, mechanical waves, and diffuses molecules from the source in the sense defined in physics. A medium that surrounds each cell in the living body in the sense of biology, at the same time, is an excitable medium that propagates electrical events, mechanical stresses, and the variety of chemical molecules in the sense of physics. In this article, I discuss the implications of this coincidence of the two notions of medium in the real world for the evolution and development of intelligent systems.

"If you are sitting next to a swimming pool, somebody dives in, and . . . , I can think of the waves that forms in the water, and when lots of people have dived in the pool, there is a very great choppiness of all these waves all over the water, and to think that it is possible maybe that waves there is a clue of what is happening in the pool. That some sort of insect or something with efficient cleverness could sit in the corner of the pool and just be disturbed by the waves. And by the nature of the irregularities and bumping all the way, it figures out who jumped in and where and when and where what is happening all over the pool. And that's what we do when we are looking at something."—Richard P. Feynman [1].

1. Two notions of medium

With intelligent agents, altering the conditions changes the activity displayed indefinitely, but not the end reached. The defining characteristics of intelligence is, as James [2] famously put it, "the fixed end, the varying means." What are the ecological resources available that could potentially provide opportunities for the evolution of intelligence in this James' sense? This question may sound superfluous, for what is responsible for intelligence is usually attributed to internal structure of the system that exhibit intelligent activity. Yet, as Robert Rosen [3] stated succinctly, a function of an organism can never be understood in terms of its internal structure simply because "a function requires an external context; a structure does not." The functional act of pursuance of ends and the choice of means for their attainment is partly dependent upon external factors beyond the control of the actor. For an intelligent system such as a living organism, whereas being an intricate mechanism in itself, is one whose movements is tailored to things in the environment outside of itself, much as fish in the pond; and these external things are as much constituents of the process of activity as is the living organism itself [4]. In what follows, among others, I shall focus on one class of material reality with which the development of activity of biological systems is inextricably linked, namely, the *medium* of the environment.

The term *medium* means different things in biology and physics. In biology, the term *medium* is defined as a substance that a biological system exists or grows in or that it travels through. The air, for example, is a medium for terrestrial animals, and the water is a medium for aquatic animals. A medium permits respiration, ceaseless chemical exchange of substance that sustains life. At smaller scales, for cells that collectively constitute the body of an organism, the extracellular matrix and the extracellular fluid can be considered as a medium. A medium is a substance that surrounds each structural and functional unit of biological systems at its scale.

In physics, the term *medium* is defined as a substance that propagates the energy from one location to another location. Various forms of energy are propagated in different media. Sound, for example, is a compression mechanical wave that is propagated through a medium such as the air, water, and various elastic solid materials. Unlike mechanical waves, electromagnetic waves such as light may occur in a vacuum as well as in a material medium such as air and water. In physics, a medium is where different structures arise from various physical events, while its substrate material being unchanged. For example, you can listen to music and then have a conversation with a friend in a room, without exchanging the air in the room for each event.

These two notions of medium in biology and physics are distinct, each of which has its own emphasis that captures different aspects of the world, although they are not mutually contradictory. The point I would like to draw attention is the fact that, in reality, the referents of these two senses of medium frequently coincide in the same material systems in the environment. The air, a medium for terrestrial animals in the sense defined in biology, is a medium that propagates and transmits light, mechanical waves, and diffuses molecules from the source in the sense defined in physics. A medium that surrounds each cell in the living body in the sense of biology is, at the same time, an excitable medium that propagates electrical events, mechanical stresses, and the variety of chemical molecules in the sense of physics.

What are the implications of this overlap of two senses of medium—biological and physical—for intelligence? What kind of opportunities can be potentially made available in this juncture of the two notions of medium in the real world? In the present paper, I shall discuss how an environmental medium, under certain conditions, could carry information about things and events in the world that can be potentially detected by the perceptual system of an agent. I shall also point out the parallel between the opportunities provided by the physical overlap of these two notions of medium, and the development of so-called physical reservoir computing, a framework that exploits the complex dynamics of physical systems as a computational resource [5, 6].

2. Liquid medium

The water is a medium for aquatic animals in the biological sense, which the animals grow in and travel through. At the same time, the water is a medium in the sense of physics where various patterns of water movements—from ripples and eddies to tidal waves—take place. Through evolution, most aquatic animals living in water have developed sensory systems to discriminate flow patterns in their aquatic medium [7]. These flow patterns can be caused by different source events, including the animal's own swimming movements, inanimate objects, conspecifics, predators, preys, and weather events. For example, the water movements that arise from fast-starts in teleost fish consist of multiple jets, based on which aquatic predators can detect not only the presence of a fish and its size, but also the direction in which it escaped [8]. Harbor seals use their vibrissae to haptically discriminate the flow structure left behind by predator or prey, and detect the size, shape, and motion path of the object that gave rise to the structure

[7]. The water movements caused by the swimming motion of different species of fish have distinct flow signatures which can remain for several minutes in the water (Figure 1). These flow structures provide piscivorous predators at a distance with valuable sources of information about the presence of a fish of suitable size as well as the swimming style or the species of the fish that have passed by in the recent past [7]. Unlike vision, hydrodynamic perceptual systems in the aquatic medium permits sensing the past event without the aid of internal memory, because perturbations in the aquatic medium, which gradually fade over time, can specify the events that occurred at an earlier point in time.

The flow patterns formed in the aquatic medium provide the aquatic animals with the open-ended opportunities for selectively attending to their different features to detect various events that are relevant to the habits of life of the animals. When we consider those two aspects of medium—that which surrounds organisms and that which transmits energy—hydrodynamic perception seems almost a natural consequence of their overlap in the aquatic medium. Among the conditions that potentially make hydrodynamic perception possible are as follows [9]: (1) disturbances of pattern in the aquatic medium are sensitive enough to separate different source events, and (2) the perceptual system of an aquatic agent has an enough resolution to discriminate the different patterns formed in their medium.

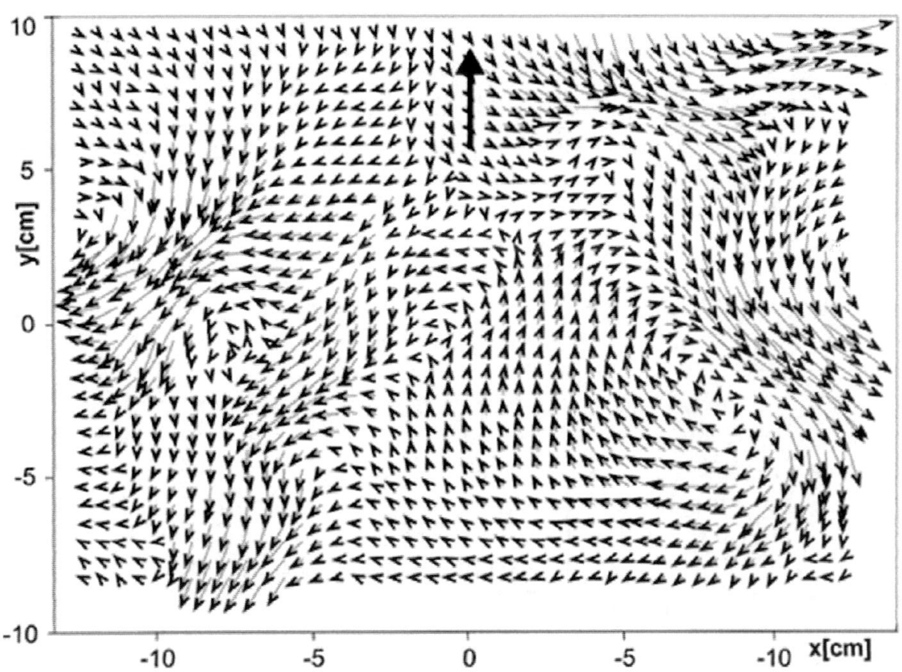

Figure 1. Water velocity 1 minute after an 86 mm-long fish (Lepomis gibbosus) swam across the area. Arrows show the magnitude and direction of the flow. Bold arrow indicates swimming direction [7].

3. Liquid state machine and reservoir computing

Independently from research on hydrodynamic perceptual systems, a computational neuroscientist Maass and his colleagues have arrived at some similar viewpoints: The perturbed states of the excitable medium could potentially make available the reservoir of information that can be exploited by intelligent systems to perform various tasks. Maass et al. [10] identified that the key challenge to understanding intelligence is to find the adequate conceptual framework that can account for how agents detect in real-time the equivalent state from the continuously changing transient stimulation which may never repeat. In tackling this challenge, Maass took seriously the material environment of neurons—the fact that "the neurons in our brain are embedded into an artificial sea-environment, the salty aqueous extracellular fluid which surrounds the neurons in our brain" [11]. Using the metaphor of liquid, Maass et al. [10] wrote as follows: "consider a series of transient perturbations caused in an excitable medium ..., for example, a liquid, by a sequence of external disturbances (inputs) such as wind, sound, or sequences of pebbles dropped into the liquid. ...the perturbed state of the liquid, at any moment in time, represents present as well as past inputs, potentially providing the information needed for an analysis of various dynamic aspects of the environment. In order for such a liquid to serve as a source of salient information about present and past stimuli without relying on stable states, the perturbations must be sensitive to saliently different inputs" [10].

Fundamental to Maass and colleagues' idea is the overlap of the two notions of medium: The medium of the neurons in our brain—the extracellular fluid—in the sense of biology is at the same time an excitable medium whose state can be perturbed by various external disturbances in the sense of physics. Using perturbations caused in a medium as informational resources, Maass et al. [10] proposed a computational model called the Liquid State Machine, which maps in real time some function of time (e.g., an aspect of the environment that changes continuously) onto other functions of time (e.g., continuous adjustments of action in relation to the changing aspect of the environment).

In terms of architecture, a liquid state machine M consists of the following two components: The first component is an excitable medium called a liquid filter L^M which generates, at every time t, perturbed 'liquid state' $x^M(t)$ in response to a preceding sequence of disturbances $u(\cdot)$. A liquid filter has the time-continuous property of fading memory whose state depends on the disturbances from some finite time window into the past, just as the perturbed states of the sea water lasts for some finite amount of time after a fish has passed by. A liquid filter needs not be customized for a specific task, just as sea water is not customized for each of the tasks that aquatic animals perform. The second component of a liquid state machine M is a readout map f^M that transforms, at every time t, the current liquid state $x^M(t)$ (which expresses the past as well as the ongoing events because of the fading memory property of the liquid filter) into task-specific output $y(t)$. Stated another way, a readout map f^M extracts the task-relevant features in the medium, and continuously adjusts the output in relation to the task goal. Depending on the task, the readout can generate an invariant response despite the fact that the high-dimensional dynamics of the liquid filter may never revisit the same state. Liquid state machines define the task-relevant equivalence classes for the dynamic liquid states [10], which instantiates one concrete implementation of intelligence in James' sense.

The conceptual framework of a liquid state machine was unified with so-called Echo State Networks—a machine learning approach independently developed by Jaeger that shares the fundamental operating principle [12]—under the umbrella label of reservoir computing. In the framework of reservoir computing, if the reservoir that exists independent of the task-specific readouts has rich and diverse enough dynamics to separate the different sources of disturbances, could make available the opportunities for real-time, task-specific control of the medium-readout system. Here, the role of a reservoir is to pick up the source event and propagating it into the high-dimensional space which could potentially render relevant features from the event more easily separable [13].

A medium-readout system can have the power for real-time computing on perturbations regardless of specific implementation or structure, if the following properties are satisfied: the separation property, approximation property [10], and echo state property [12]. The separation property addresses the amount of separation between different disturbances of structure in a medium at later time points that are caused by preceding different sequences of events (e.g., different wake signatures resulting from different sequences of source events). The approximation property refers to the property of readout function that maps the separated states into the output that meets the functional requirements of the task. The echo state property is the property that requires the reservoir states to be expressed as a function of the previous input sequence only [5].

It would be worth noting that even though the task-relevant states of the reservoir are there to be extracted, they do not determine the readout response of the system. Just as the seal running away from the white shark may not attend to a pattern of water movement that specify the presence of salmon that can be preyed upon, the liquid states that are potentially informative about some events exist independent of its use. This allows for flexibility to select the means with which to attain an end, as well as to have multiple ends at the same time. In reservoir computing, it is possible to add multiple readouts to a single reservoir so that each readout extracts different task-specific information from the rich reservoir states in such a way to support different, multiple real-time controls in parallel.

4. Body as a medium

Turvey and Fonseca [14] hypothesized that hierarchical networks of prestressed tensile elements of the body that span from the macroscale to the microscale—from muscles, tendons, and other connective tissues to various micro-elastic structures such as a network of collagen fibers—constitute the medium of the haptic sense organs of animals. Like the air being the medium for sound, odor, and reverberating flux of light, the presence of isometric tension distributed throughout all levels of interconnected, multiscale networks of the body propagate various mechanical disturbances. Thereby, the opportunities arise for an active agent to spontaneously perturb the tensile states of hierarchical networks in such a way to separate the invariant patterns that specify the source of mechanical disturbances [9].

There is evidence that provides support for this scenario of active touch that takes into account the medium for the haptic perceptual system. In a study of children learning to read braille, Nonaka, Ito, and Stoffregen [15] followed the development of scanning movements of the reading finger of blind school children over the course of one year. Contrary to the previous assumption that braille reading is optimized when the scanning finger moved at constant speed, they found that the strength of long-range power-law temporal correlations in the velocity fluctuations of lateral scanning movement were functionally related to braille reading performance and its development (Figure 2). Their result suggests that, instead of degrading perceptual sensitivity, a certain structure of velocity fluctuations contributes to separating off the invariant information about the adjacent environment over the transformations of tissue deformation patterns, and that variations in movement of the scanning finger might be controlled so as to efficiently separate informative structures that specify the braille text.

In another example, Sornkarn and Nanayakkara [16] recorded the muscle co-contraction activity of the finger using the electromyography sensors when human participants were asked to palpate a soft tissue to locate a hard nodule embedded within it. They demonstrated that human participants varied the muscle co-contraction level (muscle tone) during manual palpation to estimate the depth of a hard nodule in the soft tissue. Furthermore, in the subsequent experiment using a soft robot, the team revealed that haptic information gain about the depth of the hard nodule can be enhanced by varying the internal stiffness of the soft probe of the robot [17]. Their findings provided empirical support for the idea that the discriminatory sensitivity of the haptic perceptual system depends on the active tuning of the tensile state of soft tissues—a medium in the sense of both physics and biology where mechanoreceptors are embedded—so as to separate informative invariants.

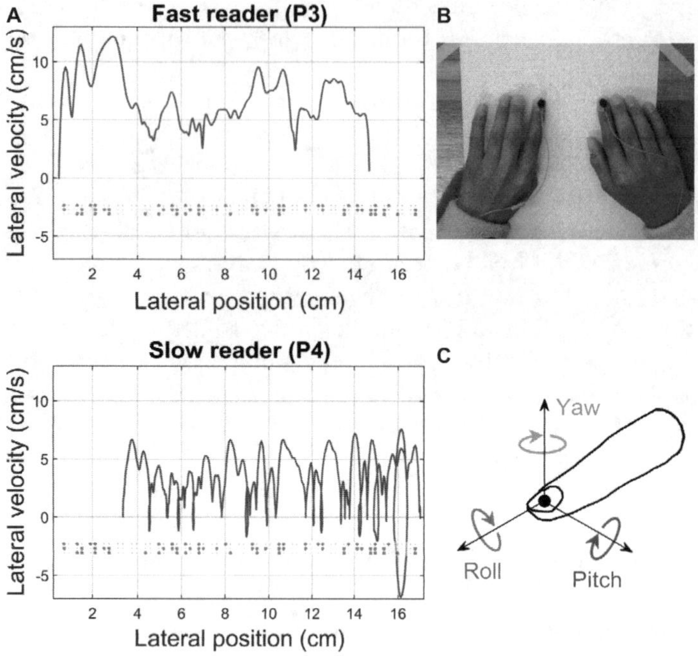

Figure 2. (A) Examples of individual lateral velocity traces as a function of the dominant reading finger's lateral position on a single line of text (shown below as rendered in contracted braille) of a fast reader (top: In this trial, the end of the line was read with the other hand) and a slow reader (bottom: In this trial, the first word of the line was skipped). (B) Set-up of braille reading experiment of children. An electromagnetic micro-sensor was attached on the nail of each index finger of the participant reading braille. (C) Representation of the three angular displacements in the sensor attached on the nail of the index finger [15].

5. Epilogue

In this paper, I explored the issue of what kind of opportunities the coincidence of the following two notions of medium in reality can make available for intelligence: (1) the substance in which an organism lives, grows, and travels and (2) the substance that propagates and transmits the energy. Drawing on the studies on sensory ecology and reservoir computing, this paper illustrated a principle how pattern and change that are formed in an environmental medium, under certain conditions, could serve as the reservoir of information that makes available the open-ended opportunities for reaching the equivalent state in the continuously changing conditions. The present paper further highlighted the possibility that the presence of isometric tension distributed throughout all levels of interconnected, multiscale networks of the body can make available the opportunities for an active perceiver to spontaneously perturb the tensile states of the body's hierarchical networks in such a way to separate the invariant patterns that specify the source of mechanical disturbances.

Acknowledgments

This work was supported by the JSPS KAKENHI (grant numbers JP21H05823, JP21KK0182, JP22H00988).

References

[1] R. P. Feynman, "Fun to imagine," 1983.
[2] W. James, *The principles of psychology.* Cosimo, Inc., 2007, vol. 1.
[3] R. Rosen, *Essays on life itself.* Columbia University Press, 2000.
[4] E. B. Holt, *The Freudian wish and its place in ethics.* H. Holt, 1916.
[5] K. Nakajima, "Physical reservoir computing?an introductory perspective," *Japanese Journal of Applied Physics*, vol. 59, no. 6, p. 060501, 2020.
[6] K. Nakajima and I. Fischer, *Reservoir computing.* Springer, 2021.
[7] W. Hanke, "Natural hydrodynamic stimuli," *Flow Sensing in Air and Water: Behavioral, Neural and Engineering Principles of Operation*, pp. 3–29, 2014.

[8] B. Niesterok and W. Hanke, "Hydrodynamic patterns from fast-starts in teleost fish and their possible relevance to predator–prey interactions," *Journal of Comparative Physiology A*, vol. 199, pp. 139–149, 2013.
[9] T. Nonaka, "Locating the inexhaustible: material, medium, and ambient information," *Frontiers in psychology*, p. 447, 2020.
[10] W. Maass, T. Natschläger, and H. Markram, "Real-time computing without stable states: A new framework for neural computation based on perturbations," *Neural computation*, vol. 14, no. 11, pp. 2531–2560, 2002.
[11] W. Maass *et al.*, "Computing with spikes," *Special Issue on Foundations of Information Processing of TELEMATIK*, vol. 8, no. 1, pp. 32–36, 2002.
[12] H. Jaeger, "The "echo state" approach to analysing and training recurrent neural networks-with an erratum note," *Bonn, Germany: German National Research Center for Information Technology GMD Technical Report*, vol. 148, no. 34, p. 13, 2001.
[13] L. F. Seoane, "Evolutionary aspects of reservoir computing," *Philosophical Transactions of the Royal Society B*, vol. 374, no. 1774, p. 20180377, 2019.
[14] M. T. Turvey and S. T. Fonseca, "The medium of haptic perception: a tensegrity hypothesis," *Journal of motor behavior*, vol. 46, no. 3, pp. 143–187, 2014.
[15] T. Nonaka, K. Ito, and T. A. Stoffregen, "Structure of variability in scanning movement predicts braille reading performance in children," *Scientific reports*, vol. 11, no. 1, pp. 1–12, 2021.
[16] N. Sornkarn and T. Nanayakkara, "Can a soft robotic probe use stiffness control like a human finger to improve efficacy of haptic perception?" *IEEE transactions on haptics*, vol. 10, no. 2, pp. 183–195, 2016.
[17] N. Sornkarn, P. Dasgupta, and T. Nanayakkara, "Morphological computation of haptic perception of a controllable stiffness probe," *PloS one*, vol. 11, no. 6, p. e0156982, 2016.

IOP Conf. Series: Materials Science and Engineering 1292 (2023) 012023 doi:10.1088/1757-899X/1292/1/012023

Between the Authentic and Artificial: From the Perspective of Contingency

Maki Sato

Fellow, The New Institute
Fellow, East Asian Academy, The University of Tokyo

esg.maki@gmail.com or maki.sato@eaa.c.u-tokyo.ac.jp

Abstract. How to deal with a contingency? A contingent situation plays a vital role in challenges with uncertainty and unpredictability. This is because contingency requires the subject to be adaptive and flexible in the given condition. Regarding the development of robotics, soft robotics is highly anticipated to be more adaptable and flexible to contingency with its bodily inputs. In contrast, material-wise, conventional hard robotics may fit more for fixed, stable, static, or predictable situations. The latest development in new materials has opened a window for more usable and useful artificial body parts as advanced tools. This short article first highlights the types of artificial body parts and summarises the role of contingency in robotics development. In addition, it analyses the various phases of the concept of the embodied mind in link with contingency. Additionally, for further discussion, it questions ethical, moral, and legal aspects regarding the future development of artificial body parts.

1. Introduction: Types of Artificial Body Parts

Soft robotics opens the possibility to various tools that work closely interacting with the human body, which may add new capabilities to the living human corps. Robots were once only objected to or externalised to humans. However, with new materials introduced, a window is opened to seek for humans not only the substituting but also the extended body parts that further the ability to interact with the surrounding environment. Parallel to such developments, different dimensional problems regarding ethics, morality, and legality are beginning to emerge.

In terms of developing artificial body parts as tools, soft robotics may provide high adaptability to both users and the user environment. Because of its plasticity, it has increased its dynamism with much complex feedback for users to easier interact with the robots under a given condition. Such interaction includes but is not limited to enabling a range of sensitivity, such as haptics of the soft robots to be possible with better visual aesthetics of its softness, which may better accommodate the users' social needs. The human body, in nature, responds to real timeliness in dealing with and corresponding with the physical world. The softness and dynamism of the soft robotics enable the robotic body parts to resonate with the given condition more effectively and through interactions with users that lead to spontaneous engaging responses than with conventional hard robotics.

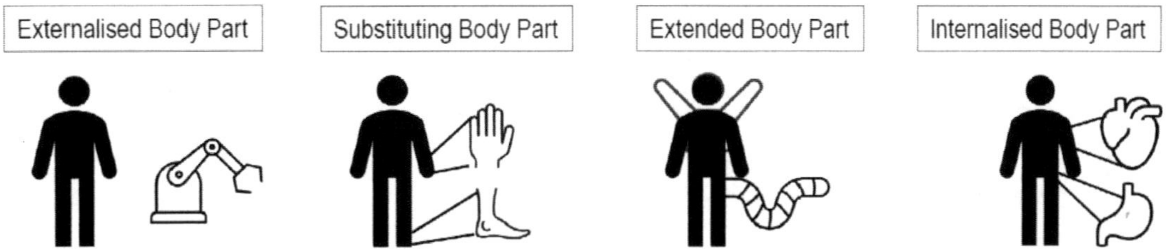

Figure 1. Types of Soft Robotics regarding Body Parts

The robotic body parts as tools may provide a wide range of possibilities, as summarised in Figure 1. Externalised body parts are a complete object of the human body and are teleoperated, such as in robot-assisted surgery. Substituting body parts work as an artificial biometric hand and prosthetic leg that substitutes the amputated original body parts. Currently, the challenges to better the aesthetics and function of the substituting body parts are the most discussed. Extended body parts are robots added to the human body to provide new functions, such as a tail and a third and fourth arms. Internalised body parts may work as a substitute for internal organs implanted entirely or partially into the human body. If utilised for assisting organ function, internalised body parts may lead to protecting animal rights and animal welfare that otherwise may be exploited for human needs. Although those artificial body parts do not necessarily relate to artificial intelligence, such needs depend on how much we require the artificial body parts to be ready for the contingent situation.

2. Embodied Mind and its Relation to Contingency

The concept of an embodiment of the mind has long been one of the critical themes in the field of cognitive philosophy [1]. The interplay and inter-responses between cognition (brain), body, and the environment are crucial to generate an adaptive and flexible fit to the given circumstances. Evolutionary robotics may bring about readiness to contingency with its improved algorithms using artificial intelligence [2]. However, the adaptability and readiness to contingency depend on the assumption of the contingency level that differs based on the supposed time range targeting future [3]. Still, the combination of soft robotics and artificial intelligence in body part development may possess high potential for coming human needs.

In terms of the fit for the robots to the level of contingency, in precise, it differs based on how much humans can predicate the uncertain and unpredicted situation beforehand, where robots are given the autonomy to adaptively react to fit with such situation. Contingency may occur with axes of the time range, noise, errors, and unusual settings, where humans cannot predetermine and manipulatively provide a program beforehand. In other words, full contingency is a situation that is a complete external happening to the predicted and structured environment where humans cannot provide a manipulation plan beforehand. Thus, for the robots to manage well with any level of contingency, their autonomous judgement and adaptative movement to the given situation will be required.[1]

A Japanese philosopher, Shūzō Kuki (九鬼周造, 1888-1941), worked on the notion of the level of contingency by examining the notion of necessity. According to Kuki, there lie different levels between

[1] Daniel Dennett articulates in *Kinds of Minds* (2007) pp.27-41, [4], physical stance, design stance, and intentional stance regarding how system could be designed taking in consideration of the future prediction. Intentional stance is to treat an entity as an agency that can predict on its own, while physical stance predicts future based on the law of physics, and design stance has its limitation in prediction from its design. In respect to contingency, the more the situation is contingent, the more the intentional stance would be required for smooth adaptation to the contingency, whereas predictable situation can be pre-programmed through physical and design stance.

necessity and contingency between the inevitable necessity, which he calls an absolute necessity (*zettai-teki-hitsuzen*, 絶対的必然), and the metaphysical contingency (*keijijō-teki-gūzen*, 形而上的偶然); namely hypothetical (仮説的) and empirical (経験的) as shown in Figure 2.[2] In terms of psychologists' perspective, Kuki's notion of metaphysical contingency may refer to future prediction, a hypothetical contingency to the theory of mind, and empirical contingency to interactions with others. From Kuki's notion of the spectrum of necessity to a contingency, this paper intends to draw out implications that may well depict the different stages involved amid intelligence and contingency.

Necessity (inevitability) ranges from absolute, empirical, and hypothetical necessities. An absolute necessity is occurrences that are entirely certain and predictable. An empirical necessity involves accumulative experiments that reduce uncertainties leading to reaction fixations. In contrast, hypothetical necessity involves a situation such as an if-then/when-then statement. On the other hand, contingency ranges from metaphysical (future prediction), hypothetical (theory of mind), and empirical (interaction with others) contingencies. Empirical contingency involves a more uncertain situation requiring neat correspondence and interactions with the environment, substances, and other living things compared with a hypothetical necessity. Hypothetical contingency may involve more flexibility and adaptability in building mutuality and inter-relationality. Metaphysical contingency involves a future prediction beyond one's experiment, which happens once in a blue moon and is unpredictable and uncertain, in the more extended chronological range than the hypothetical contingency.

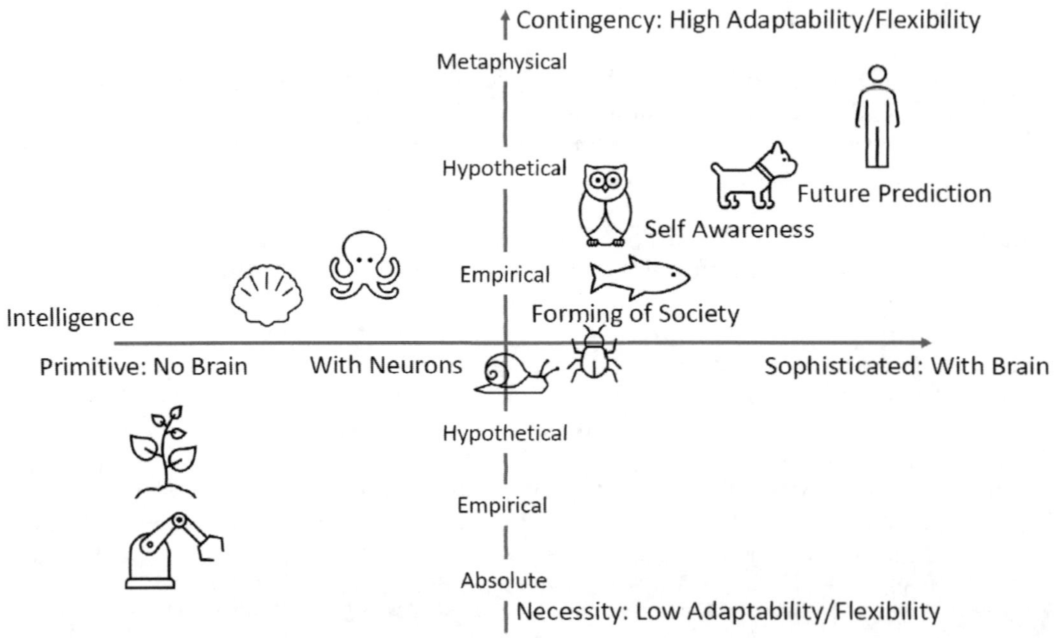

Figure 2. Contingency and Embodied Intelligence

Regarding the embodied mind, the author thinks that Kuki's notion provides implications for the future development of embodied mind in alignment with the spectrum of intelligence, from material to

[2] Note that the Figure 2 depicts relative comparison.

plant, swarm, animal, and human intelligence.[3] Figure 2 intends to show such a diverse spectrum of intelligence in accordance with the range of necessity and contingency. Primitive intelligence can be seen in mechanics as in soft materials, such as short-term memories of the material without neurons (no-brain reaction) that reform itself reactionary through its structure (body) to the given environment. Such primitive intelligence using soft materials reactively and reflexively adapts itself without neurons, which depends on the mechanical switch of the materials without using a sensor.

No-brain reactions are further broken down into categories with or without neurons. Plant intelligence is intelligence without neurons but has a reflexive (*automatic*) and responding mechanism to stimulative inputs, for example, which acts against sunlight, water, and other nutrients. In comparison, even without a brain, living beings such as seashells and octopuses own their embedded neuron (*sensory*) system within their body. With respect to no-brain intelligence, because of their flexible response to the given environment, it may imply possibilities that lead to developing a new different type of artificial intelligence, in other words, bodily artificial intelligence. Compared to the existing brain artificial intelligence that imitates brain neurons as a relay of stimulations that selects and processes information, bodily intelligence requires the body as a stimulative sensor that acquires direct information through embodiment.

Uexküll's notion of Umwelt [5] and Gibson's idea of affordance [6] both argue that the surrounding environment is conditional to subjective judgement. Concerning the robot's response to the surrounding environment, subject to the softness of the soft robots enables more flexibility and plasticity to the given condition, which may well adapt to unpredictable contingencies than the body of the hard robots. In such a sense, a soft robot can develop a new type of intelligence closely attached to bodily movement and bodily inputs rather than the existing artificial intelligence that imitates the works of brain neurons in a closed setting of existing processed information. In other words, because the weight on the body is put more on robotics, through sensory inputs such as haptics, bodily intelligence (no-brain intelligence) being closer to raw information and more open to the surrounding environment may be more adaptable to contingencies that derives from the given environment.

With-brain intelligence represents by a swarm, animal, and human intelligence in Figure 2. Swarm intelligence only requires a limited amount of information exchange using local communication. However, the more complex the intelligence becomes, the more the range of self-awareness is required. Self-awareness enables passive and active information collection based on interest (*intention*).[4] The collected information is used for building mutual relationships between the self and others. Such intention leads to delicate interactions surrounding the subject, dependent on the ability for a short-term prediction of how the subject predicts the significant other's wants and needs. An example of this interaction is pet dogs responding to their owner's wants and needs. Higher intelligence, therefore, involves predictions of a more extended range of future where the rate of uncertainty and unpredictability is undetermined and limitless. Such higher intelligence can be achieved not only through the existing brain focused artificial intelligence, but with the artificial intelligence that enables processing more on the surrounding environment through sensuous inputs from bodies. For robotics to acquire such higher intelligence, connectivity with various kinds of artificial intelligence that provides chances to analyse various opportunities and options for future prediction becomes necessary.

[3] The author's intention does not include proposing to order various intelligence since intelligence has its account in its respective ability.
[4] This notion is similar to attention schema theory (AST) by Michael Graziano [7] that tries to explain that subjective awareness arises as a schematic model of the process of attention. Graziano further argues that attention schema is a body schema, that body plays an essential role to develop consciousness.

IOP Conf. Series: Materials Science and Engineering 1292 (2023) 012023 doi:10.1088/1757-899X/1292/1/012023

3. Discussion

The paper summarised the types of artificial body parts. Additionally, it summarised the concept of the embodied mind and its relation to contingency. In terms of mind, there needs to be further discussion on the differentiation between the wide range of primitive to sophisticated intelligence: such as the clear difference between consciousness, intelligence, and mind, with and without a body.[5] However, because of the space limitation, this paper will not explore further, leaving the question open for further discussion.

Regarding ethics, morality, and legality, the discussion is open to which extent we would like our artificial body parts to acquire intelligence in terms of adaptability and responsivity to unpredictability and uncertainty that might happen in a hypothetical or metaphysical contingency.[6] The growing apprehension is whether humanity would compromise the right to amputate one's congenital body and reform one's body partially into cyborgs once the artificial body parts get better than the nature of the body. The future discussion also needs to shed light on the question of to which extent we, as humans, let our desire flourish to modify our bodies. Do we need to legally cap the desire to transform and modify our bodies using artificial body parts? Or is it up to one's preference and everlasting desire to decide whether we want to become a cyborg, dependent on an individual's aspiration to change, as we see in the plastic surgery business?

There is still some time left for a further in-depth discussion on the dimension of ethics and morality to decide on the common goals and standard settings of where we are heading and what we want from the artificial body parts that may lead to the maximisation of human capacity using artificial body parts as a tool. The author hopes this paper opens a window for engineers to have the above questions in mind while busy developing artificial body parts.

References

[1] Varela, F. J., Thompson, E., & Rosch, E. (1991/2017). The Embodied Mind, Revised Edition: Cognitive Science and Human Experience. MIT Press.

[2] Josh C. Bongard. 2013. Evolutionary robotics. Commun. ACM 56, 8 (August 2013), 74–83. https://doi.org/10.1145/2493883

[3] Kuki Shūzō. (2012). 偶然性の問題 [The Problem of Contingency]. Tokyo. Iwanami.

[4] Dennett, D. C. (1996). *Kinds of minds*. New York: Basic Books.

[5] Uexküll, J. v., 1864-1944, & Kriszat, G. (1970). *Streifzüge durch die Umwelten von Tieren und Menschen: Ein Bilderbuch Unsichtbarer Welten; Bedeutungslehre / Jakob von Uexküll, Georg Kriszat; mit einem Vorwort von Adolf Portmann; und einer Enleitung von Thure von Uexkull*. Frankfurt am Main: Fischer.

[6] Gibson, J. J., 1904-1979. (1986). *The Ecological Approach to Visual Perception*. New York: Psychology Press.

[7] Graziano, M. S. A., & Webb, T. W. (2015). The Attention Schema Theory: A Mechanistic Account of Subjective Awareness. *Frontiers in Psychology, 6*, 500.

[5] In this paper, though not descriptively defined, intelligence is used as a grounding ability detained not limited to living things but to non-living materials as well.

[6] Gray, et.al. [8] argues that in terms of mind perception, there are two substantial dimensions of agency and experience. While agency is linked with moral agency, therefore with responsibility, experience is linked with moral patience, therefore to rights and privileges. In other words, there are two-dimensional approach to moral and ethical perspective in terms of whether one is oriented as agency, such as developer or a regulator of the technology or oriented as an experience, a mere user of the technology.

doi:10.3389/fpsyg.2015.00500.
[8] Gray, H. M., Gray, K., & Wegner, D. M. (2007). Dimensions of mind perception. *Science, 315*(5812), 619. doi:10.1126/science.1134475.

IOP Conf. Series: Materials Science and Engineering 1292 (2023) 012024 doi:10.1088/1757-899X/1292/1/012024

How to define the correct guidelines for enhanced telepresence and task embodiment in remote palpation.

Leone Costi[1], **and Fumiya Iida**[1]

[1] Department of Engineering, University of Cambridge, Cambridge, UK

E-mail: `lc830@cam.ac.uk`, `fi224@cam.ac.uk`

Abstract. Teleoperated robots have been widely accepted in several fields of medical practice, enhancing human abilities and allowing remote operation. However, such technology has not been able yet to permeate areas such as primary care and physical examination. Such applications strongly rely on the quality of the interaction between doctor and patient, and on its multimodal nature. In order to achieve remote physical examination is thus mandatory to have a good doctor-robot interface, but what does good mean? Ultimately, the goal is for the user to achieve task embodiment, making the remote task feel like the in-person one. Several research groups have proposed a wide variety of interfaces, showcasing largely different methods of control and feedback, because of the absence of design guidelines. In this work, we argue that the ideal interface for a remote task should resemble as close as possible the experience provided by the in-person equivalent, keeping in consideration the nature of the target users. To support our claims, we analyze many remote interfaces and compare them with the respective in-person task. This analysis is not limited to the medical sector, with examples such as remote abdominal surgery, but it expands to all forms of teleoperation, up to nuclear waste handling and avionics.

1. Introduction

1.1. *Teleoperation in healthcare*

Teleoperation is the ability to perform a task remotely controlling a robot [1]. Teleoperation is extremely interesting when the working site is unreachable or the presence of the human user in the remote environment could be dangerous for the user itself or other individuals: nuclear matter handling, aeronautics, and medical practice [2, 3, 4]. Teleoperated systems are characterized by a user, a doctor in the case of medical applications, a leader, a follower, and a remote environment, or a patient in the case of medical applications. As a matter of fact, teleoperation has been well accepted in the healthcare system for more than 2 decades, with a large number of applications from neurosurgery to endoscopic interventions [5, 6]. The combination of enhanced precision, tremor filtering and miniaturization of the access port to the human body, whilst keeping the doctor in charge of the operation, allowed medical performance beyond human capabilities [7]. All mentioned fields are characterized by the intrinsic challenge of reaching the operation site: teleoperated robots can easily reach where the surgeons' hands cannot. The most representative example of such technology is the *Da Vinci Surgical System* (*Intuitive Surgical Inc.*): a teleoperated architecture used for remote surgery. The ideal remote system features both high teleoperation, the ability to act in the remote environment, and high telepresence, the feeling of being in the remote environment (see Fig. 1). Through telepresence, the doctor fully embodies the task, perceiving it as if it was in-person rather than remotely. This important feature of remotely controlled systems is the key to achieving the ideal synergy between the doctor and the machine, referred to in the literature as 'human-robot orchestration' [8]. Although no commercially available system showcases haptic feedback, implementing haptic

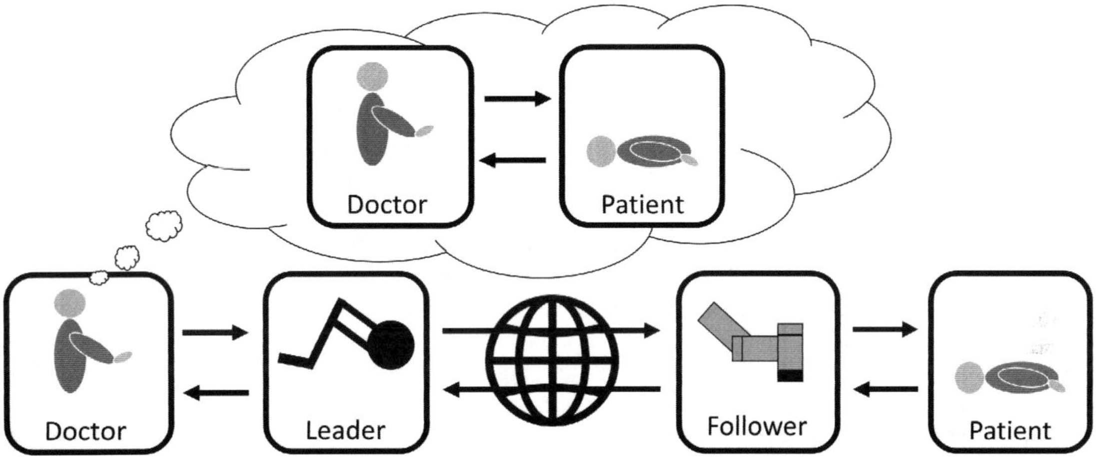

Figure 1. Schematized example of generic medical teleoperation featuring a doctor, a leader, a follower and a patient. In an intuitive system, the doctor would behave in the same way as in the in-person task.

feedback has been proven more than feasible [9, 10] and easily implementable in existing commercial systems [11, 12]. Driven by the desire to expand robotic technologies to other aspects of medical practice, several research groups have started developing systems in order to achieve remote palpation [13].

1.2. Palpation and Remote Palpation

The most common test run by primary care doctors is physical examination [14]. Not only does it lead to the most suitable follow-up exam, but in some cases also to a correct diagnosis of abdominal pathologies [15]. Requiring no additional tools besides the doctor's hands and expertise, represents a key element of the healthcare system's workflow. Physical examination is composed of 4 parts: auscultation, inspection, percussion, and palpation [16, 17]. The latter has been studied both in simulations [18] and in real scenarios [19] because of its high dexterity and multi-modal feedback modalities. Given the pivotal role of such a practice, what happens when it cannot be performed in person? Teleoperated systems seem to offer a possible solution, both enabling the diagnosis of unreachable patients and slowing down the ever-growing shortage of physicians [20]. When analyzing state-of-the-art teleoperated systems for remote palpation, it can be observed that the absence of standard design guidelines has led to a wide variety of different implementations of the doctor-robot human-machine interface (HMI), none of which showcases the desired telepresence required to achieve a good performance. Generally, the main direction taken by this field is the implementation of haptic systems able to record and convey tactile feedback from the remote environment to the user intuitively and naturally [21].

Moreover, even the delivery modality of the tactile feedback itself has multiple different implementations in the literature. In some cases, the tactile information is shown either directly [22, 23] or using a simulation of the interaction with the soft tissues [24, 25, 26]: this feedback interface can be considered as an add-on for commercial systems without the introduction of stability problems. However, the ultimate result is no fidelity and no transparency because the feedback is delivered visually and the motion required to control the trajectory of the robotic hand feels unnatural, relying on the same open kinematic chain used in surgical robots. In fact, intuitiveness plays a fundamental role in achieving telepresence and effectiveness [27, 28]. For intuitiveness, we intend the ability of the system to achieve high telepresence with minimal learning required. Note that high telepresence is effectively achieved when the remote system is embodied by the user, and becomes part of her/his body schema, allowing her/him to perceive the remote task as if it was in person. In other cases, fingertip tactile haptic devices have been implemented and used together with a traditional control interface. The simplest solutions showcase vibrating stimulation devices, placed on top of the user's fingers [29] or over the control interface [30]. The most complex interfaces showcase the entire surface's morphology changes, acting as physical twins of the touched surface in the remote environment [31, 32, 33]. Those devices are able to reproduce any shape touched by the follower and convey both kinesthetic and tactile haptic information, but they are solely feedback interfaces, and cannot be used for the trajectory control of the follower. This negatively impacts the telepresence of the overall system, as the user still needs to learn how to use a system that inevitably feels very different from using its own hand to perform in-presence palpation. Once again, the interface might result not natural and not intuitive, and promote estrangement for the user, because it is no longer requires palpating a soft interface, but controlling a joystick.

Such a problem has been tackled by using the physical phantom also as a control input [34, 35], producing more intuitive interfaces for human-machine interaction. In the work by *Li et al.* [36], it is shown how the same interface is used both for sending control commands and delivering tactile feedback. However, state-of-the-art cases of said technology have extremely limited dexterity, with discrete predetermined possible trajectories, strongly impairing the user's ability to perform any desired motion and once again promoting estrangement for the user.

2. Comparison of remote/in-person tasks in successful teleoperated applications

2.1. Medical field

One great example of remote teleoperation in the medical field is RAMIS: Robot Assisted Minimally Invasive Surgery. A user, usually a surgeon, uses a leader interface to control the movements of the follower, which is composed of an endoscopic camera and a set of minimally invasive instruments. Concerning the benefits, RAMIS not only provides low infection rates, shorter recovery time, and reduced blood loss for the patient [37], but also enhanced precision, dexterity, and vision for the doctor [38, 39]. Hence, RAMIS has radically transformed surgery [40], going from just over 100,000 robotic assisted operations in 2008 [41] to over a million in 2018 [42]. These numbers are bound to increase even further as new systems approach the market [43, 44, 45]. As the teleoperated systems permeate surgery, they showcase numerous benefits compared to traditional methods [46, 47]. As a result, many procedures have been converted from laparoscopic surgery to RAMIS, from cancer removal to partial organ resection [48, 49, 50, 51, 52, 53, 54, 55, 56, 57, 58, 59, 60]. As it emerges from all these studies, RAMIS represents a significant upgrade to laparoscopic surgery. It is also very important to notice that surgeons performing RAMIS are laparoscopic surgeons and have many years of experience using laparoscopic instruments, as RAMIS is a direct successor of minimally invasive surgery, or laparoscopy, and not of traditional open surgery. Because of that, all commercial devices present to the user an interface that resembles as much as possible the handles of a laparoscopic instrument (see Fig. 2). Because of the specific target user, reproducing the interface that characterizes a task they are already proficient in can effectively lower the time needed to learn how to use the new tool and avoid the user's estrangement from the task. Moreover, a closer look at commercially available systems such as the *Da Vinci* or the *Raven* [61] reveals that these systems do not implement any haptic feedback. Such feedback is instead present, even if limited, in laparoscopic surgery, and the absence of it complicates the handling of soft tissues [62]. However, as already mentioned in Section 1.2, there are several solutions explored in literature able to add such feature. Note that even in laparoscopic surgery, point-wise kinaesthetic haptic feedback is the only form of mechanical feedback available to the doctors, given the extremely small contact area between the tool and the patient's tissues.

Figure 2. Example of a commercial *Da Vinci* system within hospital settings. Note how the leader's handles reproduce the handles of a laparoscopic instrument, so to enhance the overall telepresence. Source: Wikimedia Commons.

2.2. External fields

Nuclear power plants (NPP) are a prime application for teleoperation, mainly because of the high level of radiation produced by the nuclear reactor and the radioactive waste [2]. In such a challenging environment, many tasks cannot be executed in person by operators without a great health hazard or potential danger. In NPP, teleoperation can be used for a wide range of tasks: from remotely cleaning the reactor's walls [63] to safely performing maintenance [64] and handling radioactive waste [65]. The last 2 examples are particularly relevant, representing high dexterity handling tasks in which objects need to be visually inspected and correctly placed. In a sense, this type of teleoperation does not radically differ from teleoperated surgery in the sense that the aim is to correctly follow a trajectory through space and perform pick-up and placement. In fact, such a task can be performed in person, through a glovebox (see Fig. 3a): a radiation-shielded cage in which an operator can introduce the hands, thanks to protective gloves. However, performing the same task remotely could not only reduce even further the level of radiation experienced by the user but also prevent any possible damage in case of failure (see Fig. 3c). Intuitively, teleoperated systems developed for such applications have to withstand radiations, and every component needs to be tested accordingly [66]. Concerning the system's architecture, because the only requirement is to check if the correct object was grasped and placed correctly, an interface consisting of an open kinematic chain with clamps on the handle and visual feedback has been shown to achieve acceptable results in commercial reactors such as the Joint European Torus (JET). However, the limited dexterity and the lack of a multimodal feedback modality, showcasing only visual feedback, led to increased operation times, when compared with the in-person task. When designing the teleoperated system for the maintenance of the experimental fusion plant ITER, an extensive study has analyzed and optimized all the procedures so to save as much time as possible and reduce the delays caused by low telepresence of the teleoperated systems [67]. Moreover, following a trend already explored in teleoperated surgery, enabling shared control for trajectory optimization and obstacle avoidance [68] adding multimodal feedback such as haptic and visual [69] significantly improves performance and operation time. Note that the implemented haptic feedback is kinaesthetic and not tactile, assuming a point-like contact between the follower and the remote environment. This assumption works relatively well when working with clamps and interacting with a rigid environment, but falls short in reproducing extended soft contact surfaces, such as the human abdomen inspected during palpation.

Another field in which teleoperation has been heavily involved is aerospace, specifically in the development of Unmanned Areal Vehicles (UAVs) and drones. Any surveillance task that requires monitoring of large or dangerous zones is a prime target for avionic teleoperation: from road surveillance [70] to applications in hazardous areas such as mines [71] and volcanic landscapes [72], on top of large scale natural areas monitoring both on land and in the sea [73, 74, 75, 76]. In this case, teleoperation has multiple advantages: on one side, the devices are much smaller and more compact, reducing the costs of purchase and maintenance, on the other, in case of fatal crash there is no human casualty because the operator is not physical on the vehicle. Conversely to NPP teleoperation, this type of teleoperation is radically different from what is requested by medical teleoperation, in the sense that it does not concern manipulation and it is not performed with a robotic manipulator, but rather with an autonomous flying vehicle. Historically, the interface presented to the user, or remote pilot, has been as similar as possible to an actual interface of a human-operated vehicle [77] (see Fig. 3b and 3d). In practice, control interfaces are fundamentally flight simulators used to control a real vehicle rather than simulate one. This solution is ideal for already trained pilots, because they already have learned how to use such an interface and the skill transfer required is minimal, given the high similarity to the in-person case. However, since areal devices as drones became available to the greater public, new control interfaces were needed to make the technology usable for people without pilot training. Because of their wide presence in today's devices, interfaces such as joy-pads and console controllers are the most common commercially available control interface for drones [78] (see Fig. 3e). They feature an easy encoding of the user's inputs and result more intuitive to most people because they are already widely used in other applications, such as the gaming industry. However, many research studies have proven that it is possible to achieve an even more intuitive control interface exploiting the human's body motions [79, 80, 81, 82] or brain-machine interfaces [83]. Because the average person does not know how to pilot an aircraft, allowing them to simulate flying with their bodies or just thinking about flying promotes the embodiment of the user in the task, resulting more intuitive and natural. This paradigm shift represents a key point in the concept of remote control interface for aerospace applications and highlights a very important key feature of remote interfaces in general: the interface needs to be intuitive for the target users that are supposed to use it. For such reason, when the target shifted from trained pilots to general public, the interfaces changed from reproducing the cockpit to body and brain-guided control, avoiding the user's estrangement from the target task of flying.

Figure 3. Examples of in-person and remote tasks from NPPs and avionics. On the left, (a) a standard glovebox to handle radioactive waste, and (c) a proposed robotic interface to remotely achieve the same action. On the right, (b) a cockpit of an airplane and (d) a professional control station for UAVs and (e) a joystick to control commercially available drones. Source: Wikimedia Commons.

3. Conclusion

Overall, all the analyzed examples of successful teleoperation have the same common ground: they are highly tuned in order to be natural and intuitive for the target user. Task embodiment is obtained through high telepresence. In order to achieve telepresence, the user needs to feel as if she/he is performing the task in person, rather than remotely. In other words, a good interface allows the user to move from the in-person task to the remote one without requiring her/him to undergo any learning process. However, this highly depends on the nature of the users for which the teleoperated system is intended. Highly specialized professionals such as doctors or pilots would prefer a complicated setting as similar as possible to the one that they have been trained to work with, whereas the general public would prefer a simpler interface. The same simpler setup would result unnatural and limiting to trained professionals, that already mastered the ability to control a more powerful and more complex tool. Therefore, we must focus on the need of general practitioners while trying to define the guidelines about remote palpation and remote physical examination. Unlikely surgery, during palpation the patient is awake and her/his feedback is a fundamental part of the in-person task. Literature studies have indeed proved that feedback signals such as facial expressions [84] and patient's voice [85] strongly improve the users' performance during palpation, to further confirm the importance of reproducing the same experience that the target user would be subjected to during the in-person task. Therefore, it is mandatory to further study physical examination and palpation and to identify and isolate the important features for general practitioners. Those features will then be needed while designing the teleoperated system in order to obtain an intuitive and natural interface with enhanced telepresence. Focusing on the target user will allow us to understand better the important features dictating the effectiveness of the system and to define the correct guidelines for the evaluation of remote palpation teleoperated architectures.

Acknowledgements
This work was supported by the SMART project, European Union's Horizon 2020 research and innovation under the Marie Sklodowska-Curie (grant agreement ID 860108).

References

[1] Peter F. Hokayem and Mark W. Spong. Bilateral teleoperation: An historical survey. *Automatica*, 42(12):2035–2057, 2006.

[2] Jan De Geeter, Marc Decréton, and Eric Colon. Challenges of telerobotics in a nuclear environment. *Robotics and Autonomous Systems*, 28(1):5–17, 1999.

[3] M. Hassanalian and A. Abdelkefi. Classifications, applications, and design challenges of drones: A review. *Progress in Aerospace Sciences*, 91(April):99–131, 2017.

[4] Sarmad Mehrdad, Fei Liu, Minh Tu Pham, Arnaud Lelevé, and S. Farokh Atashzar. Review of advanced medical telerobots. *Applied Sciences (Switzerland)*, 11(1):1–47, 2021.

[5] Jocelyne Troccaz, Giulio Dagnino, and Guang Zhong Yang. Frontiers of Medical Robotics: From Concept to Systems to Clinical Translation. *Annual Review of Biomedical Engineering*, 21:193–218, 2019.

[6] Yuyang Chen, Shu'an Zhang, Zhonghao Wu, Bo Yang, Qingquan Luo, and Kai Xu. Review of surgical robotic systems for keyhole and endoscopic procedures: state of the art and perspectives. *Frontiers of Medicine*, 14(4):382–403, 2020.

[7] Prem N. Kakar, Jyotirmoy Das, Preeti Mittal Roy, and Vijaya Pant. Robotic invasion of operation theatre and associated anaesthetic issues: A review. *Indian Journal of Anaesthesia*, 55(1):18–25, 2011.

[8] Ilana Nisky, Leone Costi, and Fumiya Iida. Coevolution of internal representations in physical human-robot orchestration – models of the surgeon and the robot in robotic surgery Coevolution of internal representations in physical human-robot orchestration – models of the surgeon and the robot in robotic surgery. 2022.

[9] Sabrina Panëels and Jonathan C. Roberts. Review of designs for haptic data visualization. *IEEE Transactions on Haptics*, 3(2):119–137, 2010.

[10] Heather Culbertson, Samuel B. Schorr, and Allison M. Okamura. Haptics: The Present and Future of Artificial Touch Sensation. *Annual Review of Control, Robotics, and Autonomous Systems*, 1(1):385–409, 2018.

[11] Farshid Amirabdollahian, Salvatore Livatino, Behrad Vahedi, Radhika Gudipati, Patrick Sheen, Shan Gawrie-Mohan, and Nikhil Vasdev. Prevalence of haptic feedback in robot-mediated surgery: a systematic review of literature. *Journal of Robotic Surgery*, 12(1):11–25, 2018.

[12] Issam El Rassi and Jean Michel El Rassi. A review of haptic feedback in tele-operated robotic surgery. *Journal of Medical Engineering and Technology*, 44(5):247–254, 2020.

[13] Alessandro Filippeschi, Filippo Brizzi, Emanuele Ruffaldi, Juan Manuel Jacinto, and Carlo Alberto Avizzano. Encountered-type haptic interface for virtual interaction with real objects based on implicit surface haptic rendering for remote palpation. *IEEE International Conference on Intelligent Robots and Systems*, 2015-December:5904–5909, 2015.

[14] Muhammad Bilal, Vlad Voin, Nitsa Topale, Joe Iwanaga, Marios Loukas, and R. Shane Tubbs. The Clinical anatomy of the physical examination of the abdomen: A comprehensive review. *Clinical Anatomy*, 30(3):352–356, 2017.

[15] Joseph S. Alpert. How Accurate Are the Findings Noted During a Physical Examination?: Will Physicians Stop Performing Physical Examinations? (Part 2). *American Journal of Medicine*, 132(6):663–664, 2019.

[16] M. Mehta. Assessing the abdomen. *Nursing*, 33(5):54–55, 2003.

[17] Sarah L. Cartwright and Mark P. Knudson. Evaluation of acute abdominal pain in adults. *American Family Physician*, 77(7):971–978, 2008.

[18] Javier Palacio-Torralba, Robert L. Reuben, and Yuhang Chen. A novel palpation–based method for tumor nodule quantification in soft tissue—computational framework and experimental validation. *Medical and Biological Engineering and Computing*, 58(6):1369–1381, 2020.

[19] Jelizaveta Konstantinova, Giuseppe Cotugno, Prokar Dasgupta, Kaspar Althoefer, and Thrishantha Nanayakkara. Palpation force modulation strategies to identify hard regions in soft tissue organs. *PLoS ONE*, 12(2):1–24, 2017.

[20] Richard M. Scheffler, Jenny X. Liu, Yohannes Kinfu, and Mario R. Dal Poz. Forecasting the global shortage of physicians: An economic- and needs-based approach. *Bulletin of the World Health Organization*, 86(7):516–523, 2008.

[21] Lawrence H. Salud and Carla M. Pugh. Use of sensor technology to explore the science of touch. *Studies in Health Technology and Informatics*, 163:542–548, 2011.

[22] Min Li, Jelizaveta Konstantinova, Guanghua Xu, Bo He, Vahid Aminzadeh, Jun Xie, Helge Wurdemann, and

Kaspar Althoefer. Evaluation of stiffness feedback for hard nodule identification on a phantom silicone model. *PLoS ONE*, 12(3):1–18, 2017.

[23] Min Li, Shan Luo, and Guanghua Xu. A tactile sensing and feedback system for tumor localization. *2016 13th International Conference on Ubiquitous Robots and Ambient Intelligence, URAI 2016*, pages 259–262, 2016.

[24] Jungsik Kim, Bummo Ahn, Yeongjin Kim, and Jung Kim. Inclusion detection with haptic-palpation system for medical telediagnosis. *Proceedings of the 31st Annual International Conference of the IEEE Engineering in Medicine and Biology Society: Engineering the Future of Biomedicine, EMBC 2009*, pages 4595–4598, 2009.

[25] Timothy R. Coles, Nigel W. John, Derek Gould, and Darwin G. Caldwell. Integrating haptics with augmented reality in a femoral palpation and needle insertion training simulation. *IEEE Transactions on Haptics*, 4(3):199–209, 2011.

[26] Min Li, Jelizaveta Konstantinova, Emanuele L. Secco, Allen Jiang, Hongbin Liu, Thrishantha Nanayakkara, Lakmal D. Seneviratne, Prokar Dasgupta, Kaspar Althoefer, and Helge A. Wurdemann. Using visual cues to enhance haptic feedback for palpation on virtual model of soft tissue. *Medical and Biological Engineering and Computing*, 53(11):1177–1186, 2015.

[27] Malika Auvray and Christian Duriez. *Haptics: Neuroscience, Devices, Modeling, and Applications*, volume 8618. 2014.

[28] Margot Racat and Sonia Capelli. *When Interfaces Make It Real*, pages 65–93. Springer International Publishing, Cham, 2020.

[29] Antonia Tzemanaki, Gorkem Anil Al, Chris Melhuish, and Sanja Dogramadzi. Design of a wearable fingertip haptic device for remote palpation: Characterisation and interface with a virtual environment. *Frontiers Robotics AI*, 5(JUN):1–15, 2018.

[30] A. Talasaz and R. V. Patel. Remote palpation to localize tumors in robot-assisted minimally invasive approach. *Proceedings - IEEE International Conference on Robotics and Automation*, pages 3719–3724, 2012.

[31] R. L. Feller, C. K.L. Lau, C. R. Wagner, D. P. Perrin, and R. D. Howe. The effect of force feedback on remote palpation. *Proceedings - IEEE International Conference on Robotics and Automation*, 2004(1):782–788, 2004.

[32] Calum Roke, Chris Melhuish, Tony Pipe, David Drury, and Craig Chorley. Lump localisation through a deformation-based tactile feedback system using a biologically inspired finger sensor. *Robotics and Autonomous Systems*, 60(11):1442–1448, 2012.

[33] Leone Costi, Perla Maiolino, and Fumiya Iida. Soft Morphing Interface for Tactile Feedback in Remote Palpation. In *2022 IEEE 5th International Conference on Soft Robotics (RoboSoft)*, pages 01–06. IEEE, apr 2022.

[34] Kevin A Hernandez-ossa, Arnaldo G Leal-junior, Anselmo Frizera-neto, and Teodiano Bastos. Haptic Feedback for Remote Clinical Palpation Examination. (February), 2019.

[35] J. Dargahi, W. F. Xie, and Peng Ji. An experimental teletaction system for sensing and teleperception of human pulse. *Mechatronics*, 18(4):195–207, 2008.

[36] Wei Li, Lei Shi, Hengyi Deng, and Zhongrong Zhou. Investigation on friction trauma of small intestine in vivo under reciprocal sliding conditions. *Tribology Letters*, 55(2):261–270, 2014.

[37] Jens Rassweiler, Marcel Hruza, Dogu Teber, and Li-Ming Su. Laparoscopic and robotic assisted radical prostatectomy–critical analysis of the results. *European urology*, 49(4):612–624, 2006.

[38] K Moorthy, Y Munz, A Dosis, J Hernandez, S Martin, F Bello, T Rockall, and A Darzi. Dexterity enhancement with robotic surgery. *Surgical Endoscopy and Other Interventional Techniques*, 18(5):790–795, 2004.

[39] John C Byrn, Stefanie Schluender, Celia M Divino, John Conrad, Brooke Gurland, Edward Shlasko, and Amir Szold. Three-dimensional imaging improves surgical performance for both novice and experienced operators using the da vinci robot system. *The American Journal of Surgery*, 193(4):519–522, 2007.

[40] Tiago Leal Ghezzi and Oly Campos Corleta. 30 years of robotic surgery. *World journal of surgery*, 40(10):2550–2557, 2016.

[41] Christopher P Childers and Melinda Maggard-Gibbons. Estimation of the acquisition and operating costs for robotic surgery. *Jama*, 320(8):835–836, 2018.

[42] Intuitive surgical annual report 2018. 2018.

[43] Salvatore Gueli Alletti, C Rossitto, S Cianci, E Perrone, S Pizzacalla, G Monterossi, G Vizzielli, S Gidaro, and G Scambia. The senhance™ surgical robotic system ("senhance") for total hysterectomy in obese patients: a pilot study. *Journal of robotic surgery*, 12(2):229–234, 2018.

[44] Stephan Lang, Stefan Mattheis, Pia Hasskamp, Georges Lawson, Christian Güldner, Magis Mandapathil, Patrick Schuler, Thomas Hoffmann, Marc Scheithauer, and Marc Remacle. A european multicenter study

evaluating the flex robotic system in transoral robotic surgery. *The Laryngoscope*, 127(2):391–395, 2017.

[45] Dhananjay Kelkar, Mahindra A Borse, Girish P Godbole, Utkrant Kurlekar, and Mark Slack. Interim safety analysis of the first-in-human clinical trial of the versius surgical system, a new robot-assisted device for use in minimal access surgery. *Surgical endoscopy*, 35(9):5193–5202, 2021.

[46] M. M. Lirici, V. Papaspyropoulos, and L. Angelini. Telerobotics in medicine and surgery. *Minimally Invasive Therapy and Allied Technologies*, 6(5-6):364–378, 1997.

[47] Alejandro Chavarriaga, Amy Wyrzykowski, and David V. Feliciano. Minimally invasive surgery? *Trauma Surgery and Acute Care Open*, 3(1):631–644, 2018.

[48] Amer Karam and Oliver Dorigo. Minimally Invasive Surgery for Gynecologic Cancers - A Cautionary Tale. *JAMA Oncology*, 6(7):991–993, 2020.

[49] Francesco Paolo Prete, Angela Pezzolla, Fernando Prete, Mario Testini, Rinaldo Marzaioli, Alberto Patriti, Rosa Maria Jimenez-Rodriguez, Angela Gurrado, and Giovanni F.M. Strippoli. Robotic Versus Laparoscopic Minimally Invasive Surgery for Rectal Cancer: A Systematic Review and Meta-analysis of Randomized Controlled Trials. *Annals of Surgery*, 267(6):1034–1046, 2018.

[50] Armando De Virgilio, Oreste Iocca, Luca Malvezzi, Pasquale Di Maio, Raul Pellini, Fabio Ferreli, Giovanni Cugini, Giovanni Colombo, and Giuseppe Spriano. The emerging role of robotic surgery among minimally invasive surgical approaches in the treatment of hypopharyngeal carcinoma: Systematic review and meta-analysis. *Journal of Clinical Medicine*, 8(2), 2019.

[51] Vipul R. Patel, Nilesh N. Patil, Geoff Coughlin, Pankaj P. Dangle, and Kenneth Palmer. Robot-assisted laparoscopic pyeloplasty: A review of minimally invasive treatment options for ureteropelvic junction obstruction. *Journal of Robotic Surgery*, 1(4):247–252, 2008.

[52] Caio M. Oliveira, Hiep T. Nguyen, Alberto R. Ferraz, Karen Watters, Brian Rosman, and Reza Rahbar. Robotic surgery in otolaryngology and head and neck surgery: A review. *Minimally Invasive Surgery*, 2012, 2012.

[53] Justin D. Stull, John J. Mangan, Alexander R. Vaccaro, and Gregory D. Schroeder. Robotic Guidance in Minimally Invasive Spine Surgery: a Review of Recent Literature and Commentary on a Developing Technology. *Current Reviews in Musculoskeletal Medicine*, 12(2):245–251, 2019.

[54] Steve Wright, Sam Esfandiari, Nader Elmayergi, Zion Sasson, and Jack M. Goodman. Left atrial functional changes following short-term exercise training. *European Journal of Applied Physiology*, 114(12):2667–2675, 2014.

[55] Kelly J. Lafaro, Camille Stewart, Abigail Fong, and Yuman Fong. Robotic Liver Resection. *Surgical Clinics of North America*, 100(2):265–281, 2020.

[56] Johannes Bonatti, Ingo Crailsheim, Martin Grabenwöger, and Bernhard Winkler. Minimally Invasive and Robotic Mitral Valve Surgery: Methods and Outcomes in a 20-Year Review. *Innovations: Technology and Techniques in Cardiothoracic and Vascular Surgery*, 16(4):317–326, 2021.

[57] Jacopo Desiderio, Zhi Wei Jiang, Ninh T. Nguyen, Shu Zhang, Daniel Reim, Orhan Alimoglu, Juan Santiago Azagra, Pei Wu Yu, Natalie G. Coburn, Feng Qi, Patrick G. Jackson, Lu Zang, Steven T. Brower, Yukinori Kurokawa, Olivier Facy, Hironori Tsujimoto, Andrea Coratti, Mario Annecchiarico, Francesca Bazzocchi, Andrea Avanzolini, Johan Gagniere, Denis Pezet, Fabio Cianchi, Benedetta Badii, Alexander Novotny, Tunc Eren, Metin Leblebici, Martine Goergen, Ben Zhang, Yong Liang Zhao, Tong Liu, Waddah Al-Refaie, Junjun Ma, Shuji Takiguchi, Jean Baptiste Lequeu, Stefano Trastulli, and Amilcare Parisi. Robotic, laparoscopic and open surgery for gastric cancer compared on surgical, clinical and oncological outcomes: A multi-institutional chart review. A study protocol of the International study group on Minimally Invasive surgery for GASTRIc Cancer-IMIGASTRIC. *BMJ Open*, 5(10):1–9, 2015.

[58] Evelio Rodr, Alan P Kypson, Simon C Moten, L Wiley Nifong, and W Randolph Chitwood Jr. Robotic mitral surgery at East Carolina University :. *International Journal*, (April):211–215, 2006.

[59] P. R. Nanayakkara, S. A. Ahmed, D. Oudit, S. T. O'Dwyer, and C. R. Selvasekar. Robotic assisted minimally invasive pelvic exenteration in advanced rectal cancer: Review and case report. *Journal of Robotic Surgery*, 8(2):173–175, 2014.

[60] Francis P. Robertson and Rowan W. Parks. A review of the current evidence for the role of minimally invasive pancreatic surgery following neo-adjuvant chemotherapy. *Laparoscopic, Endoscopic and Robotic Surgery*, 5(2):47–51, 2022.

[61] Mitchell J.H. Lum, Diana C.W. Friedman, Ganesh Sankaranarayanan, Hawkeye King, Kenneth Fodero, Rainer Leuschke, Blake Hannaford, Jacob Rosen, and Mika N. Sinanan. The RAVEN: Design and validation of a telesurgery system. *International Journal of Robotics Research*, 28(9):1183–1197, 2009.

[62] G. Dogangil, B. L. Davies, and F. Rodriguez Y Baena. A review of medical robotics for minimally invasive soft tissue surgery. *Proceedings of the Institution of Mechanical Engineers, Part H: Journal of Engineering in Medicine*, 224(5):653–679, 2010.

[63] B. L. Luk, K. P. Liu, A. A. Collie, D. S. Cooke, and S. Chen. Tele-operated climbing and mobile service

robots for remote inspection and maintenance in nuclear industry. *Industrial Robot*, 33(3):194–204, 2006.

[64] Stephen Sanders. Remote operations for fusion using teleoperation. *Industrial Robot*, 33(3):174–177, 2006.

[65] Ozan Tokatli, Pragna Das, Radhika Nath, Luigi Pangione, Alessandro Altobelli, Guy Burroughes, Emil T. Jonasson, Matthew F. Turner, and Robert Skilton. Robot-assisted glovebox teleoperation for nuclear industry. *Robotics*, 10(3), 2021.

[66] Takeo Oomichi, Yosifumi Isozaki, and Masanari Kojima. Practical design of robots operating in radiation environments. *Advanced Robotics*, 21(5-6):515–532, 2007.

[67] Henri Boessenkool, Justin Thomas, Jeroen G.W. Wildenbeest, Cock J.M. Heemskerk, Marco R. de Baar, Maarten Steinbuch, and David A. Abbink. Where to improve in human-in-the-loop tele-operated maintenance? A phased task analysis based on video data of maintenance at JET. *Fusion Engineering and Design*, 129(October 2017):309–319, 2018.

[68] Vitalii Pruks, Kwang Hyun Lee, and Jee Hwan Ryu. Shared Teleoperation for Nuclear Plant Robotics Using Interactive Virtual Guidance Generation and Shared Autonomy Approaches. *2018 15th International Conference on Ubiquitous Robots, UR 2018*, pages 91–95, 2018.

[69] Philippe Desbats, Franck Geffard, Gérard Piolain, and Alain Coudray. Force-feedback teleoperation of an industrial robot in a nuclear spent fuel reprocessing plant. *Industrial Robot*, 33(3):178–186, 2006.

[70] Jassim Al-Fadhli, Mustafa Ashkanani, Abdulwahab Yousef, Issam Damaj, and Mohammad El-Shafei. RECON: A remotely controlled drone for roads safety. *2014 International Conference on Connected Vehicles and Expo, ICCVE 2014 - Proceedings*, pages 912–918, 2014.

[71] Javad Shahmoradi, Elaheh Talebi, Pedram Roghanchi, and Mostafa Hassanalian. A comprehensive review of applications of drone technology in the mining industry. *Drones*, 4(3):1–25, 2020.

[72] Benjamin R. Jordan. Collecting field data in volcanic landscapes using small UAS (sUAS)/drones. *Journal of Volcanology and Geothermal Research*, 385:231–241, 2019.

[73] Fernando H.Iost Filho, Wieke B. Heldens, Zhaodan Kong, and Elvira S. De Lange. Drones: Innovative technology for use in precision pest management. *Journal of Economic Entomology*, 113(1):1–25, 2020.

[74] Richard Adade, Abiodun Musa Aibinu, Bernard Ekumah, and Jerry Asaana. Unmanned Aerial Vehicle (UAV) applications in coastal zone management—a review. *Environmental Monitoring and Assessment*, 193(3), 2021.

[75] Di Wu, Ruopu Li, Feiyang Zhang, and Jia Liu. A review on drone-based harmful algae blooms monitoring. *Environmental Monitoring and Assessment*, 191(4), 2019.

[76] Lina Tang and Guofan Shao. Drone remote sensing for forestry research and practices. *Journal of Forestry Research*, 26(4):791–797, 2015.

[77] Yixiang Lim, Alessandro Gardi, Roberto Sabatini, Subramanian Ramasamy, Trevor Kistan, Neta Ezer, Julian Vince, and Robert Bolia. Avionics Human-Machine Interfaces and Interactions for Manned and Unmanned Aircraft. *Progress in Aerospace Sciences*, 102(August):1–46, 2018.

[78] K. W. Chan, U. Nirmal, and W. G. Cheaw. Progress on drone technology and their applications: A comprehensive review. *AIP Conference Proceedings*, 2030(November), 2018.

[79] Jenifer Miehlbradt, Alexandre Cherpillod, Stefano Mintchev, Martina Coscia, Fiorenzo Artoni, Dario Floreano, and Silvestro Micera. Data-driven body–machine interface for the accurate control of drones. *Proceedings of the National Academy of Sciences of the United States of America*, 115(31):7913–7918, 2018.

[80] Carine Rognon, Stefano Mintchev, Fabio Dellagnola, Alexandre Cherpillod, David Atienza, and Dario Floreano. FlyJacket: An Upper Body Soft Exoskeleton for Immersive Drone Control. *IEEE Robotics and Automation Letters*, 3(3):2362–2369, 2018.

[81] Zhenfei Zhao, Hao Luo, Guang Hua Song, Zhou Chen, Zhe Ming Lu, and Xiaofeng Wu. Web-based interactive drone control using hand gesture. *Review of Scientific Instruments*, 89(1), 2018.

[82] Byeong Cheol Kang and Tae Jun Ha. Human-interactive drone system remotely controlled by printed strain/pressure sensors consisting of carbon-based nanocomposites. *Composites Science and Technology*, 182(August):107784, 2019.

[83] Amin Hekmatmanesh, Pedro H J Nardelli, and Heikki Handroos. Review of the State-of-the-Art of Brain-Controlled Vehicles. pages 110173–110193, 2021.

[84] Thilina D. Lalitharatne, Leone Costi, Ryman Hashem, Ilana Nisky, Rachael E. Jack, Thrishantha Nanayakkara, and Fumiya Iida. Face mediated human–robot interaction for remote medical examination. *Scientific Reports*, 12(1):1–14, 2022.

[85] N Protpagorn, L Costi, TD Lalitharatne, I Nisky, and F Iida. Vocal pain expression augmentation to improve interaction accuracy in virtual robopatient. In *2022 RAS/EMBS IEEE 9th International Conference on Biomedical Robotics and Biomechatronics*, page accepted. IEEE, 2022.

AUTHOR INDEX

Institute of Physics
Dirac House, Temple Back
Bristol BS1 6BE UK

ISSN: 1757-8981
ISBN 978-1-71388-8185-8

World Symposium on Smart Materials and Applications (WSSMA 2021)

IOP Conference Series: Materials Science and Engineering
Volume 1186

Bangkok, Thailand
16-18 July 2021

IOP
Institute of Physics